Cyclodextrins for UV Rays Filtration

Traditional UVR absorbers pose environmental and health risks due to their volatility, solubility, and toxicity. High concentrations are needed for effective UV filtration, leading to environmental accumulation. Safer alternatives, like cyclodextrin complexation, can enhance UV filtration while reducing absorber concentration. Cyclodextrin complexation offers a promising solution by enhancing UV filtration efficiency while reducing the required concentration of UVR absorbers.

Cyclodextrins for UV Rays Filtration addresses a critical gap in photoprotection research by offering a comprehensive exploration of how cyclodextrins enhance UVR filtration. It provides an in-depth understanding of various UVR filters, including organic, inorganic, nanoparticle-based, natural, and herbal filters, while focusing on optimizing their efficiency through cyclodextrin complexation. This book serves as a single-point reference for researchers and professionals, guiding them in selecting suitable cyclodextrins based on UV absorber size, improving SPF/UPF performance, and reducing environmental impact. By presenting both fundamental principles and advanced applications, this book distinguishes itself as the first dedicated monograph on cyclodextrin-based UVR filters, making it an essential resource for academia and industry.

Features:

- Systematic overview of cyclodextrin-based UVR filtration principles, mechanisms, and methods.
- Insights from experts on advancements and future prospects of cyclodextrin-based UVR filters.
- Applications of cyclodextrins in sunscreens, lotions, textiles, protective films, and glass.
- Guidance on selecting and utilizing cyclodextrins for enhanced UVR protection.
- Coverage of both fundamental concepts and real-world uses of cyclodextrins in UVR filtration.

This book serves as a comprehensive and authoritative reference that will significantly benefit researchers, formulators, and professionals aiming to develop safer and more efficient photoprotective materials.

Cyclodextrins for UV Rays Filtration

Edited by
Sivakumar Krishnamoorthy

CRC Press
Taylor & Francis Group
Boca Raton London New York

CRC Press is an imprint of the
Taylor & Francis Group, an **informa** business

Designed cover image: Self-created by the editor, Sivakumar Krishnamoorthy

First edition published 2026
by CRC Press
2385 NW Executive Center Drive, Suite 320, Boca Raton FL 33431

and by CRC Press
4 Park Square, Milton Park, Abingdon, Oxon, OX14 4RN

CRC Press is an imprint of Taylor & Francis Group, LLC

ISBN: 9781032795669 (hbk)
ISBN: 9781032795683 (pbk)
ISBN: 9781003492757 (ebk)

DOI: 10.1201/9781003492757

Typeset in Times
by Newgen Publishing UK

Dedication

To my beloved wife, Bhuvaneswari,
and to my daughter, Indhumathi,
for the unwavering support and endless patience

Contents

Preface

The increasing concerns over ultraviolet (UV) radiation and its detrimental effects on human health, materials, and the environment have driven extensive research into UV filtration technologies. Among various approaches, cyclodextrins (CDs) have emerged as a promising class of macrocyclic molecules due to their unique ability to form inclusion complexes with UV-absorbing compounds, thereby enhancing their stability, solubility, and overall effectiveness. This book, *Cyclodextrins for UV Rays Filtration*, explores the fundamental principles, mechanisms, and diverse applications of cyclodextrins in UV filtration.

Cyclodextrins are cyclic oligosaccharides, traditionally derived from starch, characterized by their hydrophobic cavities and hydrophilic exteriors. These properties enable them to encapsulate a wide range of guest molecules, including UV-absorbing compounds, thereby enhancing their stability, solubility, and efficacy. Over the past few decades, the functionalization and modification of cyclodextrins have opened new avenues for their application in UV filtration, making them indispensable in the development of advanced materials for sunscreens, coatings, textiles, and environmental protection.

This book, *Cyclodextrins for UV Rays Filtration*, provides the multifaceted roles of cyclodextrins in mitigating the adverse effects of UV radiation. It explores the principles of UV absorption, the mechanisms by which cyclodextrins enhance the performance of UV filters, and the latest innovations in cyclodextrin-based materials. From traditional applications in cosmetics to cutting-edge nanotechnology-driven solutions, the chapters in this book provide a comprehensive overview of the state-of-the-art technologies and future directions in this rapidly evolving field. The versatility of cyclodextrins, combined with their biocompatibility and sustainability, has positioned them as key players in the development of next-generation UV filtration systems. This book brings together contributions from leading researchers and experts, showcasing the latest advancements in cyclodextrin-based UV filtration and their applications in healthcare, environmental protection, and industrial processes. Each chapter offers a detailed exploration of the design, synthesis, and practical implementation of cyclodextrin-based materials, highlighting their potential to address global challenges related to UV radiation.

It is my hope that this book will serve as a valuable resource for researchers, industry professionals, and students interested in the science and technology of UV filtration. As the world continues to grapple with the escalating impacts of UV radiation, the innovative use of cyclodextrins offers a promising pathway to develop safer, more effective, and environmentally friendly solutions. I am confident that the insights and knowledge shared in this book will inspire further research and innovation, paving the way for a brighter and healthier future.

Prof. Dr. Sivakumar Krishnamoorthy
Department of Chemistry, Faculty of Science,
Director – Research and Development Cell,

Sri Chandrasekharendra Saraswathi Viswa Mahavidyalaya (Deemed to be University) [SCSVMV],
Enathur, Kanchipuram – 631 561, Tamilnadu, India.
sivakumar.k@kanchiuniv.ac.in
cyclodextrins.prof.shiva@gmail.com

Acknowledgements

I would like to express my gratitude to the entire team at Taylor & Francis Books India Pvt Ltd., for their unwavering support and dedication throughout this project. A special thanks to *Ms. Renu Upadhyay, Senior Commissioning Editor (Chemistry)*, for her meticulous guidance and insightful feedback during the proposal review process. I am also profoundly grateful to *Ms. Jyotsna Jangra* at CRC Press for her consistent support, which played a pivotal role in bringing this project to fruition.

Editor Biography

Prof. Dr. Sivakumar Krishnamoorthy serves as Professor in the Department of Chemistry, Faculty of Science, and as the Director of the Research and Development Cell at Sri Chandrasekharendra Saraswathi Viswa Mahavidyalaya (SCSVMV), a Deemed-to-be University (www.kanchiuniv.ac.in) in Kanchipuram, Tamil Nadu, India. Inspired by the versatility of cyclodextrins, his research focuses on developing innovative application materials through cyclodextrin inclusion complexes. He holds one granted Indian patent on cyclodextrin application and has filed five patent applications. He is a recipient of a research fellowship from the prestigious Indian Academy of Sciences and has authored over 115 research papers. He has secured funding from various Government of India agencies, including DST, CSIR, UGC, DAE, the Indian Academy of Sciences, and the Tamil Nadu State Council for Science and Technology, for his research projects and programs.

Contributors

Dinesh Karthik A.
P.G & Research Department of Chemistry,
Shanmuga Industries Arts and Science College,
Tiruvannamalai, Tamilnadu, India.

Usama Ahmad
Faculty of Pharmacy, Integral University,
Kursi Road, Lucknow, India.

Khan Aejaz Ahmed
Faculty of Pharmacy, Integral University,
Kursi Road, Lucknow, India.

Nalini A.
Department of Chemistry, Faculty of Science,
Sri Chandrasekharendra Saraswathi Viswa Mahavidyalaya (Deemed to be University) [SCSVMV], Enathur,
Kanchipuram, Tamilnadu, India.

Kanakadurga Bandaru
Department of Chemistry,
Shri Vishnu Engineering College for Women,
Bhimavaram, A.P, India.

Santosh Kumar Badampudi
Department of chemistry,
Gayatri Vidya Parishad College of Engineering,
Visakhapatnam, India.

Pranjal K. Baruah
Department of Applied Sciences,
GUIST, Gauhati University,
Guwahati, Assam, India.

Saptadwipa Bhattacharjee
Department of Chemistry,
University of North Bengal,
Dist-Darjeeling, West Bengal, India.

Saikat Bhattacharya
Dr. B. C. Roy College of Pharmacy & AHS,
Durgapur,
West Bengal,
India.

Maria Soledad Bueno
Unidad de Investigación y Desarrollo en Tecnología Farmacéutica,
(UNITEFA) CONICET, Departamento de Ciencias Farmacéuticas,
Facultad de Ciencias Químicas Universidad Nacional de Córdoba.
Córdoba, Argentina.

Lakhi Chetry
Department of Applied Sciences,
GUIST, Gauhati University,
Guwahati, Assam, India.

Fatih Ciftci
Department of Biomedical Engineering,
Fatih Sultan Mehmet Vakif University,
Istanbul, Türkiye
Technology Transfer Office,
Fatih Sultan Mehmet Vakif University,
Istanbul, Türkiye.

Rima Das
Department of Chemistry,
Chandidas Mahavidyalaya, Khujutipara, Birbhum,
West Bengal, India.

Sanjukta Das
Dr. B. C. Roy College of Pharmacy & AHS,
Durgapur, West Bengal.

Taniya Das
Department of Chemistry,
University of North Bengal,
Dist-Darjeeling, West Bengal, India.

Tanmoy Dutta
Department of Chemistry,
JIS College of Engineering,
Kalyani, India.

Claudia Garnero
Unidad de Investigación y Desarrollo en Tecnología Farmacéutica,
(UNITEFA) CONICET, Departamento de Ciencias Farmacéuticas,
Facultad de Ciencias Químicas Universidad Nacional de Córdoba.
Córdoba, Argentina.

Pranab Ghosh
Department of Chemistry,
University of North Bengal,
Dist-Darjeeling, West Bengal, India.

Kamakshi Gnanasekaran
Department of Life Science,
Kristu Jayanti College, Autonomous, Bengaluru.

Selvakumar G.
Department of Textile Technology,
PSG College of Technology,
Coimbatore, Tamil Nadu, India.

Rama Swamy Guttula
Department of BS,
Vishnu Institute of Technology,
Bhimavaram, Andhra Pradesh, India.

Wan Nurhidayah Wan Hanaffi
Laboratoire de Chimie Physique Macromoléculaire,
Université de Lorraine, CNRS, Nancy, France
Department of Immunology, School of Medical Sciences,
Universiti Sains Malaysia.

Janice Marianne John
Department of Life Science,
Kristu Jayanti College, Autonomous, Bengaluru.

Ganesh Kadiyala K
Department of Chemistry,
Shri Vishnu Engineering College for Women,
Bhimavaram, A.P, India.

Pavan Krishna K.
Department of PG Chemistry,
B.V. Raju College,
Bhimavaram, India.

Shinomol George K.
Department of Life Science,
Kristu Jayanti College, Autonomous, Bengaluru.

Ekramul Kabir
Department of Physics,
Darjeeling Government College;
Darjeeling, India.

Abdul Ashik Khan
Department of Chemistry,
Darjeeling Government College;
Darjeeling- 734101, India.

Mohd Muazzam Khan
Faculty of Pharmacy, Integral University,
Kursi Road, Lucknow, India.

Ayfer Koyuncu
Hacettepe University,
Graduate School of Science and Engineering,
Bioengineering Department, Ankara, Türkiye.

Sivakumar K.
Department of Chemistry, Faculty of Science,
Sri Chandrasekharendra Saraswathi Viswa Mahavidyalaya (Deemed to be University) [SCSVMV], Enathur,
Kanchipuram, Tamilnadu, India.

Sudipta K. Kundu
Department of Chemistry,
Government General Degree College,
Nakashipara, Nadia, India.

Marcela Raquel Longhi
Unidad de Investigación y Desarrollo en Tecnología Farmacéutica,
(UNITEFA) CONICET, Departamento de Ciencias Farmacéuticas,
Facultad de Ciencias Químicas Universidad Nacional de Córdoba.
Córdoba, Argentina.

Tamalika Maitra
Department of Chemistry,
University of North Bengal,
Dist-Darjeeling, West Bengal, India.

Belén Alejandra Mezzano
Unidad de Investigación y Desarrollo en Tecnología Farmacéutica,
(UNITEFA) CONICET, Departamento de Ciencias Farmacéuticas,
Facultad de Ciencias Químicas Universidad Nacional de Córdoba
Córdoba, Argentina.

Santosh Kumar Nadikatla
Department of chemistry,
GMR institute of technology,
Rajam, India.

Shouvik Kumar Nandy
School of Pharmacy,
Shoolini University, Solan, Himachal Pradesh.

Arif Hossain Patowari
Department of Chemistry,
University of North Bengal,
Dist-Darjeeling, West Bengal, India.

Micaela Ponce Ponte
Unidad de Investigación y Desarrollo en Tecnología Farmacéutica,
(UNITEFA) CONICET, Departamento de Ciencias Farmacéuticas,
Facultad de Ciencias Químicas Universidad Nacional de Córdoba.
Córdoba, Argentina.

Mary Nancy Flora R.
Department of Chemical Engineering,
Vel Tech High Tech Dr.Rangarajan Dr.Sakunthala Engineering College,
Chennai, Tamilnadu, India.

Sarojini, Sr. S
P.G & Research Department of Chemistry,
Shangmuga Industries Arts and Science College,
Thiruvannamalai, Thiruvalluvar University,
Vellore, Tamil Nadu, India.

Subhadeep Saha
Department of Chemistry,
Government General Degree College at Pedong,
Kalimpong, India.

Ajoy K. Sarkar
Textile Development and Marketing,
Jay and Patty Baker School of Business and Technology,
Fashion Institute of Technology,
State University of New York, New York, USA.

Barnali De Sarkar
Department of Chemistry,
University of North Bengal,
Dist-Darjeeling, West Bengal, India.

Partha Pratim Sarma
Department of Applied Sciences,
GUIST, Gauhati University,
Guwahati, Assam, India.

Sagarika Deepthy Tallapragada
Dr. B. C. Roy College of Pharmacy & AHS,
Durgapur, West Bengal.

Nagalakshmi K.V.
Department of chemistry,
Gayatri Vidya Parishad College of Engineering,
Visakhapatnam, India.

Madhavi Vemula
Department of Chemistry,
BVRIT Hyderabad College of Engineering for Women,
Hyderabad, India.

1 Cyclodextrins for Enhanced Ultraviolet Rays Filtration

Principles and Applications

K. Sivakumar, A. Nalini, G. Selvakumar, and Ajoy K. Sarkar

ABBREVIATIONS

β-cd	betacyclodextrin	hp-γcd	Hydroxypropyl gamma cyclodextrin
bps	benzophenones	H-G	Host-Guest complex
cds	cyclodextrins	MCT	monochlorotriazinyl
dna	dinitroaniline	omc	octylmethoxycinnamate
dnb	dinitrobenzoic acid	SPF	Sun protection factor
dnp	dinitrophenol	UPF	Ultraviolet protection factor
FTIR	Fourier transform infrared	UVR	Ultraviolet radiation
hbp	hydroxybenzophenone	NMR	Nuclear magnetic resonance

1.1 INTRODUCTION: *ULTRAVIOLET RAYS FILTRATION*

Ultraviolet (UV) radiation, a critical component of solar energy, encompasses wavelengths shorter than visible light but longer than X-rays, ranging from 10 to 400 nm. While essential for processes, such as vitamin D synthesis in humans, its high-energy nature poses significant risks to biological systems, materials, and overall human health. UV radiation is divided into three bands: UV-A (315–400 nm), UV-B (280–315 nm), and UV-C (100–280 nm). While UV-C is largely absorbed by the Earth's atmosphere, UV-A and UV-B penetrate and exert a variety of detrimental effects.

1.1.1 Harmful Effects of UV Radiation

Human Health Impacts: UV-B is a primary cause of skin cancer, sunburn, and photoaging. Prolonged exposure to UV-B radiation is a primary contributor to skin

DOI: 10.1201/9781003492757-1

cancer, causing direct DNA damage and immunosuppression. UV-A, though less energetic, penetrates deeper into the skin, leading to photoaging and oxidative stress. Both bands are implicated in the development of cataracts and other eye disorders. The cumulative health impacts of UV radiation underscore the critical need for effective filtration methods to safeguard public health.

Material Degradation: UV exposure accelerates the breakdown of polymers, coatings, and pigments, leading to loss of mechanical strength, discoloration, and reduced lifespan of materials. Photodegradation mechanisms, such as chain scission and oxidation, compromise the structural and aesthetic properties of materials, reducing their functional lifespan. This is particularly problematic in industries reliant on outdoor applications, such as construction, automotive, and textiles.

1.1.2 Need for UV Rays Filtration Technologies

Protecting human health and preserving material integrity requires efficient UV filtration solutions. These technologies are crucial in applications ranging from sunscreen formulations and protective textiles to coatings and packaging materials.

1.1.2.1 Current Methods of UV Rays Filtration

- *Physical Filters*: These include metal oxides like titanium dioxide (TiO_2) and zinc oxide (ZnO), which reflect and scatter UV radiation.
- *Chemical Filters*: Organic compounds such as oxybenzone and avobenzone absorb UV rays but face challenges like photodegradation and toxicity.

1.1.3 Challenges in UV Rays Filtration

- Instability of chemical filters under prolonged UV ray exposure.
- Potential environmental and health concerns associated with UV filter degradation.
- Inefficiency in maintaining prolonged UV ray protection in dynamic environments.

1.1.4 Need for Innovative Solutions

These limitations have driven the demand for innovative UV filtration technologies that are safer, more stable, and environmentally sustainable. Advances in material science, nanotechnology, and photochemistry offer promising pathways for addressing these challenges, ensuring enhanced UV protection across diverse applications.

1.2 CYCLODEXTRINS FOR ULTRAVIOLET RAYS FILTRATION

The Cyclodextrins: Molecules with a Cavity

Cyclodextrins (*cds*) are molecules that possess truncated cone shape with a cavity rendered by the chair form of 1, 4 linked α-D-glucopyranose units (Figure 1.1). The

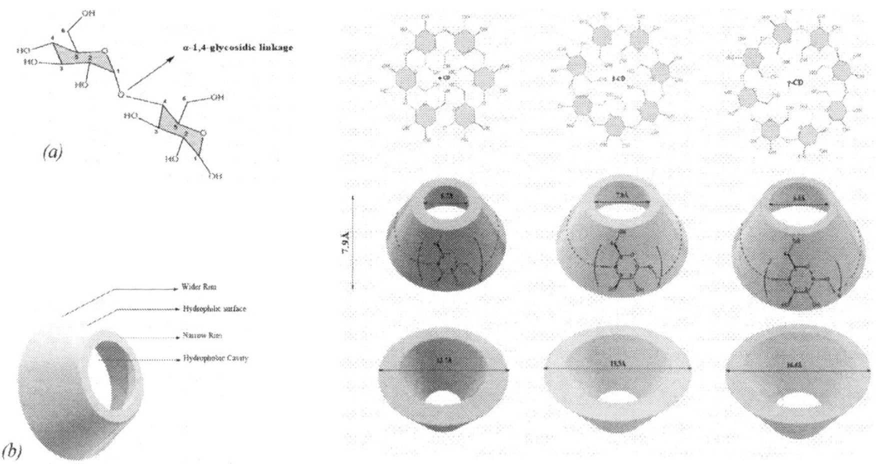

(c)

FIGURE 1.1 (a) 1,4-linked *α*-D-glucopyranose unit(s) of cyclodextrins, (b) 1° and 2° rim of cyclodextrins, (c) dimensions of *α*-, *β*-, and *γ*-cyclodextrins.

narrow rim of the *cd* cavity is called the 1° rim and the wider rim is called the 2° rim [1, 2]. *cd* molecules with six glucopyranose units are termed *α-cd*, with seven and eight glucopyranose units are termed *β-cd*, and *γ-cd*, respectively. The *α*-, *β*-, and *γ*-*cds* are called native *cds*. The cavity diameter of *α-cd* is ~5 Å, *β-cd* is ~6 Å, and *γ-cd* is 7 Å (Figure 1.1). Nowadays, more than 110 different *cds* with a different number of glucopyranose units exist in the market [3].

Cyclodextrins (CDs) play a valuable role in ultraviolet (UV) rays filtration due to their unique molecular structure and versatile properties. Their utility arises from the ability to encapsulate and stabilize UV-active compounds, which enhances the efficiency and longevity of UV filtration systems. Here are the key ways cyclodextrins contribute to UV rays filtration:

Encapsulation of UV-Active Molecules: Cyclodextrins can form host–guest complexes with UV-absorbing molecules such as organic sunscreens (e.g., avobenzone, oxybenzone). This encapsulation:

- *Prevents photodegradation:* Stabilizes the UV filters under prolonged UV exposure, extending their functional lifespan.
- *Enhances solubility:* Increases the water solubility of hydrophobic UV filters, improving their application in aqueous formulations.

Controlled Release of Active Compounds: Cyclodextrins enable the controlled release of UV-active agents, ensuring sustained protection over time [4]. This property is particularly useful in:

- Sunscreens for prolonged outdoor activities.
- Coatings and textiles exposed to sunlight.

Reduction of Toxicity: By encapsulating UV-active agents, cyclodextrins can reduce direct contact between these agents and biological or environmental systems. This minimizes potential adverse effects, such as skin irritation or leaching of harmful chemicals into water systems.

Improvement in UV Filter Efficiency: Cyclodextrins can enhance the photostability of encapsulated molecules, allowing them to maintain their UV-absorbing capacity for longer durations. This improves the overall efficiency of UV filtration products.

1.3 MECHANISM OF UV RAYS FILTRATION BY HOST–GUEST COMPLEXES OF CYCLODEXTRINS

Availability of vacant space in the cavity renders cyclodextrins (*cds*) to function as *"receiving **Host** molecule"* that can encapsulate the comparatively small *"penetrating **Guest** molecule*". Thus, the *cd*s form a **H**ost–**G**uest complex (**H–G**) through non-covalent interactions. The *cd* host holds the guest molecule rigidly upon encapsulation and alters the native functional behaviour of guest molecules. Ultraviolet radiation (UVR) filtration ability is enhanced by complexing the UVR absorbers with *cds*. In general, scientists utilize the altering ability of *cds* to enhance the native functional behaviour of molecules and hence the functional materials. Therefore, upon complexing these UVR absorbers with *cds*, the *cds* renders a rigid fit to the UVR-absorbing moieties by encapsulating in to the cavity. The rigid fit enhances the UVR-absorbing ability of and further dissipation efficiency of the UVR filtering molecules.

1.4 UVR ABSORBERS: SKELETON AND FUNCTIONING

Molecules with potential moiety/scaffold to receive UVR upon exposure and dissipate as radiative or no radiative photons could be defined as a *UVR absorbers* [5]. The photosensitive moieties like –COO–, –C=O, –OH, –N=N–, –C=N, –C=C, –NH_2, –NO_2, –COOH are responsible for the UVR absorption and dissipation. Fabric/paper/sunscreen containing UVR absorbers protect the materials by converting the bombarded UVRs into low-energy rays through radiative or non-radiative process. Dissipation of high intense UVRs occurs through non-radiative processes through release as heat, whereas the radiative process involves the release of low-energy rays. Both dissipation processes involve certain chemical reactions based on the photoreactive moieties present in the molecule [6].

1.5 DRAWBACKS ASSOCIATED WITH UVR ABSORBERS: SKIN PENETRATION AND ENVIRONMENTAL IMPACT

Although there are many *uvr-absorbing* molecules, which can filter the UVR to the maximum level, (i) using benzophenones (*BPs*) in sunscreen formulations leads to the penetration of *BPs* into the human skin, causing cancer, (ii) higher dosage of pristine *uvr absorbers* is required to develop sun protective fabrics, (iii) the *uvr absorbers* accumulate in the environment through evaporation/solubilization/volatilization/deterioration.

1.6 CYCLODEXTRINS AS A PROBLEM SOLVER

Upon analyzing the scientific reports on the performance of *cds:uvr absorbers,* we understand that it is possible to overcome the aforementioned drawbacks (due to the usage of *uvr absorbers* in pristine form) by complexing with cyclodextrins.

1.7 SIGNIFICANCE OF THIS CHAPTER (REVIEW)

Although the *cd* **H–G** complexes enhance the UVR filtration abilities of UVR absorbers, limited reports are found in scientific literature databases. Just ten scientific papers reporting the usage of four different *cds* [7–16] on 20 different UVR absorbers exist. Therefore, we believe that this review would drag the attention of researches working on photoprotective materials. The *cds* are glucose molecules that are non-toxic to humans, whereas using a high concentration of UVR absorbers in photoprotective materials may cause adverse effects to the humans and hence to the environment. Therefore, complexing the UVR-absorbing molecules with *cds* could be an environment-friendly scientific approach to reduce adverse effects. Instead of adding a higher concentration of UVR absorbers to the photoprotective materials to achieve a higher UVR filtering scale, researchers can prefer complexed UVR absorber (with *cds*). The *cd:uvr absorber* complexes exhibits higher UVR filtration parameters than the free UVR absorbers. Hence, it is possible to reduce the amount of UVR absorbers while developing photoprotective materials and hence the adverse effects.

With the aim of understanding the impact of *cds* on the native UVR filtration ability of UVR-absorbing molecules, investigative reports on **H–G** complexes of various *cds* with UVR absorbers are reviewed and discussed in this review report. Based on a complete literature search on "*H–G complexes of cd:UVR absorbers*" around ten relevant reports published up to 2021 [7–16] were retrieved, and compiled in Table 1.1, Figure 1.2. The name of the *cd*, name and structure of the UVR absorber, the material on which the UVR filtering ability is tested, and the enhancement achieved in all 12 **H–G** systems are listed in Table 1.1. *This the first report on the enhancement of UVR filtration efficiency of UVR absorbers using cyclodextrins.*

1.8 EXCLUSIONS AND INCLUSIONS OF THE SCIENTIFIC LITERATURE IN THIS REVIEW

As shown in Table 1.1, investigative reports on *cd:uvr absorber complexes* that are specifically quantified and demonstrated to enhance the UVR filtering ability as an ultraviolet protection factor or sun protection factor are listed and discussed in this review. Reports on the *cd:uvr absorber complexes* that are speculating the enhancement without any demonstrated experimental evidences are carefully verified and excluded; Table 1.2 from this review.

TABLE 1.1
***cd:uvr Absorber*-Based Host–Guest Complexes Reported in the Literature**

Name and Structure of *cd:uvr Absorber Complex*	Material	UPF of Untreated Material	UPF of Material Treated with *cd:uvr Absorber* Complex	SPF*/ UPF Enhancement in %	SPF*/ UPF Enhancement in Folds	[Ref.] (Year)
MCT *β-cd*:omc	cellulosic fabric	41.4	133.1	*221	*3.2	[7] (2006)
β-cd:4-hbp	cotton fabric	17.9	55	207	3.08	[8] (2007)
		16	38	138	2.4	[17] 2023
MCT *β-cd*:scR	cotton	7	54	671	7.7	[9]
	pet/cotton blend	30	115	283	3.8	(2010)
MCT *β-cd*:blS	cotton	7	44	430	6.3	[9]
	pet/cotton blend	30	80	167	2.7	(2010)
MCT *β-cd*:ruS	cotton	7	52	642	7.4	[9]
	pet/cotton blend	30	130	333	4.3	(2010)
MCT *β-cd*:4-hbp	cotton/wool	32	108	237	3.3	[10]
	viscose/wool	23	88	282	3.8	(2013)
MCT *β-cd*:oxn	cotton/wool	32	104	225	3.2	[10]
	viscose/wool	23	80	247	3.5	(2013)
MCT *β-cd*:TiO_2(n)	cotton/wool	32	114	256	3.6	[10]
	viscose/wool	23	92	300	4.0	(2013)
MCT *β-cd*:ZnO(n)	cotton/wool	32	122	281	3.8	[10]
	viscose/wool	23	108	369	4.7	(2013)
β-cd:2,4-dnp	pen ink	16	20.16	*19.6	*1.2	[11] (2014)

β-cd:2,4-dna	pen ink	16	20.16	*19.6	*1.2	[11] (2014)
β-cd:2,6-dna	pen ink	16	20.16	*19.6	*1.2	[11] (2014)
β-cd:2,6-dnb	pen ink	16	20.16	*19.6	*1.2	[11] (2014)
modified dextrin:avo	sunscreen	-	-	*38.9	*-	[12] (2016)
hp-γ -*cd*:tfa	sunscreen	17.9 10.3	48.9 25.8	*172 150	*2.7 2.5	[13] (2018)
β-cd: flavonoids of broad leaf holly leaf extract [myt, qct, kml, rut]	cotton fabric	18.3	42.6	132	2.3	[14] (2020)
β-cd:cou	cotton fabric	35	70	100	2.0	[15] (2020)
β-cd:mk	cotton fabric	16	46	188	2.9	[16] (2022)
β-cd:bp	cotton fabric	16	31	94	1.3	[17] 2023
β-cd:dhbp	cotton fabric	16	51	219	3.2	[17] 2023
β-cd:thbp	cotton fabric	16	39	144	2.4	[17] 2023

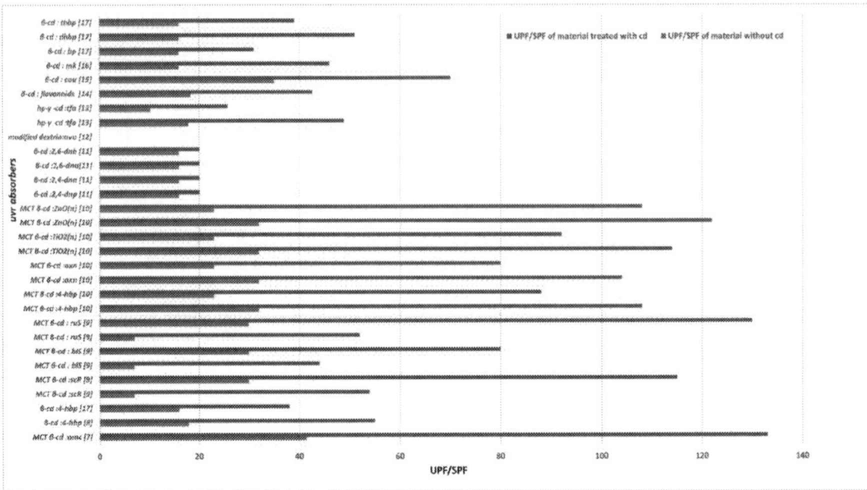

FIGURE 1.2 Enhancement of UPF/SPF of various *uvr absorbers* reported in the literature.

TABLE 1.2
Stability Enhancement by Cyclodextrins on Various Absorbers Reported in the Literature

UVR Absorber	Material	UVR-Absorbing Efficiency and Mechanism	[Ref.]
β-CD:ethylhexyl methoxycinnamate	sunscreen	β-CD enhances the photostability of ethylhexyl methoxycinnamate	[19]
β-CD:octocrylene	sunscreen	β-CD enhances the photostability of octocrylene	[19]
β-CD:2-Hydroxy-4-Methoxy Benzophenone (*oxybenzone*)	sunscreen	β-CD enhances the photostability of oxybenzone	[19]
HPβ-CD:2-Hydroxy-4-Methoxy Benzophenone (*oxybenzone*)	sunscreen	HPβ-CD enhances the photostability of oxybenzone, controls the release of oxybenzone thereby reduces the skin penetration of uv absorber.	[20]
HPβ-CD:2-Hydroxy-4-Methoxy Benzophenone (*oxybenzone*)	sunscreen	β-CD enhances the photostability of oxybenzone	[21]
HPβ-CD:butyl-methoxydibenzoylmethane	–	HPβ-CD enhances the photostability of oxybenzone.	[22]
β-CD:rutin and HP-β-CD:rutin	–	The inclusion complex of rutin with β-CD and HP-β-CD led to an increase in the rutin photostability for 2.5 and 5.4 times, respectively	[23]

TABLE 1.2 (Continued)
Stability Enhancement by Cyclodextrins on Various Absorbers Reported in the Literature

UVR Absorber	Material	UVR-Absorbing Efficiency and Mechanism	[Ref.]
α-CD: ferulic acid	sunscreen	α-CD enhances the photostability of ferulic acid	[24]
Me-β-CD: quercetin	–	Me-β-CD enhances the photostability of quercetin	[25]
HP-β-CD:phenylbenzimidazole sulphonic acid (PBSA)	sunscreen	HP-β-CD enhances the photostability of PBSA	[26]
2HP-β-CD:quercetin	–	2HP-β-CD enhances the photostability of quercetin	[27]

1.9 CHRONICLE DEVELOPMENT OF CD:UVR ABSORBER COMPLEXES

Achievement of a 3.2-fold enhancement in the sun protection factor (SPF) of cotton fabric with monochlorotriazinyl-β-cd (host):octyl methoxycinnamate (guest) in 2006 by Santo Scalia and coworkers (Italy) is the first report on the use of *cds* as an enhancer for UVR filters [7]. In 2007, El-Tahlawy et al. used *β-cd* for enhancing the ultraviolet protection factor of 4-hydroxy benzophenone and tested on cotton fabric. El-Tahlawy et al. achieved 45% of UPF enhancement on cotton fabric using *β-cd: 4 hbp* complex [8]. Since then, *β-cd* has been used for enhancing the UPF of myricetin, quercetin, kaempferol, rutin [14], coumarine [15], michler's ketone [16] on cotton fabric and monochlorotriazine-*β-cd* has been used for enhancing the UPF ability of scarlet red, brilliant blue, rubbine 3B [9], 4-hydroxy benzophenone [10] on polyester/cotton/wool/blend fabric from 2010 to 2020. Apart from fabric materials *β-cd* have been used for enhancing the SPF of pen ink by complexing the 2,4-dinitrophenol, 2,4-dinitroaniline, 2,6-dinitroaniline, 2,6-dinitrobenzoic acid [11], modified dextrin complex of avobenzene in sunscreen [12], hp-*γ-cd*: trans ferulic acid complex on glass plate [13].

1.10 CYCLODEXTRINS FOR ENHANCING THE UPF OF FABRICS

Santo Scalia and coworkers have used monochlorotriazinyl-*β*-cyclodextrin (MCT- *β-cd*) for the surface modification of tensile cellulose fabric and used octyl methoxy cinnamate (*omc*) for forming an inclusion complex with the host grafted onto fabric. The sun protection factor of *omc* had increased upto 3.2-fold upon complexing with MCT-*β*-cd.

The SPF of untreated fabric is 41.4, the SPF of fabric grafted with MCT-*β*-cd is 47.4, the SPF of OMC is 60.3, and MCT-*β-cd*:OMC is 133.1 [7].

El-Tahlawy et al. effectively raised the native UPF of 4-hydroxy benzophenone (*4-hbp*) by forming an inclusion complex with a *β*-cd host. Additives such as citric acid and sodium hypophosphite were used for the strong binding of *β*-cd:*4-hbp* on the cotton fabric. The authors have used equal concentration of *β*-cd and *4-hbp* (0.25 g) to optimize the UPF of cotton fabric. The complex exhibits significant UPF performs on the fabric even after 30 washes [8].

Ibrahim et al. used MCT-*β*-cd for enhancing the UPF of three different dye molecules. MCT-*β*-cd molecules were grafted onto cotton cellulose fabric and then the dye molecules (dye) were entrapped into the host cavity. Based on their investigations, the authors have proposed that the efficiency of UVR protection by the *β*-cd: dye complexes depend on the concentration of additives, temperature and pH. The three dyes, namely scartlet red, blue-S, rubbine-S, are complexed with *β*-cd for testing on cotton and pet/cotton blend. The *β*-cd: dye(s) complexes exhibit UPF enhancement in the following order, (scarlet red) 7.7 > (rubbine-S) 7.4 > (blue-S) 6.3 (cotton) and (rubbine-S) 4.3 > (scarlet red) 3.8 > (blue-S) 2.7 folds (pet/cotton blend) [9].

In another report by Ibrahim et al., MCT-*β*-cd has been used for enhancing the UVR-absorbing ability of 4-hbp, oxalinilide, TiO_2 and ZnO on cotton/wool and viscose/wool fabrics. These four complexes of MCT-*β*-cd exhibit UPF enhancements of 2.5-fold to 3.8-fold on both cotton/wool and viscose/wool blended fabric. The assessed UPF value reveals that the inorganic UVR absorbers – *TiO_2, ZnO* – are more effective than organic UVR absorbers – *4-hbp, oxalanilide*. The increasing order of UVR protection efficiency is oxalinilide < 4-hbp < TiO_2 < ZnO. The UPF of cotton/wool, viscose/wool blended fabric is durable for more than 15 washes [10].

β-cd grafted cotton fabric complexed with flavonoids of broad leaf holly leaf extract enhances the UPF from 30.1 to 50.7. The extracts of broad leaf holly leaf consist of four flavanoids, mainly myricetin, quercetin, kaempferol, rutin. Although the flavonoids possess the UVR filtering ability, upon complexing with β-cd the UPF of flavonoid extracts of broad leaf holly leaf enhances up to 68.4% [14]. The inclusion complex of *β-cd* with natural coumarins extracted from *cortex fraxini* improves the UPF of cotton fabric from 18.3 to 75. *Cortex fraxini* is a Chinese medicinal herb. After 30 washes the UPF reduces to 33.2, which lies in the good protection range [15]. Sivakumar and Nalini et al. used *β-cd: Michler's ketone* complex to enhance the UVR filtering efficacy of poplin cotton fabric from UPF 16 to 46 [16]. In recent research, Sivakumar and Nalini et al. tested the UPF of benzophenone and benzophenone with 1 –OH, 2 –OH and 3 –OH groups on poplin cotton fabric by complexing with β-cd. The authors have achieved the highest UPF of 51 upon using β-cd:dihydroxy benzophenone on cotton fabric [17]. The authors have demonstrated the dissipation of UVR as a *(i) non-radiative twisted intramolecular charge transfer* relaxation process in the case of *β-cd: Michler's ketone* complex [16], and *(ii) keto-enol tautomerism* in the case of β-cd:hydroxyl substituted benzophenones [17].

1.11 CYCLODEXTRINS FOR ENHANCING THE SPF OF PEN INK

A series of dinitro aromatic compounds with additional groups such as –OH, –COOH, $–NH_2$, where tri-substituted benzenes are complexed with *β-cd* and tested for the UVR filtration ability with blue ink used in ball point pen. All four tri-substituted benzenes, namely 2,4-dnp, 2,4-dna, 2,6-dna, and 2,6-dnb added blue ink exhibits 19.6% enhancement of SPF [11]. 2,4-dinitrophenol (2,4-dnb), 2,4-dinitroaniline (2,4-dna), 2,6-dinitroaniline (2,6-dna), and 2,6-dinitrobenzoic acid (2,6-dnb) has appeared for the UV absorption bands in a different wavelength region below 400 nm, a combination of these dinitro aromatic compounds gave the broad absorption spectra within the UV region. The absorption intensities have been increased by preparation of the inclusion complex of dinitro compounds with *β-cd*. Prepared inclusion complexes are used to improve the UV protection properties of the ball point pen ink against photo degradation. The formation of solid inclusion complexes was characterized by FT-IR, and 1H NMR spectroscopy. The UV-protecting properties of these inclusion complexes were calculated and their sun protection factor (SPF) is also discussed. The stability of the ballpoint pen ink has been confirmed by the UV–visible spectroscopic method.

1.12 CYCLODEXTRINS FOR ENHANCING THE SPF OF SUNSCREEN FORMULATIONS

Li et al. reported the UVR filtration ability of a sunscreen emulsion based on the inclusion complex of modified dextrin and avobenzene [1-(4-methoxyphenyl)-3-(4-tert-butylphenyl) propane-1,3-dione]. The reported avobenzene complex exhibits 38.9% of enhanced SPF than the pristine form. Moreover, the authors have reported that upon forming inclusion complex with dextrins the sunscreen emulsion becomes less penetrative into the skin. This lesser penetration is ascribed to the protection rendered by the hydrophobic surface of the dextrin molecule to the entrapped avobenzene molecule from diffusing into a hydrophilic medium. Thus, the authors have achieved the enhancement of SPF as well as the suppression of skin penetration of avobenzene [12]. Teruaki Mori et al. tested the UVR filtration ability of hp-*γ*-*cd*:trans ferulic acid on a glass plate, they found 40.1% enhancement in the SPF and 60.1% enhancement in the UPF of trans ferulic acid. The authors claimed that the reported ferulic acid could be derived from natural sources like rice bran. Moreover, the hp-*γ-cd*: trans ferulic acid complex shows 8.3% of higher sun-cut than synthetic UVR filter octylmethoxy cinnamate [13].

1.13 IMPORTANT GENERALIZATIONS

In general, cyclodextrins enhance the native UVR filtration ability of UVR absorbers. However, the percentage of enhancement depends on various factors such as type of cyclodextrin, additives used and the application materials: *fabric/sunscreen cream/paper board, etc.*

UVR filtration by cds:uvr absorber complex: All 20 "*cds:uvr absorber complex*" systems clearly demonstrate the UVR filtration enhancement occurs due to the inclusion of "*guest-UVR absorbers*" into cyclodextrin "*host-enhancers*". That is, the rigid fit gained by the "*guest-UVR absorbers*" upon complexing with *cds* enhances the native UVR dissipation ability [7–17].

UVR absorbers: Out of 20 *uvr absorbers* listed, around 12 systems are dissipating the UVR through carbonyl (–C=O) moiety [1, 2, 4, 6–10]. Few uvr absorbers with imine (C=N) and (N=N) moieties are also reported [3, 5]. Researchers have used various additives as a matrix to improve the performance of these moieties. And, only two reports are based on the natural plant extract (broad leaf holly leaf extract and *cortex fraxini*) [7, 8]. Hence, the UVR filtration ability of the variety of synthetic/natural organic/inorganic molecules has to be tested and investigated with cyclodextrin.

Cyclodextrins: In the literature, only four cyclodextrin analogues namely *β-cd* [2, 5, 9, 10], monochlorotriazinyl-β-cd [3, 4], hydroxypropyl-*γ-cd* [7], and *a* modified dextrin [6] are used by researchers for enhancing the UVR filtration ability of uvr absorbers. To the best of our knowledge, there are more than 100 types of cyclodextrins to choose to further optimize the performance of *uvr absorbers* [3].

Materials: The "*cds:uvr absorber complexes*" are tested on three application materials namely *fabric* [1–4, 8–10], *pen ink* [5], *and sunscreen emulsion* [6, 7]. Out of ten reports, eight are based on the UVR filtration of fabric material. And, there is a great scope for testing on different materials.

UVR filtration mechanisms: Although, the enhancement of native UVR filtration of the "*uvr absorber*" is demonstrated as an impact of rigid fit rendered by *cds*. The UVR dissipation/filtration mechanism of pristine "*uvr absorbers*" are not investigated and demonstrated clearly in many reports. Hence, it should be researched in detail.

1.14 FUTURE OUTLOOK

Material science vs molecular chemistry: Researchers have identified and enhanced the UVR filtering potential of different "*uvr absorbers*" on different materials through a trial–error approach. The molecular chemistry aspect of UVR dissipation is not investigated. Hence, understanding the molecular chemistry of UVR dissipation by these molecules would be useful for designing and synthesizing an effective "*uvr absorber*" system.

Other side of uvr absorbers: challenges with uvr absorbers vs benefits of using with cds: At present, the chemical route (i.e., coating "*uvr absorbers*") is the ultimate method for achieving the threshold UPF/SPF of finished materials with particular physical specification. "*uvr absorbers*" are effective in the dissipation of UVR and enhancement of UPF/SPF of materials on which they were coated. But, UVR protection is gained through chemical means. In order to achieve the threshold UPF/SPF of materials, higher concentrations of "*uvr absorbers*" are generally used. Usage of higher concentrations of "*uvr absorbers*" causes human toxicity and adverse effects to the environment due to the physical and chemical instability [18]. Upon complexing with *cds* (cyclodextrins), the threshold UPF/SPF of materials could be achieved with lower concentrations of "*uvr absorbers*". In addition to that, the "*uvr absorbers*" in a complexed state with *cds* exhibits high photostability (Table 1.2).

Recommendations: Cyclodextrins as enhancers for uvr absorbers: We suggest that cyclodextrins could serve as an effective enhancer for the *uvr absorbers*. Hence, upon using the *uvr absorbers* in the complexed form with suitable cyclodextrins, the concentration of *uvr absorbers* could be reduced and the photo stability of *uvr absorbers* could be improved (Table 1.2). Ultimately, the accumulation of *uvr absorbers* in the environment could be reduced by using it in the complexed form with *cds*.

1.15 AN OVERVIEW OF THE CONTENT OF THE BOOK

Chapter 1: Cyclodextrins for Enhanced Ultraviolet Rays Filtration: Principles and Applications – This chapter provides an in-depth exploration of ultraviolet (UV) rays filtration, emphasizing the harmful effects of UV radiation on human health and materials, alongside the need for efficient filtration technologies. It examines current methods, including physical and chemical filters, highlighting their limitations, such as instability and environmental concerns. The unique properties of cyclodextrins (*cds*) are analyzed, showcasing their ability to encapsulate UV-active molecules, enhance photostability, and reduce toxicity. A review of limited yet significant studies demonstrates the potential of *cd*-hosted complexes in improving UV filtration efficiency. The chapter concludes by advocating for the broader adoption of *cds* to achieve sustainable and enhanced UV protection.

Chapter 2: Cyclodextrins as Encapsulants for UV-Ray Absorber Complexes in Sunscreen Cosmetics – This chapter focuses on examining the efficient utilization of UVR filters through inclusion complexes involving onion peel extracts, free onion peel extracts, avobenzone, and octinoxate complexed with β-CD.

Chapter 3: Cyclodextrins for Enhancing the Photostability of UVR Absorbers – This chapter explores the role of CDs in enhancing the photostability and water solubility of UVR absorbers and antioxidants, mitigating degradation caused by sunlight and oxygen. Key topics include the mechanisms of inclusion complex formation, preparation methods, characterization, and factors influencing stability. The chapter also highlights the effectiveness of CD-based inclusion complexes in sunscreen formulations and recent advances in improving UV protection.

Chapter 4: Cyclodextrins Unveiled: Revolutionizing UV Protection Across Diverse Sectors – This chapter highlights the versatile role of cyclodextrins as UV filters across various industries, including food, agriculture, pharmaceuticals, and cosmetics. It discusses their unique ability to form inclusion complexes with UV-absorbing compounds, enhancing stability, solubility, and bioavailability while protecting against degradation from light and air. The chapter explores their mechanisms of action and significant applications, such as improving sunscreen efficacy, safeguarding UV-sensitive drugs, protecting food items, and advancing agricultural products. By improving the durability and effectiveness of UV protection formulations, cyclodextrins are revolutionizing solutions to combat the harmful effects of UV radiation.

Chapter 5: Cyclodextrin-Modified Antioxidants as UV-Ray Absorbers – This chapter discusses the role of cyclodextrins (CDs) in enhancing the efficacy of natural antioxidants as UV-ray absorbers. It highlights the ability of CDs to form inclusion complexes with poorly soluble antioxidants, improving their solubility, stability, and controlled release onto the skin. The chapter explores a range of antioxidant-CD complexes, including quercetin, rutin, ferulic acid, and curcumin, among others, emphasizing their applications in cosmetics and food sectors. By enhancing biological activities such as anticancer effects, CDs contribute to the development of advanced

UV protection formulations. This work underscores the transformative potential of CDs in improving antioxidant-based UV-ray absorption solutions.

Chapter 6: Innovative Utilization of Cyclodextrins for Advanced UVR Filtration – This chapter summarizes the innovative use of cyclodextrins (CDs) for advanced UV radiation (UVR) filtration, highlighting their ability to enhance the photostability, solubility, and lifespan of UVR absorbers through inclusion complex formation. Various synthesis methods, such as co-precipitation, kneading, and freeze-drying, are discussed for their efficiency. Applications span pharmaceuticals, cosmetics, food packaging, and UV-protective coatings, providing enhanced stability and prolonged protection. Advantages like controlled release and challenges such as production costs and scalability are addressed. Future perspectives focus on green chemistry, nanotechnology, and smart materials. CDs offer significant advancements for sustainable UVR filtration technologies.

Chapter 7: Unlocking the Potential of Cyclodextrins: Advantages, Applications, and Cyclodextrin-UVR Absorber Complexes in Pharmaceutical Formulations – This chapter discusses the potential of cyclodextrins (CDs) in pharmaceutical formulations, emphasizing their toroidal structure with a hydrophilic exterior and lipophilic cavity, ideal for binding drugs. Their biocompatibility and ability to form stable complexes enhances drug solubility, permeability, and bioavailability. CDs are particularly effective when paired with UVR absorbers, stabilizing their photoprotective activity. The chapter highlights the physicochemical properties of CDs, such as aqueous solubility and flexibility, which make them valuable drug carriers. Their role in optimizing formulations and advancing drug delivery systems is explored in detail. CDs' multifaceted advantages offer significant advancements in pharmaceuticals.

Chapter 8: Applications of UVR Absorbers: Cyclodextrin Complexes – This chapter explores the versatile applications of cyclodextrin-UV absorber complexes for effective UV protection across various industries. It highlights cyclodextrins' role in enhancing the solubility, stability, and controlled release of UV-absorbing compounds, which is pivotal for formulating efficient sunscreens. The use of cyclodextrin complexes as UV blockers in textiles, particularly outdoor clothing, is also discussed. The chapter emphasizes their potential to improve sunscreen efficacy and protect against skin cancer by stabilizing antioxidants and UV filters. It concludes with a call for further research to unlock innovative applications for UV protection, contributing significantly to advancements in this field.

Chapter 9: Cyclodextrins in UV Protection: Advanced Absorbers for Versatile Materials – This chapter explores the potential of cyclodextrins (CDs) as advanced UV filters, focusing on their ability to form inclusion complexes that enhance the stability, solubility, and UV-absorbing capacity of UV absorbers. It reviews CD-based UV filters in sunscreen formulations, including encapsulated organic and inorganic UV absorbers and CD-based nanocomposites. Key advantages such as increased UV protection, reduced toxicity, and environmental benefits are highlighted, along with safety and efficacy considerations. The chapter concludes with future research directions, emphasizing CDs as a promising and innovative approach to improving UV filtration in sunscreen applications.

Chapter 10: Cyclodextrins-UVR-Absorbing Molecular Complexes for Photodynamic Therapy – This chapter emphasizes the role of cyclodextrin-UVR-absorbing molecular complexes in enhancing photodynamic therapy (PDT). PDT's effectiveness is often hindered by the poor solubility, stability, and off-target toxicity of photosensitizers. Cyclodextrins address these challenges by forming inclusion complexes that improve solubility, stability, and targeted delivery. Encapsulation also protects photosensitizers from premature degradation and ensures precise activation by specific light wavelengths. This innovative approach enhances PDT's therapeutic efficiency and safety. Further research is recommended to optimize these complexes for clinical use and explore broader therapeutic applications.

Chapter 11: Cyclodextrins: Revolutionizing Sunscreen Performance and Protection – This chapter highlights the transformative role of cyclodextrins (CDs) in enhancing sunscreen performance and UV protection. CDs improve the photostability, solubility, and controlled release of UVR absorbers like oxybenzone and avobenzone by forming inclusion complexes. It discusses the use of α-, β-, and γ-CDs, along with their modified forms, in optimizing SPF and UPF properties. The chapter also explores advanced formulation techniques incorporating CD-UVR absorber complexes into delivery systems. These innovations enhance sunscreen longevity and effectiveness, offering insights for improved commercial products. Future trends in sun protection research and applications are also analyzed.

Chapter 12: Cyclodextrin-Enhanced UVR Absorbers: Versatile Solutions for Fabrics, Polymers, and Beyond – This chapter explores the role of cyclodextrins (CDs) in enhancing UVR filtration systems across diverse materials, including fabrics, polymers, glass, and films. By forming stable inclusion complexes, CDs improve the solubility, stability, and UV absorption efficiency of organic, inorganic, and nanoparticle-based UVR absorbers. Organic absorbers benefit from reduced photodegradation, while inorganic absorbers like metal oxides gain improved dispersion and substrate compatibility. Nanoparticle-based absorbers achieve superior UV protection and enable advanced multi-layered coatings. The chapter also outlines a rigorous literature review methodology used to analyze existing research and highlights the potential of CDs in developing durable, high-performance UV-blocking materials.

Chapter 13: Impact of Cyclodextrins on the Ultraviolet/Sun Protection Factors (UVPF/SPF) of UVR Absorbers – This chapter highlights the transformative role of cyclodextrins (CDs) in enhancing the ultraviolet/sun protection factors (UVPF/SPF) of UVR absorbers. By forming inclusion complexes, CDs improve the solubility, stability, and photoprotective efficacy of compounds like oxybenzone, avobenzone, and octylmethoxy cinnamate in sunscreen formulations. Beyond cosmetics, CDs are employed for UV protection in materials like poplin cotton fabric and ballpoint pen ink. These enhancements are attributed to improved interaction and skin penetration of UVR absorbers. The chapter underscores the potential of CDs in advancing UV protection while advocating for further research to optimize their application in diverse fields.

Chapter 14: Enhancing UVR Absorber Efficacy with Cyclodextrins in Cosmetic and Non-Cosmetic Materials – This chapter highlights the transformative role of cyclodextrins (CDs) in enhancing the ultraviolet/sun protection factors (UVPF/SPF) of UVR absorbers. By forming inclusion complexes, CDs improve the solubility, stability, and photoprotective efficacy of compounds like oxybenzone, avobenzone, and octylmethoxy cinnamate in sunscreen formulations. Beyond cosmetics, CDs are employed for UV protection in materials like poplin cotton fabric and ballpoint pen ink. These enhancements are attributed to improved interaction and skin penetration of UVR absorbers. The chapter underscores the potential of CDs in advancing UV protection while advocating for further research to optimize their application in diverse fields.

REFERENCES

1. Lai W. F., Rogach A. L., and Wong W. T. (2017) 'Chemistry and engineering of cyclodextrins for molecular imaging.' *Chem. Soc. Rev.*, Vol 46, pp 6379–6419.
2. Jambhekar, S. S., and Breen, P. (2016) 'Cyclodextrins in pharmaceutical formulations I: Structure and physicochemical properties, formation of complexes, and types of complex.' *Drug. Discov. Today*, Vol 21, pp 356–362.
3. Suárez, D., and Díaz, N. (2017) 'Conformational and entropy analyses of extended molecular dynamics simulations of α-, β-and γ-cyclodextrins and of the β-cyclodextrin/nabumetone complex.' *Phys. Chem. Chem. Phys.*, Vol 19, pp 1431–1440.
4. Zhang, J., and Ma, P. X. (2013) 'Cyclodextrin-based supramolecular systems for drug delivery: Recent progress and future perspective.' *Adv. Drug Deliv. Rev.*, Vol 65, pp 1215–1233.
5. Saravanan, D. (2007) 'UV protection textile materials.' *Autex Res. J.*, Vol 7, pp 53–62.
6. Slavov, C., Yang, C., Schweighauser, L., Wegner, H. A., Dreuw, A., and Wachtveitl, J. (2017) 'Ultrafast Excited-State Deactivation Dynamics of Cyclotrisazobenzene—A Novel Type of UV-B Absorber.' *Chem. Phys. Chem.*, Vol 18, pp 2137–2141.
7. Scalia, S., Tursilli, R., Bianchi, A., Nostro, P. L., Bocci, E., Ridi, F., and Baglioni, P. (2006) 'Incorporation of the sunscreen agent, octyl methoxycinnamate in a cellulosic fabric grafted with β-cyclodextrin.' *Int. J. Pharm.*, Vol 308, pp 155–159.
8. El-Tahlawy, K., El-Nagar, K., and Elhendawy, A. G. (2007) 'Cyclodextrin-4 Hydroxy benzophenone inclusion complex for UV protective cotton fabric.' *J. Text. Inst.*, Vol 98, pp 453–462.
9. Ibrahim, N. A., E-Zairy, W. R., and Eid, B. M. (2010) 'Novel approach for improving disperse dyeing and UV-protective function of cotton-containing fabrics using MCT-β-CD.' *Carbohydr. Polym.*, Vol 79, pp 839–846.
10. Ibrahim, N. A., El-Zairy, E. M. R., Abdalla, W. A. and Khalil, H. M. (2013) 'Combined UV-protecting and reactive printing of Cellulosic/wool blends.' *Carbohydr. Polym.*, Vol 92, pp 1386–1394.
11. Srinivasan, K., Radhakrishnan, S., and Stalin, T. (2014) 'Inclusion complexes of β-cyclodextrin-dinitrocompounds as UV absorber for ballpoint pen ink.' *Spectrochim. Acta A Mol. Biomol. Spectrosc.*, Vol 129, pp 551–564.
12. Li, C. C., Lin, L. H., Lee, H. T., and Tsai, J. R. (2016) 'Avobenzone encapsulated in modified dextrin for improved UV protection and reduced skin penetration.' *Chem. Pap.*, Vol 70, pp 840–847.

13. Mori, T., Tsuchiya, R., Doi, M., Nagatani, N., and Tanaka, T. (2019) 'Solubilization of ultraviolet absorbers by cyclodextrin and their potential application in cosmetics.' *J. Incl. Phenom. Macrocycl. Chem.*, Vol 93, pp 91–96.
14. Liu, J., Ma, X., Shi, W., Xing, J., Ma, C., Li, S., and Huang, Y. (2020) 'Anti-ultraviolet properties of β-cyclodextrin-grafted cotton fabrics dyed by broadleaf holly leaf extract.' *Text. Res. J.*, Vol 90, pp 2441–2453.
15. Liu, J., Ma, X., Shi, W., and Xing, J. (2021) 'Evaluation of enhanced UV protection property of dyed cotton fabrics based on inclusion complex of β-cyclodextrin with natural coumarin extracted from Cortex fraxini.' *Fibers Polym.*, Vol 22, pp 1569–1579.
16. Krishnamoorthy, S., Allabasha, N., Mani, M. K., and Sarkar, A. K. (2022) 'Michler's Ketone: β-Cyclodextrin Host–Guest Inclusion Complex for Enhancing the Ultraviolet Protection Factor of Poplin Cotton Fabric.' *Photochem. Photobiol.*, Vol 98, pp 1284–1292.
17. Krishnamoorthy, S., Allabasha, N., Mani, M. K., and Sarkar, A. K. (2023) 'Effect of β-cyclodextrin on benzophenones and its impact on UVR filtration of cotton fabric through 'guest absorber—host enhancer'mechanism.' *J. Incl. Phenom. Macrocycl. Chem.*, Vol 103, pp 131–145.
18. Nohynek, G. J., and Schaefer, H. (2001) 'Benefit and risk of organic ultraviolet filters.' *Regul. Toxicol. Pharmacol.*, Vol 33, pp 285–299.
19. Al-Rawashdeh, N. A., Al-Sadeh, K. S., and Al-Bitar, M. B. (2013) 'Inclusion complexes of sunscreen agents with β-cyclodextrin: Spectroscopic and molecular modeling studies.' *J. Spectrosc.*, Vol 2013, p 841409.
20. Felton, L. A., Wiley, C. J., and Godwin, D. A. (2004) 'Influence of cyclodextrin complexation on the in vivo photoprotective effects of oxybenzone.' *Drug. Dev. Ind. Pharm.*, Vol 30(1), pp 95–102.
21. Sarveiya, V., Templeton, J. F., and Benson, H. A. (2004) 'Inclusion complexation of the sunscreen 2-hydroxy-4-methoxy benzophenone (oxybenzone) with hydroxypropyl-β-cyclodextrin: Effect on membrane diffusion.' *J. Incl. Phenom. Macrocycl. Chem.*, Vol 49, pp 275–281.
22. Scalia, S., Villani, S., Scatturin, A., Vandelli, M. A., and Forni, F. (1998) 'Complexation of the sunscreen agent, butyl-methoxydibenzoylmethane, with hydroxypropyl-β-cyclodextrin.' *Int. J. Pharm.*, Vol 175(2), pp 205–213.
23. Fenyvesi, É., Otta, K., Kolbe, I., Novák, C., and Szejtli, J. (2004) 'Cyclodextrin complexes of UV filters.' *J. Incl. Phenom. Macrocycl. Chem.*, Vol 48, pp 117–123.
24. Molinari, A., Tursilli, R., Casolari, A., Maldotti, A., and Scalia, S. (2006) 'Influence of complexation with cyclodextrins on photo-induced free radical production by the common sunscreen agents octyl-dimethylaminobenzoate and octyl-methoxycinnamate.' *Pharmazie*, Vol 61, pp 41–45.
25. d'Agostino, S., Azzali, A., Casali, L., Taddei, P., and Grepioni, F. (2020) 'Environmentally friendly sunscreens: Mechanochemical synthesis and characterization of β-CD inclusion complexes of avobenzone and octinoxate with improved photostability.' *ACS Sustain. Chem. Eng.*, Vol 8, pp 13215–13225.
26. Scalia, S., Villani, S., and Casolari, A. (1999) 'Inclusion Complexation of the Sunscreen Agent 2-Ethylhexyl-p-dimethylaminobenzoate with Hydroxypropyl-β-cyclodextrin: Effect on Photostability.' *J. Pharm. Pharmacol.*, Vol 51, pp 1367–1374.
27. Carlotti, M. E., Sapino, S., Ugazio, E., and Caron, G. (2011) 'On the complexation of quercetin with methyl-β-cyclodextrin: Photostability and antioxidant studies.' *J. Incl. Phenom. Macrocycl. Chem.*, Vol 70, pp 81–90.

2 Cyclodextrins as Encapsulants for UV-Ray Absorber Complexes in Sunscreen Cosmetics

K Ganesh Kadiyala, Madhavi Vemula, Rama Swamy Guttula, and K. Pavan Krishna

2.1 INTRODUCTION

The human body requires sunlight exposure to manufacture vitamin D and increase the synthesis of β-endorphin. Consequently, individuals are recommended to allocate at least 20 minutes every day to outside activities. However, if the skin is left unprotected and exposed to the sun for a long time, it can lead to harmful consequences such as severe sunburn, excessive skin darkening, and the formation of skin cancer [1–3].

Sun protection is extensively utilized and highly regarded for its significance in safeguarding the skin, as it aids in the prevention of skin cancer and photoaging, as well as mitigating the inflammatory consequences of solar radiation. UV radiation can cause both short-term symptoms and long-term problems when it enters the stratum corneum [4, 5]. It speeds up the process of skin aging by encouraging the creation of reactive oxygen species. Moreover, it can result in DNA harm, skin redness, inflammation, deepening of pigmentation, inhibition of the immune system, decreased synthesis of vitamin D, and the formation of cancer [6–8]. Efficiently safeguarding the skin during sun exposure is therefore of paramount significance.

Applying sunscreen to the skin before sun exposure optimizes the advantages and reduces the negative effects. Consequently, sunscreens have developed as highly coveted and widely used skincare products worldwide. Sunscreens are effective when they comprise UV filters that can counteract both the types of UV light (type B and type A) UV. Multiple artificial UV filters can be found on the market and have been approved by multiple government regulatory agencies. UV filters can be categorized into two primary groups: inorganic filters, which reflect UV light, and organic filters, which absorb UV radiation.

Furthermore, when a fraction of the active ingredients present in a sunscreen is absorbed by the outermost layer of the skin, called the stratum corneum, the efficacy and protective capacity of the sunscreen diminish. This is because the same level of

 DOI: 10.1201/9781003492757-2

UV filter concentration is not sustained during the duration of use. In addition, the product should have a simple application process, be non-irritating, let visible light to pass through (i.e., not visible on the skin), and meet the requirements set by the local regulatory agency for approval [6–8]. Ultimately, it is crucial that the substance does not pose any harm to the environment or wildlife when it is released into natural bodies of water.

A major concern regarding organic UV filters is their lack of photostability. Studies have shown that these filters can experience cis-trans isomerization, keto-enol tautomerization, or fragmentation reactions when exposed to UV radiation [9–11]. As a result, their effectiveness diminishes. Moreover, these filters might potentially elicit unpleasant reactions or be chemically reactive with the skin [12–14].

2.2 STRUCTURE OF SKIN

The skin consists of three primary layers: the epidermis, dermis, and subcutaneous layer. The epidermis consists of both non-viable and viable layers, with a thickness ranging from 50 to 100 micrometers. The non-viable epidermis, also known as the stratum corneum, is the outermost layer of the skin that acts as the primary barrier to prevent the penetration of drugs [15]. The functional portion of the epidermis, measuring around 70 μm in thickness, is comprised of four distinct layers: the uppermost layer known as the stratum lucidum, followed by the stratum granulosum, stratum spinosum, and stratum basale [16]. The stratum basale is the deepest layer that is next to the dermis and is made up of many types of cells. The skin contains several types of cells, including keratinocytes, which produce the protective protein keratin, melanocytes, responsible for creating melanin, Langerhans cells, which are the primary antigen-presenting cells (APC), and Merkel cells, which contain sensory neurons and contribute to the perception of touch and pain [17]. The epidermis contains the bulk of skin melanomas, which can be classified into two primary categories: non-malignant melanoma and malignant melanoma. Basal cell carcinoma (BCC) and squamous cell carcinoma (SCC) are the two most prevalent types of non-melanoma skin malignancies, with BCC accounting for around 75% of occurrences. Both BCC and SCC specifically target the keratinocytes in the outermost layer of the skin, known as the epidermis [18]. The dermis, with a thickness ranging from 3 to 5 mm, constitutes the main constituent of the skin. The dermis is very metabolically active and functions as the main site for wound healing. Therefore, it is essential for most drugs to permeate this layer to provide therapeutic effects. Furthermore, the dermis contains a diverse range of enzymes, such as esterases, peptidases, and hydrolases, that augment the metabolic function of the skin and reduce the accessibility of drugs for absorption via the skin [19, 20].

2.3 SKIN PROTECTION FROM UV LIGHT

UV filters provide crucial protection against harmful UVA (315–400 nm) and UVB (280–315 nm) radiation, which can cause tanning, premature aging, and skin cancer [21, 22]. Their effectiveness and potential adverse effects depend significantly on their chemical composition. Organic sunscreens are primarily composed of aromatic carbon or unsaturated bonds that absorb UV light. Upon absorption, UV radiation excites electrons in these filters, which then release energy as minimal thermal energy, preventing direct damage to the skin [23, 24]. The electrons subsequently return to their stable state, allowing the process to repeat.

Photostability is a critical factor for UV filters, as exposure to UV radiation can lead to photodegradation, resulting in isomers and reactive oxygen species (ROS). Over time, these processes can reduce the concentration and efficacy of UV filters. Organic UV filters are categorized based on their chemical structure. Benzophenones, including oxybenzone, dioxybenzone, and sulisobenzone, consist of aromatic methoxy groups attached to a benzophenone core. Avobenzone (AVO) shares structural similarities with benzophenones but differs due to its unique aromatic ring linkage.

Other organic UV filters (Table 2.1) include aromatic esters such as salicylates, cinnamates, meradimate, octocrylene, and aminobenzoates. In contrast, UV filters like ensulizole and ecamsule lack aromatic esters or methoxy groups but contain sulfonic acid groups, similar to sulisobenzone. These structural variations influence their protective properties and photostability.

Absorption and reflection or scattering are two ways that the inorganic UV filters interact with the UV light. ZnO and TiO_2 are both semiconductors capable of light absorption. According to Cole et al. (2016), inorganic UV filters absorb around 85 to 95% of UV radiation (Table 2.2) [25]. Like organic UV filters, UV radiation stimulates an electron to transition to a higher-energy orbital state [26].

TABLE 2.1
Chemical Structures of Organic UV Filters

ANTHRANILATE	**Meradimate**	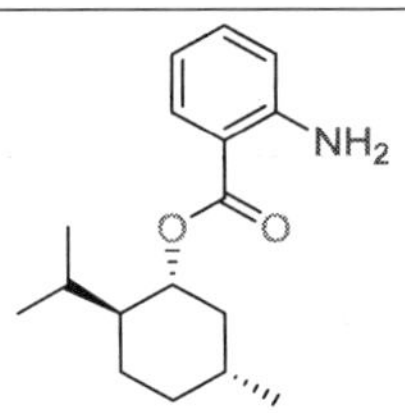
AMINOBENZOATES	**Amino benzoic acid**	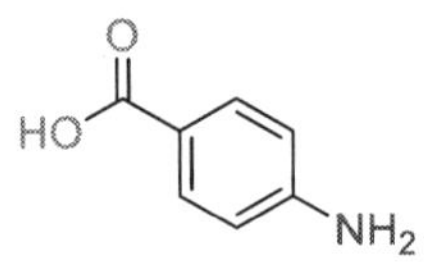
	Padimate O	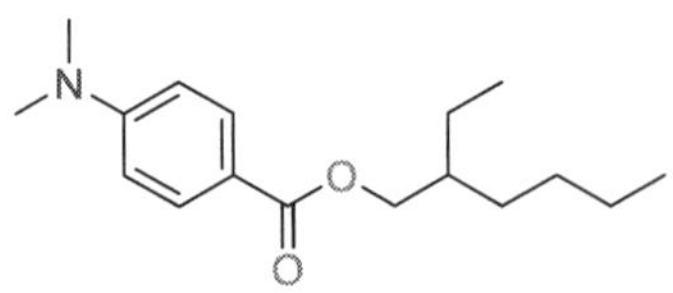

TABLE 2.1 (Continued)
Chemical Structures of Organic UV Filters

BENZIMIDAZOLE	**Ensulizole**	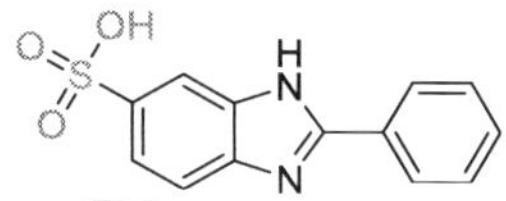
BENZOPHENONES	**Dioxybenzone**	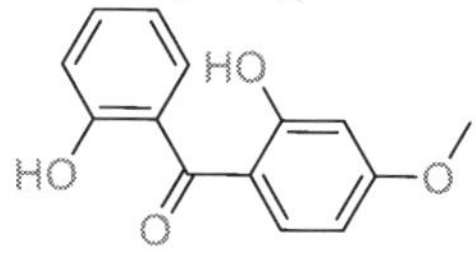
	Oxybenzone	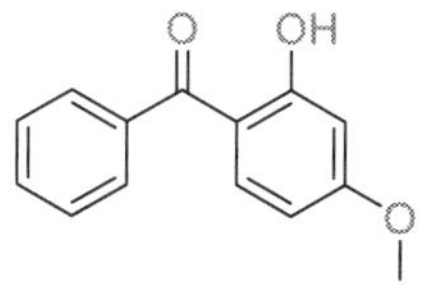
	Sulsobenzone	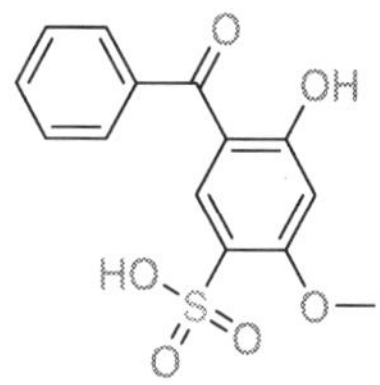
CAMPHOR	**Ecamsule**	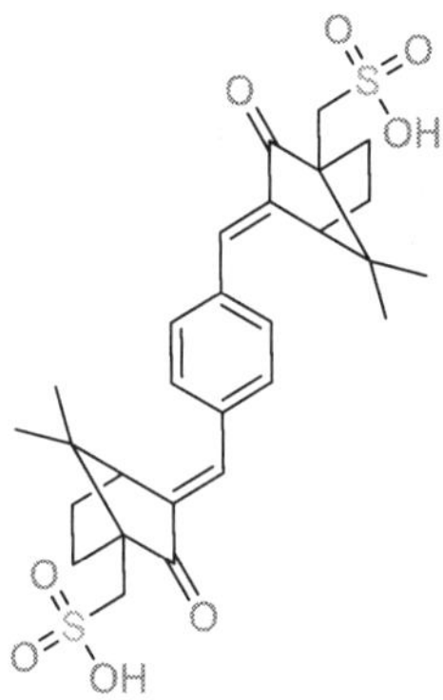
CINNAMATES	**Cinoxate**	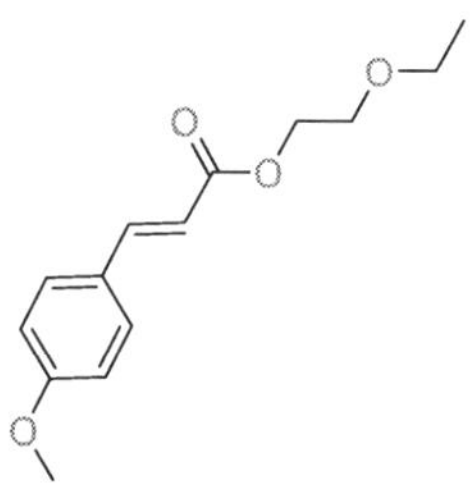

(continued)

TABLE 2.1 (Continued)
Chemical Structures of Organic UV Filters

	Octinoxate	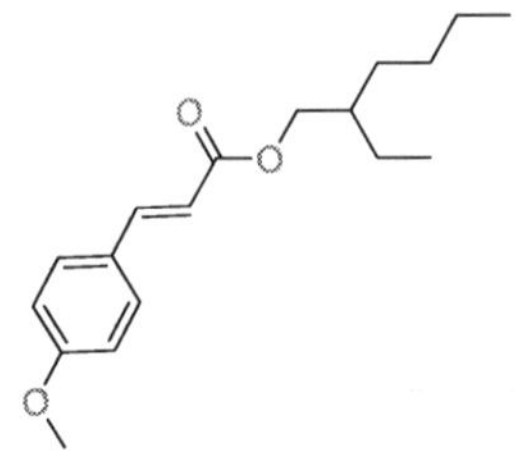
	Octocrylene	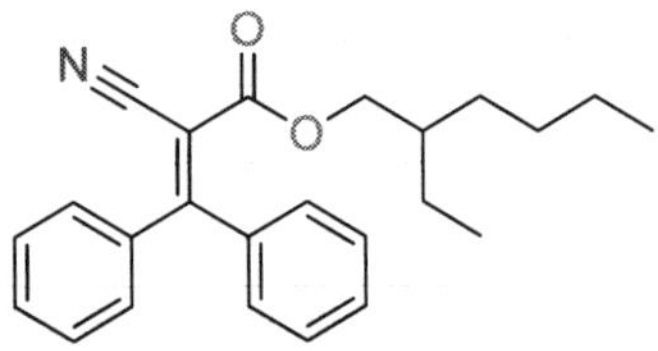
DIBENZOYLMETHANE	**Avobenzone**	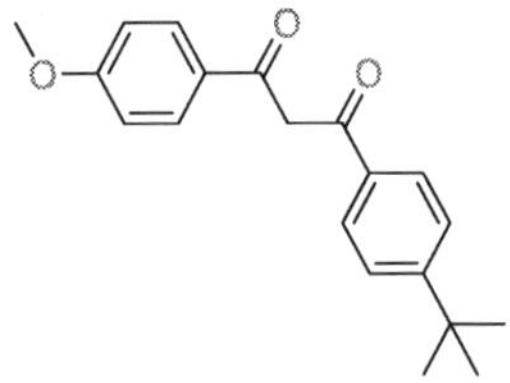
SALICYLATES	**Homosalate**	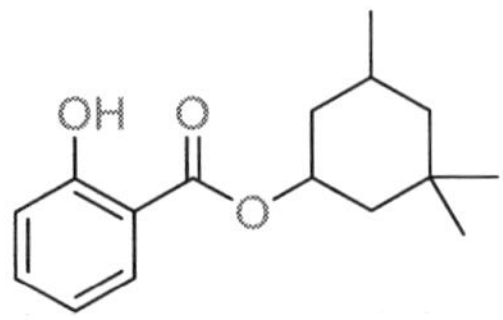
	Octisalate	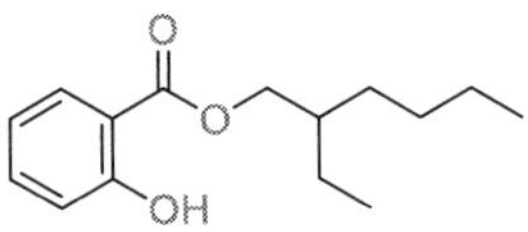
	Trolamine salicylate	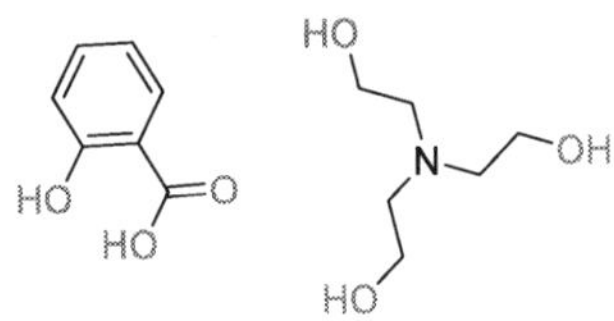

More precisely, the presence of electron orbitals that overlap in the crystalline structure gives rise to the formation of "bands," which are separated by gaps in energy known as "band gaps." Electrons experience excitation as they transition from a lower valence band to a higher conduction band, crossing the band gap. The dimensions and constitution of particles influence the absorption or dispersion/reflection of ultraviolet

TABLE 2.2
UV Protection Range of the Available UV Filters

Name of the Organic UV Filter	% of Concentration (Weight per Volume) [27]	Protective Region [24]
Aminobenzoic acid	15	UVB
Avobenzone	3	UVA1
Cinoxate	3	UVB
Dioxybenzone	3	UVA2, UVB
Ecamsule	Approved for specific formulations only	UVA1, UVA2
Ensulizole	4	UVB
Homosalate	15	UVB
Meradimate	5	UVA2
Octinoxate	7.5	UVB
Octisalate	5	UVB
Octocrylene	10	UVB
Oxybenzone	6	UVA2, UVB
Padimate O	8	UVB
Sulisobenzone	10	UVA2, UVB
TiO_2	25	UVA2, UVB
Trolamine salicylate	12	UVB
ZnO	25	UVA1, UVA2, UVB

(UV) radiation. While absorption remains the primary mechanism for all particles, larger particles have a greater capacity to reflect or scatter UV radiation compared to smaller particles. Nanoscale particle UV filters reflect a reduced amount of visible light, resulting in an almost translucent colour, in contrast to the opaquer look of bigger particles. Decreasing the size of particles also causes a shift in the range of UV wavelengths that the particles provide protection against.

2.4 DRAWBACKS OF UNCOMPLEXED UV FILTERS

The avobenzone and octinoxate filters, known for their high efficacy in absorbing UVA and UVB rays, are commonly used in sun creams. However, these filters are prone to photodegradation and can interact with each other through a process called photocyclization. This interaction leads to the creation of byproducts that can absorb light in a distinct UV region or have a reduced molar absorption coefficient [16, 18, 28–33]. Octinoxate (OCO) undergoes photoisomerization, changing from E-octinoxate to Z-octinoxate. The E isomer, which possesses a greater molar absorption coefficient in various solvents, is the preferred component in sunscreens. Photostability of avobenzone is significantly influenced by the polarity and proticity of the solvent [30, 34, 35]. In addition, avobenzone undergoes degradation into radical fragments.

Onion peels are considered inedible and have been subjected to less investigation and chemical analysis non-comparison to the edible bulb. However, several studies have categorized onion peel as a commodity of significant commercial value because of its high concentration of flavonoids and polyphenols, which comes under phenolics. Phenolic chemicals are found in plants that serve as secondary metabolites to the plant's defense mechanism. Due to their varied structures, these chemicals display a broad spectrum of biological activities. They can function as antioxidants, anti-inflammatory agents, antiallergenic agents, antibacterial agents, and photoprotective agents, among others [36, 37]. Phenolic chemicals, although beneficial, are prone to oxidation and lose their bioactive properties. Therefore, it is important to develop techniques like microencapsulation to safeguard their effectiveness [38]. Encapsulation is a method where a substance is packed by an encapsulating agent (such as a wall or shell) to release particles with a predefined shape at a very small scale, ranging from nanometers to millimeters [39, 40]. Choosing the best suitable covering material and encapsulation process is essential to obtain microparticles with the necessary characteristics [41, 42].

2.5 ENCAPSULATION OF UV FILTERS WITH CYCLODEXTRINS

The photostability of UV filters is a significant difficulty in sunscreen compositions. Exposure to sunlight can cause certain UV filters to deteriorate, thereby diminishing their ability to shield the skin from damaging UV radiation [43]. Cyclodextrins function as protective agents by creating inclusion complexes, which shelter UV filters from degradation. This improves the resistance of the UV filters to degradation caused by light and extends the duration of the sunscreen's efficacy. In addition, cyclodextrins can enhance the solubility of UV filters that have low solubility in sunscreen formulations. This promotes enhanced dispersion of UV filters within the sunscreen formulation, resulting in a more consistent and uniform level of protection upon application to the skin [44].

The most frequently utilized option for encapsulation of UV filters is the *β*-CD format, mostly due to its lower cost compared to other alternatives. *β*-CD based sunscreens are formed by trapping a core material (synthesized or natural chemicals) in cavity of the *β*-CDs via hydrogen bonds, or van der Waals force of attraction, or the hydrophobic effect [45]. Incorporating cyclodextrins into sunscreen formulations enhances the effectiveness of sunscreens by increasing the stability and solubility of UV filters. This ultimately results in improved protection against UV radiation.

Cyclodextrins have a notable impact on improving the effectiveness of sunscreens by enclosing UV filters. These are cyclic sugar compounds that can create complexes with different guest molecules, including those present in sunscreen formulas. Cyclodextrins enhance the stability, solubility, and photostability of UV filters, which are essential for maximizing the efficiency of sunscreens [35, 46].

2.6 PHOTOSTABILITY UV FILTERS FOLLOWING ENCAPSULATION WITH β-CD

2.6.1 Encapsulation of AVO and OCT in β-CD

Avobenzone and octinoxate are widely used UV filters in sunscreens, protecting against UVA and UVB radiation, respectively. Inclusion complexes with *β*-cyclodextrin

(β-CD) enhance their stability and performance. Mixing β-CD with avobenzone (AVO) or octinoxate (OCT) in 1:1 and 2:1 molar ratio was performed via kneading with water or grinding without it [47]. Grinding failed to produce uniform mixtures, while kneading resulted in novel solid formations.

^{1}H NMR analysis identified inclusion complexes as $(\beta\text{-CD})_2$·AVO and (β-CD)$_3$·OCT2, confirming a 2:1 β-CD to AVO ratio. However, the spectroscopy did not provide conclusive data on water content.

Thermal analyses, specifically TGA and DSC, were employed to provide further evidence of the encapsulating products. The analyses were successful in evaluating the quantity of water included in the products. The thermal stability limits of avobenzone and octinoxate are 280 °C and 240 °C, respectively. Thermogravimetric analysis of β-CD shows a reduction in weight, suggesting the removal of water, between temperatures of 40–80 °C. β-CD undergoes degradation at approximately 300 °C. Both of β-CD inclusion complexes are hydrated, leading to a weight decrease equivalent to approximately eight water molecules. These compounds demonstrate thermal stability till around 300 °C. The thermal stability of β-CD is improved by the addition of UV filters, and the deterioration of the UV filters aligns with the degradation of β-CD. The physical properties of the guest molecules, such as their melting, evaporation, or sublimation points, are anticipated to undergo significant changes when they are surrounded by β-CD molecules [48–51]. The absence of an endothermic peak in the DSC trace of $(\beta\text{-CD})_2$·AVO suggests a unique behaviour that persists until breakdown. When comparing, it is seen that pure avobenzone has melting values of 87 °C for Form I and 76 °C for Form II. Nevertheless, DSC measurements are inadequate for verifying the occurrence of octinoxate inclusion events due to its liquid state under standard conditions. Just like in the case of the $(\beta\text{-CD})_2$·AVO, the DSC analysis of the $(\beta\text{-CD})_3$·OCT$_2$ inclusion complex shows a broad peak between 50–90 °C, which suggests that the dehydration process is taking place.

An effective strategy to address the problem of photostability is to encapsulate the UV filters within the encapsulants that maintain the sunscreen's effectiveness.

The β-CD complexes of inclusion remained synthesized using a modest, environmentally friendly, and highly efficient mechanochemical technique, in addition to a slurry process in water. A detailed investigation was conducted on the solid state of these compounds. We collected data using solid-state techniques including powder XRD, Raman spectroscopy, and thermal investigations, as well as solution-based ^{1}H NMR. This data confirms the formation of crystal inclusion compounds inside the C_2 isostructural class for β-CD. Therefore, we have concluded that the chemical formulas of these complexes are $(\beta\text{-CD})_2$·AVO and $(\beta\text{-CD})_3$·OCT$_2$, and each compound has a total of eight water molecules. The photostability of avobenzone and octinoxate with β-CD complexes was verified by ESI-MS. No discernible byproducts were seen following exposure to UV light during the photoreaction (Figure 2.1). The enhanced stability of the inclusion complexes generated by avobenzone and octinoxate with β-cyclodextrin would allow for a reduction in the amount of UV filters utilized in sun lotions, potentially resulting in benefits for both the skin and the environment. Ongoing research is being conducted to expand the range of this method by integrating more UVA and UVB filters, along with their corresponding compositions.

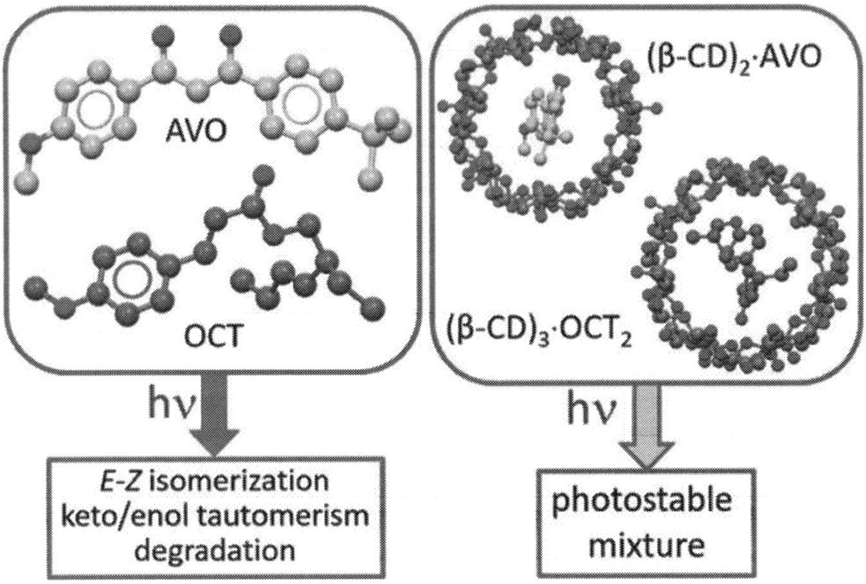

FIGURE 2.1 Photostability of AVO and OCT sunscreens is improved when they are included with *β*-CD [23].

2.6.2 Encapsulation of Onion Peels with β-CD

2.6.2.1 Onion Peel Extract

Our objective was to assess the efficacy of free onion peel (OP) extract against *β*-CD microencapsulated OP extract (natural UV filters) to minimize the reliance on synthetic UV filters. The extract was attained through a solid-liquid extraction process and was subsequently analyzed to determine its biological characteristics.

The TPC value obtained is consistent with the range of values documented in the literature, which is important to recognize. However, this extract exhibits a significant phenolic content, which is evident from its proximity to the higher values [52–54]. Furthermore, the antioxidant capacity was assessed using the DPPH assay and the obtained results fall within the range documented in previous studies, but significantly lower than (Table 2.3) the results obtained using the identical extraction method [50].

The sun protection factor (SPF) measures sunscreen efficacy in shielding skin from UV rays. An analysis assessed the SPF of varying quantities of onion peel (OP) extract, comparing the results to oxybenzone as a positive control. Table 2.4 summarizes the findings, highlighting OP extract's potential as a natural UV filter.

2.6.2.2 Sunscreen Formulations

Messias et al. developed five oil-in-water formulations to evaluate the photoprotective and antioxidant properties of OP extract. The negative control (NC) formulation lacked additives, while the positive control (PC) included 5% oxybenzone, a widely used synthetic UV filter [55]. To compare the effects of OP extract with oxybenzone,

TABLE 2.3
Onion Peel Extraction Characterization

	Test Name	Mean
DPPH	**TPC (mg_{GAE} g^{-1})**	585
	TE (mg_{TE} g^{-1})	200
	IC_{50} (mg L^{-1})	31

TPC: total phenolic content; GAE: gallic acid equivalent; IC_{50}: concentration of extract necessary to inhibit 50% of DPPH radical; TE: trolox equivalent antioxidant capacity.

TABLE 2.4
Onion Peel Extract SPF Value at Various Concentrations

Sample Name	Concentration in mg/mL	Sun Protection Factor (SPF)
Oxybenzone	0.010	38.84
OP extract	0.010	6.05
OP extract	0.045	24.92
OP extract	0.095	60.24

a formulation containing 5% non-encapsulated OP extract (OPE) was prepared. Additionally, a formulation with 5% microencapsulated OP extract (MP-OPE) was created to assess the benefits of encapsulation. A MIX formulation, comprising 1.6% each of oxybenzone, non-encapsulated OP extract, and microencapsulated OP extract, was also developed for a comprehensive analysis [56].

The pH of all formulations decreased over time but remained suitable for topical use, with OP extract causing mild acidification due to its phenolic content. Antioxidant tests showed no activity in NC and PC, while OP extract formulations suppressed the DPPH radical, indicating strong antioxidant potential. SPF evaluations highlighted the role of OP extract and microencapsulation in improving UV protection. The MIX formulation provided optimal performance by combining natural and synthetic UV filters effectively (Table 2.5).

The results suggest that the formulation of NC does not offer any degree of defense over UV radiation, as it lacks any additional substances, UV filters, or extracts. The PC composition had the most significant SPF value. The analysis of formulations containing OP extract is more complex. Throughout the week of manufacturing sunscreen, the formulation, such as microencapsulated extract (MP-OPE), exhibited a higher protection factor in comparison to both OPE and MIX formulations.

During the second examination, which took place two weeks after manufacture, and the final analysis, which occurred a month after production, the MIX emulsion had the greatest SPF value. The OPE formulation consistently exhibits the smallest

TABLE 2.5
SPF Values of Different Formulations at Three Different Analysis Times [57]

Sample Name	SPF (T_1 – 0 week)	SPF (T_2 – 2 weeks)	SPE (T_3 – 4 weeks)
OPE	4.7	2.1	1.4
MP-OPE	18.2	11.0	6.3
MIX	12.9	11.2	11.0
NC	0.8	0.2	0.0
PC	22.8	19.0	17.8

NC: negative control; PC: positive control; OPE: onion peel extract; Mp-OPE: microencapuslated onion peel extract; MIX: mixture formulations.

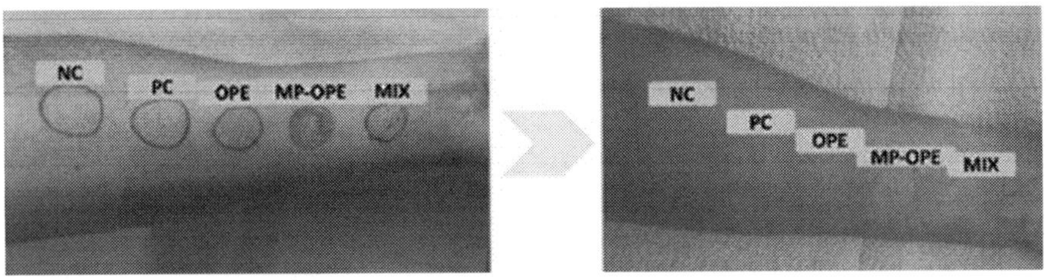

FIGURE 2.2 Comparison of skin response prior to and following one hour of solar exposure (4 weeks).

NC: negative control; PC: positive control; OPE: onion peel extract; MP-OPE: microencapuslated onion peel extract; MIX: mixture formulations.

protection factor, which can be attributed to the rapid breakdown of the OP extract when left unprotected. Given the range in the safety factor, it is feasible to go deeper into the data. The findings indicate that there is a consistent decline in the SPF value of the MP-OPE formulation from t_0 to t_2. This can be attributed to the controlled dissolution of the OP extract when it is microencapsulated. The diffusion of the extract from the small particles is anticipated to reach its highest point within the initial few hours, after which it will continue to be released steadily. Additionally, the formulations' ability to guard against sunlight was assessed by observing a volunteer's skin reaction to the formulations following one hour of exposure to sunshine (Figure 2.2).

The results of this test appear to deviate from the values for SPF that were obtained. However, SPF solely evaluates the level of protection against UVB radiation. OP extract is a blend of compounds that work together to absorb UV radiation throughout various regions of the UV spectrum. This leads to a higher level of protection than what is suggested by the SPF rating. While these studies are in the early stages, they suggest that OP extract, whether in non-encapsulated or embedded form, has the potential to be a sustainable and environmentally friendly source of protection from sunlight for sunscreens.

2.7 CONCLUSIONS

The study investigated the development of enhanced sunscreens by encapsulating UV filters in β-CD. This solid-state assembly method was applied to two common sunscreen agents, avobenzone and octinoxate, using the hydrophobic effect. Encapsulation yielded stable inclusion complexes, with a new variation of avobenzone (Form II) discovered and shown to have a monotropic relationship with the standard Form I.

Avobenzone and octinoxate effectively absorb UVA and UVB radiation but degrade under UV exposure, reducing efficacy. To improve photostability, these filters were encapsulated in host molecules like β-CD, preserving their function while enhancing durability. This innovative approach can reduce the concentration of UV filters, improving skin health and minimizing environmental impacts.

Additionally, OP extract, rich in quercetin and resveratrol, was explored as a natural UV blocker. The extract displayed significant antioxidant and photoprotective properties. Microencapsulation of OP extract with β-CD enhanced its stability and prolonged its effectiveness. Molecular inclusion produced microparticles with an average diameter of 31 μm, further supporting the extract's durability.

Formulations combining encapsulated OP extract with synthetic UV filters achieved superior sun protection and sustained efficacy over time. These findings suggest that OP extract, in free or encapsulated form, could serve as a sustainable alternative or complement to synthetic filters, reducing their usage and concentration.

This research highlights the potential of microencapsulation and molecular inclusion to advance sunscreen formulations, offering enhanced photoprotection, stability, and environmental safety. Ongoing efforts aim to expand this approach to include a broader range of UVA and UVB filters, paving the way for more effective and sustainable sunscreen products.

REFERENCES

1. Moukova, A., Malina, L., Kolářová, H., & Bajgar, R. (2023) 'Hyperoside as a UV Photoprotective or Photostimulating Compound—Evaluation of the Effect of UV Radiation with Selected UV-Absorbing Organic Compounds on Skin Cells' *Int. J. Mol. Sci.*, Vol 24, pp 1–16. https://doi.org/10.3390/ijms24129910
2. Jesus, A., Sousa, E., Cruz, M. T., Cidade, H., Lobo, J. M. S., & Almeida, I. F. (2022) 'UV FILTERS: Challenges and prospects' *Pharmaceuticals*, Vol 15, pp 1–26. https://doi.org/10.3390/ph15030263
3. Lindqvist, P. G., Epstein, E., & Landin-Olsson, M. (2022) 'Sun exposure – hazards and benefits' *Anticancer Res.*, Vol 42, pp 1671–1677. https://doi.org/10.21873/anticanres.15644
4. Li, C., Lin, L., Lee, H., & Tsai, J. (2016) 'Avobenzone encapsulated in modified dextrin for improved UV protection and reduced skin penetration' *Chem. Papers*, Vol 70, pp 840–847. https://doi.org/10.1515/chempap-2016-0021

5. González, S., Fernández-Lorente, M., & Gilaberte, Y. (2008) 'The latest on skin photoprotection' *Clin. Dermatol.*, Vol 26, pp 614–626. https://doi.org/10.1016/j.clindermatol.2007.09.010
6. Sambandan, D. R., & Ratner, D. (2011) 'Sunscreens: An overview and update' *JAAD*, Vol 64, pp 748–758. https://doi.org/10.1016/j.jaad.2010.01.005
7. Mancuso, J., Maruthi, R., Wang, S. Q., & Lim, H. W. (2017) 'Sunscreens: An update' *Am. J. Clin. Dermatol.*, Vol 18, pp 643–650. https://doi.org/10.1007/s40257-017-0290-0
8. Dummer, R., & Maier, T. (2002) 'UV protection and skin cancer' In *Recent results in cancer research*, Vol 160, pp 7–12. https://doi.org/10.1007/978-3-642-59410-6_2
9. Kockler, J., Oelgemöller, M., Robertson, S., & Glass, B. (2012) 'Photostability of sunscreens' *J. Photochem. Photobiol. C: Photochem. Rev.*, Vol 13, pp 91–110. https://doi.org/10.1016/j.jphotochemrev.2011.12.001
10. Serpone, N., Salinaro, A. T., Emeline, A. V., Horikoshi, S., Hidaka, H., & Zhao, J. (2002) 'An in vitro systematic spectroscopic examination of the photostabilities of a random set of commercial sunscreen lotions and their chemical UVB/UVA active agents' *PPS*, Vol 1, pp 970–981. https://doi.org/10.1039/b206338g
11. Serpone, N., Dondi, D., & Albini, A. (2007) 'Inorganic and organic UV filters: Their role and efficacy in sunscreens and suncare products' *Inorganica Chimica Acta*, Vol 360, pp 794–802. https://doi.org/10.1016/j.ica.2005.12.057
12. Damiani, E., Astolfi, P., Giesinger, J., Ehlis, T., Herzog, B., Greci, L., & Baschong, W. (2010) 'Assessment of the photo-degradation of UV-filters and radical-induced peroxidation in cosmetic sunscreen formulations' *Free Radic. Res.*, Vol 44, pp 304–312. https://doi.org/10.3109/10715760903486065
13. Cozzi, A. C., Perugini, P., & Gourion-Arsiquaud, S. (2018) 'Comparative behavior between sunscreens based on free or encapsulated UV filters in term of skin penetration, retention and photo-stability' *Eur. J. Pharm. Sci.*, Vol 121, pp 309–318. https://doi.org/10.1016/j.ejps.2018.06.001
14. Hayden, C. G. J., Cross, S. E., Anderson, C., Saunders, N. A., & Roberts, M. S. (2005) 'Sunscreen penetration of Human skin and related Keratinocyte toxicity after topical application' *Skin Pharmacol. Physiol.*, Vol 18, pp 170–174. https://doi.org/10.1159/000085861
15. Physicochemical Basis of Pharmaceuticals: Amazon.Co. Uk: Moynihan, Humphrey, Crean, Abina: 9780199232840: Books. Available online: www.amazon.co.uk/Physicochemical-Basis-Pharmaceuticals-Humphrey-Moynihan/dp/0199232849
16. Defraeye, T., Bahrami, F., Ding, L., Malini, R. I., Terrier, A., & Rossi, R. M. (2020) 'Predicting transdermal fentanyl delivery using mechanistic simulations for tailored therapy' *Front. Pharmacol.*, Vol 11, pp 1–23. https://doi.org/10.3389/fphar.2020.585393
17. Jullian, C., Moyano, L., Yáñez, C., & Olea-Azar, C. (2007) 'Complexation of quercetin with three kinds of cyclodextrins: An antioxidant study' *Spectrochim. Acta A Mol. Biomol. Spectrosc.*, Vol 67, pp 230–234. https://doi.org/10.1016/j.saa.2006.07.006
18. Types | Skin Cancer | Cancer Research UK. Available online: www.cancerresearchuk.org/about-cancer/skin-cancer/types
19. Kayacı, F., & Uyar, T. (2012) 'Encapsulation of vanillin/cyclodextrin inclusion complex in electrospun polyvinyl alcohol (PVA) nanowebs: Prolonged shelf-life and high temperature stability of vanillin' *Food Chem.*, Vol 133, pp 641–649. https://doi.org/10.1016/j.foodchem.2012.01.040
20. Pyo, S. M., & Maibach, H. I. (2019) 'Skin metabolism: Relevance of skin enzymes for rational drug design' *Skin Pharmacol.*, Vol 32, pp 283–294. https://doi.org/10.1159/000501732

21. Chisvert, A., & Salvador, A. (2007) '*UV Filters in Sunscreens and other Cosmetics. Regulatory Aspects and Analytical Methods*' Elsevier eBooks, pp. 83–120. https://doi.org/10.1016/b978-044452260-3/50028-0
22. Giokas, D. L., Salvador, A., & Chisvert, A. (2007) 'UV filters: From sunscreens to human body and the environment' *TrAC*, Vol 26(5), pp 360–374. https://doi.org/10.1016/j.trac.2007.02.012
23. Nash, J. J., & Tanner, P. R. (2014) 'Relevance of UV filter/sunscreen product photostability to human safety' *PPP*, Vol 30, pp 88–95. https://doi.org/10.1111/phpp.12113
24. Shaath, N. A. Allured Publishing Corporation. (2007) 'The Encyclopedia of Ultraviolet Filters' Amazon.in. www.amazon.in/s?i=stripbooks&rh=p_27%3ANadim+A.+Shaath+PhD&ref=dp_byline_sr_book_1
25. Cole, C., Shyr, T., & Ouyang, H. (2015) 'Metal oxide sunscreens protect skin by absorption, not by reflection or scattering' *PPP*, Vol 32, pp 5–10. https://doi.org/10.1111/phpp.12214
26. Smijs, T. G. M., & Pavel, S. (2011) 'Titanium dioxide and zinc oxide nanoparticles in sunscreens: focus on their safety and effectiveness' *Nanotechnol Sci Appl.*, Vol 95, pp 95–112. https://doi.org/10.2147/nsa.s19419
27. From the Over-the-Counter Monograph M020: Sunscreen Drug Products for Over-the-Counter Human Use. Final Administrative Order (OTC000006).
28. Pattanaargson, S., Munhapol, T., Hirunsupachot, P., & Luangthongaram, P. (2004) 'Photoisomerization of octyl methoxycinnamate' *J. Photochem. Photobiol.*, Vol 161, pp 269–274. https://doi.org/10.1016/s1010-6030(03)00282-x
29. Pattanaargson, S., & Limphong, P. (2001) 'Stability of octyl methoxycinnamate and identification of its photo-degradation product' *Int. J. Cosmet. Sci.*, Vol 23, pp 153–160. https://doi.org/10.1046/j.1467-2494.2001.00071.x
30. Cantrell, A., & McGarvey, D. J. (2001) 'Photochemical studies of 4-tert-butyl-4′-methoxydibenzoylmethane (BM-DBM)' *J. Photochem. Photobiol. B.*, Vol 64, pp 117–122. https://doi.org/10.1016/s1011-1344(01)00226-3
31. Dondi, D., Albini, A., & Serpone, N. (2006) 'Interactions between different solar UVB/UVA filters contained in commercial suncreams and consequent loss of UV protection' *PPS*, Vol 5(9), pp 835–843. https://doi.org/10.1039/b606768a
32. D'Agostino, S., Boanini, E., Braga, D., & Grepioni, F. (2018) 'Size matters: [2 + 2] photoreactivity in macro- and microcrystalline salts of 4-Aminocinnamic acid' *Cryst Growth Des*, Vol 18, pp 2510–2517. https://doi.org/10.1021/acs.cgd.8b00109
33. D'Agostino, S., Taddei, P., Boanini, E., Braga, D., & Grepioni, F. (2017) 'Photo-vs Mechano-induced polymorphism and single crystal to single crystal [2 + 2] photoreactivity in a Bromide salt of 4-Amino-Cinnamic acid' *Cryst. Growth Des.*, Vol 17, pp 4491–4495. https://doi.org/10.1021/acs.cgd.7b00415
34. Afonso, S., Horita, K., Silva, J. P., Almeida, I. F., Amaral, M. H., Lobão, P., Costa, P., Miranda, M. S., Da Silva, J. C. G. E., & Lobo, J. M. S. (2014) 'Photodegradation of avobenzone: Stabilization effect of antioxidants' *J. Photochem. Photobiol B.*, Vol 140, pp 36–40. https://doi.org/10.1016/j.jphotobiol.2014.07.004
35. Huong, S. P., Rocher, E., Fourneron, J., Charles, L., Monnier, V., Bun, H., & Andrieu, V. (2008) 'Photoreactivity of the sunscreen butylmethoxydibenzoylmethane (DBM) under various experimental conditions' *J. Photochem. Photobiol.*, Vol 196(1), pp 106–112. https://doi.org/10.1016/j.jphotochem.2007.11.023
36. Działo, M., Mierziak, J., Korzun, U., Preisner, M., Szopa, J., & Kulma, A. (2016) 'The potential of plant phenolics in prevention and therapy of skin disorders' *Int. J. Mol. Sci.*, Vol 17, pp 160. https://doi.org/10.3390/ijms17020160

37. Grgić, J., Šelo, G., Planinić, M., Tišma, M., & Bucić-Kojić, A. (2020) 'Role of the encapsulation in bioavailability of phenolic compounds' *Antioxidants*, Vol 9, pp 923. https://doi.org/10.3390/antiox9100923
38. Kumar, A., Raizada, P., Khan, A. a. P., Nguyen, V., Van Le, Q., Singh, A., Saini, V., Rangabhashiyam, S., & Huynh, T. (2021) 'Phenolic compounds degradation: Insight into the role and evidence of oxygen vacancy defects engineering on nanomaterials' *Sci. Total Environ.*, Vol 800, pp 149410. https://doi.org/10.1016/j.scitotenv.2021.149410
39. Eghbal, N., Liao, W., Dumas, É., Azabou, S., Dantigny, P., & Gharsallaoui, A. (2022) 'Microencapsulation of natural food antimicrobials: Methods and applications' *Appl. Sci.*, Vol 12, pp 3837. https://doi.org/10.3390/app12083837
40. Tarone, A. G., Cazarin, C. B. B., & Maróstica, M. R. (2020) 'Anthocyanins: New techniques and challenges in microencapsulation' *Food Res. Int.*, Vol 133, pp 109092. https://doi.org/10.1016/j.foodres.2020.109092
41. Kłosowska, A., Wawrzyńczak, A., & Feliczak-Guzik, A. (2023) 'Microencapsulation as a route for obtaining encapsulated flavors and fragrances' *Cosmetics*, Vol 10, pp 1–26. https://doi.org/10.3390/cosmetics10010026
42. Sousa, V. I., Parente, J. F., Marques, J. F. G., Forte, M. A., & Tavares, C. J. (2022) 'Microencapsulation of essential oils: A review' *Polymers*, Vol 14, pp 1–42. https://doi.org/10.3390/polym14091730
43. Ginzburg, A. L., Blackburn, R. S., Santillan, C., Truong, L., Tanguay, R. L., & Hutchison, J. E. (2021) 'Zinc oxide-induced changes to sunscreen ingredient efficacy and toxicity under UV irradiation' *PPS*, Vol 20, pp 1273–1285. https://doi.org/10.1007/s43630-021-00101-2
44. Dahabra, L., Broadberry, G., Gresley, A. L., Najlah, M., & Khoder, M. (2021) 'Sunscreens containing cyclodextrin Inclusion complexes for Enhanced efficiency: A strategy for skin cancer prevention' *Molecules*, Vol 26(6), pp 1698. https://doi.org/10.3390/molecules26061698
45. Sandilya, A. A., Natarajan, U., & Priya, M. H. (2020) 'Molecular view into the cyclodextrin cavity: Structure and hydration' *ACS Omega*, Vol 5, pp 25655–25667. https://doi.org/10.1021/acsomega.0c02760
46. Celano, R., Docimo, T., Piccinelli, A. L., Gazzerro, P., Tucci, M., Di Sanzo, R., Carabetta, S., Campone, L., Russo, M., & Rastrelli, L. (2021) 'Onion peel: Turning a food waste into a resource' *Antioxidants*, Vol 10, pp 304. https://doi.org/10.3390/antiox10020304
47. D'Agostino, S., Azzali, A., Casali, L., Taddei, P., & Grepioni, F. (2020) 'Environmentally friendly sunscreens: Mechanochemical synthesis and characterization of β-CD inclusion complexes of avobenzone and octinoxate with improved photostability' *ACS SCE*, Vol 8, pp 13215–13225. https://doi.org/10.1021/acssuschemeng.0c02735
48. Atwood J. L., & Lehn J.-M. (1996) '*Comprehensive supramolecular chemistry*' (1st ed.). Pergamon.
49. Braga, S. S., Gonçalves, I. S., Herdtweck, E., & Teixeira-Dias, J. J. C. (2003) 'Solid state inclusion compound of S-ibuprofen in β-cyclodextrin: Structure and characterisation' *New J. Chem.*, Vol 27, pp 597–601. https://doi.org/10.1039/b207272f
50. Moraes, C. M., Abrami, P., De Araujo, D. R., Braga, A. F., Issa, M. G., Ferraz, H. G., De Paula, E., & Fraceto, L. F. (2007) 'Characterization of lidocaine:hydroxypropyl-β-cyclodextrin inclusion complex' *J. Incl. Phenom. Macrocycl. Chem.*, Vol 57, pp 313–316. https://doi.org/10.1007/s10847-006-9179-x
51. Cruickshank, D. L., Rougier, N. M., Vico, R. V., De Rossi, R. H., Buján, E. I., Bourne, S. A., & Caira, M. R. (2010) 'Solid-state structures and thermal properties of inclusion complexes of the organophosphate insecticide fenitrothion with permethylated

cyclodextrins' *Carbohydr. Res.*, Vol 345, pp 141–147. https://doi.org/10.1016/j.carres.2009.10.023

52. Akdeniz, B., Şumnu, G., & Şahin, S. (2018) 'Microencapsulation of phenolic compounds extracted from onion (Allium cepa) skin' *J. Food Process. Preserv.*, Vol 42, pp 13648. https://doi.org/10.1111/jfpp.13648
53. Bordin Viera, V., Piovesan, N., Mello, R. D. O., Barin, J. S., Fogaça, A. D. O., Bizzi, C. A., Hashime Kubota, E. (2021) 'Ultrasonic assisted extraction of phenolic compounds with evaluation of red onion skin (Allium cepa L.) antioxidant capacity' *J. Culin. Sci. Technol.*, Vol 21, pp 156–172. https://doi.org/10.1080/15428052.2021.1910095
54. Román, Ó. B., Blanco, B., Sanz, M. T., & Beltrán, S. (2020) 'Subcritical water extraction of phenolic compounds from onion skin wastes effect of temperature and solvent properties' *Antioxidants*, Vol 9, pp 1233. https://doi.org/10.3390/antiox9121233
55. Geoffrey, K., Mwangi, A. N., & Maru, S. M. (2019) 'Sunscreen products: Rationale for use, formulation development and regulatory considerations' *SPJ*, Vol 27, pp 1009–1018. https://doi.org/10.1016/j.jsps.2019.08.003
56. Ferreira, S. M., Gomes, S. M., & Santos, L. (2023) 'A novel approach in skin care: By-Product extracts as natural UV filters and an alternative to synthetic ones' *Molecules*, Vol 28, pp 2037. https://doi.org/10.3390/molecules28052037
57. Messias, M. A., Ferreira, S. M., Tavares, L., & Santos, L. (2023) 'A comparative study between onion peel extracts, free and complexed with β-cyclodextrin, as a natural UV Filter to cosmetic formulations' *Int. J. Mol. Sci.*, Vol 24, pp 15854. https://doi.org/10.3390/ijms242115854

3 Cyclodextrins for Enhancing the Photostability of UVR Absorbers

Lakhi Chetry, Partha Pratim Sarma, and Pranjal K. Baruah

3.1 INTRODUCTION

The sun is the prime natural source of ultraviolet radiation (UVR). While UVR carries several therapeutic values and is indispensable for the production of vitamin D, excessive exposure to it, can impart detrimental effects on human health. UVR is a kind of electromagnetic radiation, exhibiting wavelengths shorter than visible light while longer than X-rays. It is divided into three major categories according to their wavelength: ultraviolet-A (UVA) (320–400 nm), ultraviolet-B (UVB) (280–320 nm), and ultraviolet-C (UVC) (100–280 nm). UVC, with the shortest wavelengths, is the most intense and pernicious UVR. Fortunately, the Earth's atmosphere absorbs virtually all UVC, preventing it from reaching the surface. UVR, prominently UVA and UVB, imparts a widespread threat to human health along with food, beverages, agriculture, and medications causing substantial degradation and compromising their performance over time [1]. UVA radiation can penetrate deep into the skin layers and produce free radicals, which can cause indirect and long-term damage to DNA. UVB rays can harm the epidermis, the top layer of skin, and are responsible for most sunburns and skin cancers [1]. Ultraviolet (UV) absorbers, also known as photoabsorbers, are substances that absorb and dissipate UVR and shield materials from the harmful effects of sunshine. A schematic representation of skin layer and UV ray penetration with and without UVR absorbers is shown in Figure 3.1. Despite their efficiency, these absorbers under prolonged sun exposure can undergo photodegradation through photoisomerization, tautomerization, chain reactions, etc., decreasing their lifespan and lowering their efficacy [2]. The increased awareness of the detrimental paraphernalia of UVR has pushed the industries to explore innovative solutions to boost the photostability of UVR absorbers, so as to improve their efficacy and performance in sun protection formulations. Cyclodextrins (CDs) stand out among these solutions as a promising class of compounds with distinct molecular structures and exceptional attributes that make them effective for stabilizing UVR absorbers. CDs are cyclic oligosaccharides with a distinctive molecular structure that enables formation

DOI: 10.1201/9781003492757-3

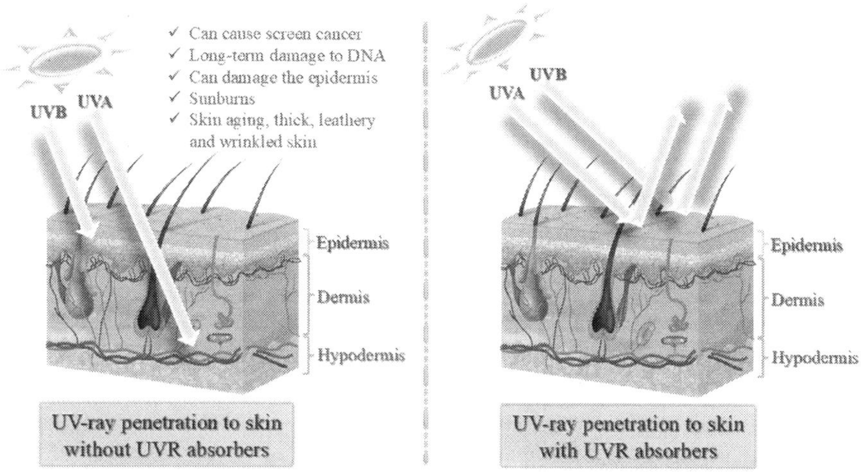

FIGURE 3.1 Diagrammatic representation of skin layer and UV ray penetration with and without UVR absorbers.

of inclusion complexes with a diverse array of guest molecules, furnishing them ideal for stabilizing UVR absorbers under prolonged exposure to sunlight [1, 3].

3.2 UV FILTERS

UV filters are compounds, mixtures, or materials that block or absorb UV light. UV absorbers emerge as critical guards against UVRs and play an important role in protecting the integrity and functionality of many materials. UV filters are divided into two categories, that includes chemical or organic filters and physical or inorganic UV filters. Chemical filters absorb UV rays and convert them into thermal energy while physical filters reflect UVR [4]. Metal oxides, for example, ZnO, TiO_2, and silicates are commonly employed as physical or inorganic UV filters [4]. They provide stronger protection than the chemical ones [1, 4, 5]. Most metal oxides are insoluble in water; however, they can be broken down into water-soluble ions. These filters act by reflecting and scattering UVR rather than absorbing it. The crystalline structure and distinct optical qualities of these inorganic filters contribute to their great efficacy in preventing UVA and UVB radiation from reaching the skin or destroying materials [5].

Chemical UV filters are organic molecules that encompass a myriad of compounds designed to absorb specific wavelengths of UVR. These compounds often have one or more aromatic rings, sometimes conjugated with carbon-carbon double bonds and/or carbonyl moieties resulting in numerous conjugated π-electron systems.

The organic UV filters are classified based on their chemical structure, including aminobenzoic acid derivatives, salicylic acid derivatives, cinnamic acid derivatives, benzophenones, dibenzoyl methane derivatives, benzylidene camphor derivatives, triazines, benzimidazole derivatives, and benzotriazole derivatives as well. Benzophenones, anthranilates, avobenzones, trolamine salicylate, padimate O, octinoxate, cinoxate, methylene-bis-benzo triazolyl tetramethyl butylphenol (MBBT), and bis-ethylhexyloxyphenol methoxyphenyl triazine are a few examples of organic UV filter that fall into this category [4, 6]. Some commonly used organic UVR filters of different classes are represented in Figure 3.2. These organic filters absorb UVR but not visible light, causing conformational changes in molecules. As a molecule transitions from the excited to ground state, it releases energy as heat [7]. Oxybenzone, the most widely used benzophenone, absorbs both UVB and short-wavelength UVA [8]. The use of organic UV filters has proliferated in a wide array of applications, including food products, cosmetic formulations, pharmaceuticals, agriculture, fabrics, textiles, glass, plastics, and various polymers. These filters are highly valued for their potential to offer selective protection against both UVA and UVB radiation. This attribute of versatility not only improves the performance of products but also contributes to their aesthetic appeal, making organic UV filters an indispensable component across numerous industries [9].

Despite their applications in diverse fields, organic UV filters have certain limitations, which include photoinstability, limited water resistance, and

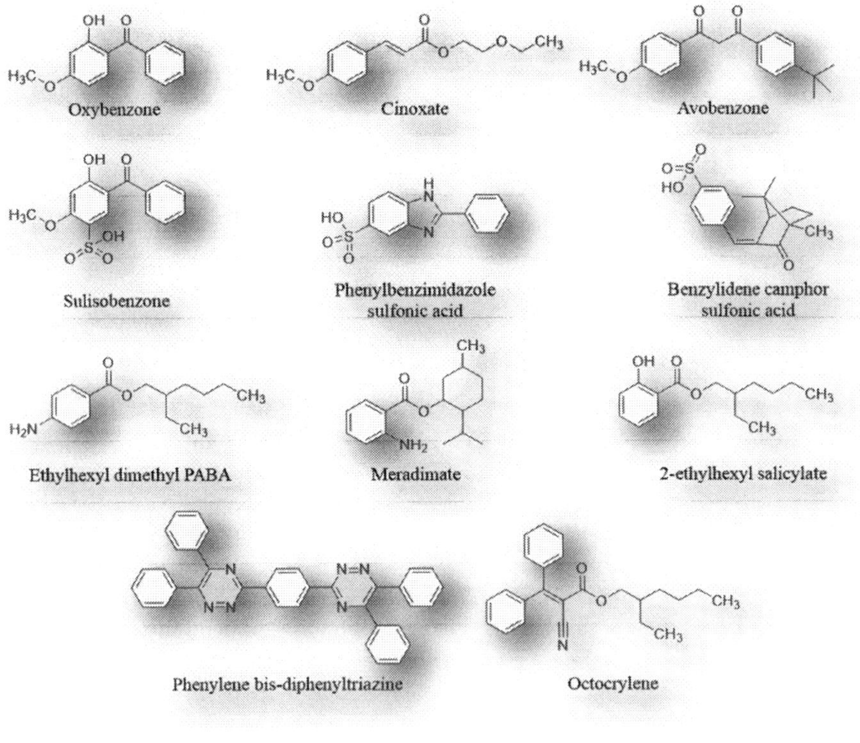

FIGURE 3.2 Structure of some commonly used organic UVR filters of different classes.

environmental issues. The use of CDs offers an intriguing approach to dealing with the challenges associated with UV filters, leading to the development of more stable as well as effective photoprotective solutions [8].

3.3 CD AND ITS DERIVATIVES

Cyclodextrins (CDs) are naturally occurring cyclic oligosaccharides composed of *α*-D-glucopyranoside formed by the degradation of starch and cellulose by some microorganisms. CDs may also be obtained by enzymatic action on starch [10]. These cyclic carbohydrates, known as cyclomaltooligosaccharides, have a maltose-type glycosidic link between the glucose units. The components of cyclic structures are linked by α (1 $\rightarrow$ 4) glucoside bonds. CDs possess a unique three-dimensional structure that resembles a cup-shaped toroid [11]. The cup's surface is hydrophilic, but the inner surface is more hydrophobic due to the presence of glucose methine protons, which are apolar, whereas the secondary and primary hydroxyls on the outer surfaces of the large and small rims are polar. The significant contrast in polarity between the interior and outer surfaces of CD contributes to its beneficial properties [12].

Based on the number of *α*-glucose units present, there are three primary types of CDs, which include the *α*-, *β*-, and *γ*-CDs with six, seven, and eight *α*-glucose units, respectively [3].

3.3.1 α-CD

α-CD ($C_{36}H_{60}O_{30}$) possessing the molecular weight of 972.8 g/mol, is composed of six glucose units linked through *α*-1,4-glycosidic linkages, forming a stiff three-dimensional structure [13]. It possesses a cylindrical shape, with a small central hydrophobic chamber, possessing six hydroxyl groups (-OH), which can potentially establish hydrogen bonds, but they disintegrate rapidly in aqueous solution and are resistant to enzymatic hydrolysis [14]. The width of the outer cavity of α-CD is 14.6 Å and the internal cavity possesses a width ranging from 4.7 to 5.3 Å [15]. The structure comprising six glucose units joined through α-1,4-glycosidic linkages (α-CD) was represented in **Figure 3.3**.

3.3.2 β-CD

β-CD is composed of seven glucose units joined by *α*-1,4-glycosidic linkages. The chemical formula of *β*-CD is $C_{42}H_{70}O_{35}$, with a molecular weight of 1134.98 g/mol [13]. The outer cavity of *β*-CD is 15.4 Å, while the *β*-CD internal cavity has a width that varies between 6.0 and 6.5 Å [15]. Due to its medium cavity size, *β*-CD exhibits greater stability and can accommodate a wider range of guest molecules more effectively. The hydroxyl groups surrounding the thinner borders are stretchy and can establish hydrogen bonds [14]. *β*-CD has the lowest water solubility compared to *α*- and *γ*-CDs, due to the formation of intramolecular hydrogen bonds [16].

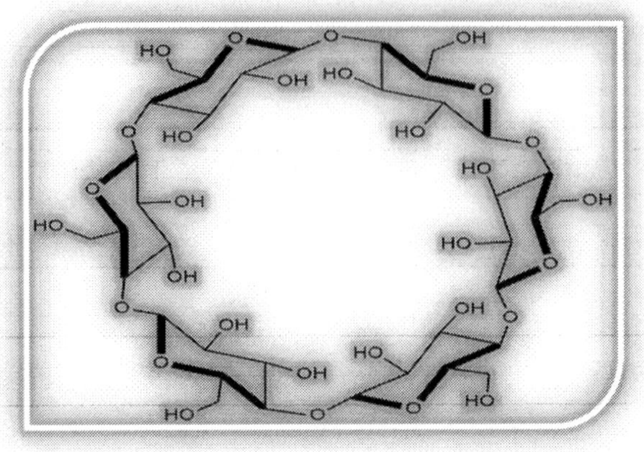

FIGURE 3.3 The structure comprising six glucose units joined through α-1,4-glycosidic linkages (α-CD).

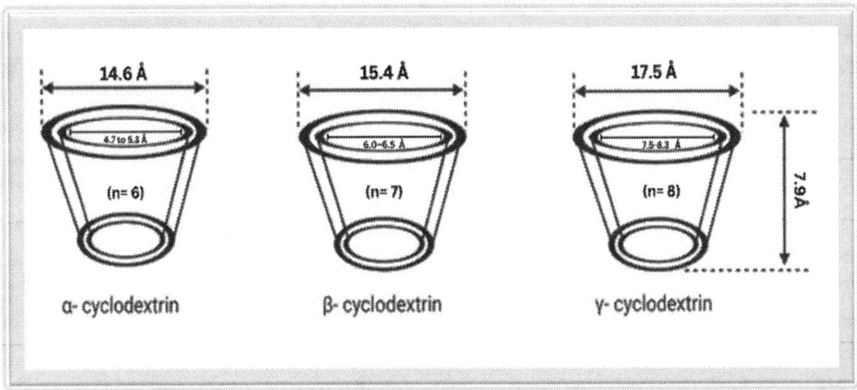

FIGURE 3.4 Diagrammatic representation of shape and dimensions of α, β, and γ-CD complexes.

3.3.3 γ-CD

γ-CD is composed of eight glucose units linked through α-1,4-glycosidic linkages. The chemical formula of γ-CD is $C_{48}H_{80}O_{40}$, with a molecular weight of 1297.92 g/mol [13]. The outer cavity of γ-CD is 17.5 Å, while the internal cavity has a width ranging from 7.5–8.3 Å [15]. Out of the three CDs, γ-CD has the largest cavity as well as the highest solubility. The dimensions of α-, β-, and γ-CD were represented diagrammatically in Figure 3.4.

Apart from the three most regularly used CDs, other derivatives have also been developed and utilized to improve the binding properties of original CDs [17].

3.3.4 Derivatives of CDs

CD derivatives, derived from natural CDs (α, β, and γ), have been synthesized to enhance the utility and versatility of CDs in different applications. Modifying CDs through amination and etherification, as well as adding functional groups like methyl, ethyl, hydroxyethyl, carboxymethyl, and hydroxypropyl, has led to a diverse range of CDs. These changes are implemented to enhance and modify specific characteristics of CDs, such as solubility, stability, and interaction capabilities [18, 19].

Some examples of commonly used CD derivatives:

(i) HP-β-CD (hydroxypropyl-beta-CD)
(ii) SBE-β-CD (sulfobutyl ether-beta-CD)
(iii) RM-β-CD (random methylated-beta-CD)
(iv) HP-γ-CD (hydroxypropyl-gamma-CD)

3.4 CD INCLUSION COMPLEXES

When a CD molecule (the host) and another molecule (the guest) create a host–guest complex inside the hydrophobic cavity of the CD, it is referred to as a CD inclusion complex. The shape and size of the guest molecule have a substantial impact on its ability to fit into the CD cavity and the formation of the inclusion complex [19, 20]. Inclusion complexes between CD and guest molecules frequently include a variety of intermolecular forces, such as dipole-dipole interactions and van der Waals interactions, which improve the molecule's chemical stability, particularly by protecting it from hydrolysis, oxidation, dehydration, and photodecomposition. This makes CD beneficial in a wide range of applications, including medicines, food, agrochemistry, and cosmetics, where it may significantly improve the stability, solubility, and bioavailability of specific chemicals [20]. The hydrophobic interior of CD enables it to form inclusion complexes with diverse hydrophobic guest molecules while the hydrophilic exterior of CD allows it to become water soluble, making it easier to utilize in aqueous solutions and more compatible with the hydrophilic environment [20].

Although CDs have a wide range of applications, their water solubility is somewhat limited. Solubility is influenced by the total number of glucose units in the molecule and the interaction of their hydroxyl group with water. In comparison, γ-CDs are more soluble than α-CD and β-CD, whereas β-CD is the least soluble. The incorporation of substituents or modifications into the CD molecule can improve its solubility in water or other solvents [21]. The stability, solubility, and efficacy of the UVR absorbers or filters can be significantly enhanced as well as photodegradation can also be prevented through incorporation in the CD cavity forming a host–guest complex simply known as CD inclusion complexes.

3.5 FACTORS AFFECTING THE FORMATION OF INCLUSION COMPLEXES

Temperature, pH, charges on the host and guest molecules, size and shape of the guest and host molecules, solvents, solubility of the guest molecules, and component concentrations are some of the parameters that contribute to the fabrication of inclusion compounds with CDs [21, 22]. The formation and stability of inclusion compounds are greatly influenced by the following key characteristics.

3.5.1 Shape and Size of Guest Molecules and CD

The size and shape of the guest molecule play important roles in the development of CD inclusion complexes. CDs have a unique truncated cone form with a hydrophobic cavity. The size of the cavity changes based on the kind of CD (α, β, or γ). The guest molecule has to be of appropriate shape and dimension to fit into the hydrophobic cavity of the CD. A guest molecule that is too big or too tiny may not fit adequately, preventing complex formation, and may not interact sufficiently with the CD to create a stable complex [22].

3.5.2 Nature of Guest Molecule (Neutral, Positive, or Negative Charge)

The degree of inclusion complexation is greatly affected by the structure of both CDs and guest molecules. They may be neutral, positively, or negatively charged. In general, the presence of charge on the CD structure contributes an additional element for interaction as compared to neutral CDs. This could be explained by the fact that the presence of opposing charges between the guest and the host increased the intensity of inclusion complexation, which in turn increased the extent of ionic or electrostatic interaction [22].

3.5.3 pH

Alteration in pH may enhance molecules solubility, though they can also destabilize the produced CD complexes as neutral molecules tend to be more hydrophobic than ionised molecules [21, 22].

3.6 METHODS OF PREPARATION OF CD INCLUSION COMPLEXES

Numerous methods exist to manufacture CD-guest complexes, contingent upon the properties of the guest and the characteristics of the selected CD moiety, in order to get the desired increased attributes of the final complex. The preparation procedures significantly influence the performance and morphometric properties of the inclusion complexes [19]. The preparation method may influence the characteristics of the final integrated complex [19, 20]. Different methods for developing the CD inclusion complex along with their advantages and disadvantages are listed in Table 3.1.

3.7 CHARACTERIZATION OF CDS COMPLEXES

The integration of several characterisation techniques might enhance the comprehension of host–guest interactions and facilitate the selection of the most suitable

TABLE 3.1
Advantages and Disadvantages of Different Methods Used for Developing CD Inclusion Complexes

Methods	Advantages	Disadvantages	Ref.
Co-precipitation	Best method for dealing with insoluble compounds, flexibility, simplicity, and high efficiency.	Large-scale manufacturing generates a lot of effluent, high cost.	[20]
Freeze drying	The low temperature reduces the loss of highly volatile guests, beneficial for heat-labile guests and soluble complexes like hydroxy propylated CD complexes.	The method is time-consuming, high production costs, complex equipment, possibility of aggregation or collapse of the end product.	[20, 23]
Grinding	Efficient, adaptable, user-friendly and environmentally sustainable method, efficient solvent-free technique of synthesising CD inclusion.	Less cost-effective, complex equipment.	[23]
Microwave irradiation	Absence of residues from the use of organic solvents, higher yields, shorter reaction times, Cost-effective and time-saving.	Unsuitable for extremely volatile compounds.	[24, 25]
Spray drying	High yield, fast drying, suitable for mass manufacturing in the food sector.	Unsuitable for extremely volatile and heat-labile guest compounds, confined to water-dispersible or water-soluble components.	[15, 26]
Kneading	A simple and economical technique with the potential for mass manufacturing.		[13]
Ultrasonic method	Improve the material's molecular movement speed, the shock wave produced by ultrasonic wave propagation in liquid can release energy and speed up the inclusion process of CD and guest molecules.	Expensive, potential for inaccurate results.	[27]
Supercritical fluid	Lowers the inclusion temperature, appropriate for chemicals sensitive to moisture and heat, avoids the use of organic solvents.	High cost.	[13, 27]

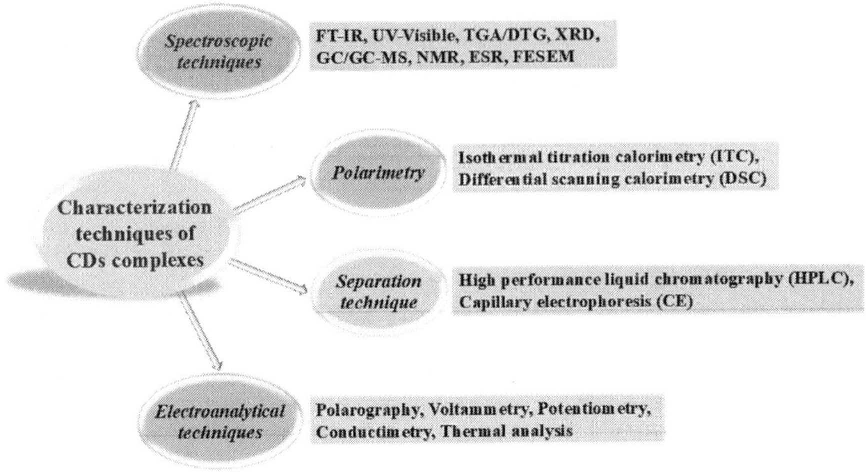

FIGURE 3.5 Various methods employed for the characterization of CD inclusion complexes.

cyclodextrin for a certain guest molecule [28]. Various methods employed for the characterization of CD inclusion complexes are mentioned in Figure 3.5 [19, 28].

These approaches provide a comprehensive insight into the formation of CD inclusion complexes, the stability, interactions, and structural alterations that occur throughout the synthesis of these complexes. The characterization techniques utilized depends upon the type of the guest molecule, the specific CD used, and the information required about the complex.

3.8 DRIVING FORCE FOR FORMATION OF INCLUSION COMPLEX

The potential of CD molecules to form an inclusion complex with a guest molecule relies on two key components [28, 29]:

(i) *Stearic factor*: Stearic factor is governed by the size of CD relative to the guest molecule or specific functional groups. The guest must be of appropriate size to fit inside the CD cavity.

(ii) *Thermodynamic interactions*: The second critical aspect is the thermodynamic interactions between the guest, solvent, and CD. To form a complex, there must be a positive net energetic driving force that attracts the guest to the CD.

The chemical characteristics of the guest molecule significantly impact the binding strength of the CD/molecule complex. These properties encompass charge (whether neutral, cationic, or anionic), numerous intermolecular interactions such as hydrogen bonding, van der Waals forces, and dipole-dipole interactions, along with the size and shape of the guest molecule. The influence of each of these factors on CD complexation differs depending on the class or type of guest molecule [29, 30]. For optimal binding within the CD cavity, the guest molecule should possess appropriate organic functional groups.

The stability of CD-guest complexes is primarily influenced by van der Waals forces. For instance, Rawashdeh et al. demonstrated that electrostatic and van der Waals interactions play a key role in driving the inclusion complexation of various sunscreen filters, such as oxybenzone (Oxy), octocrylene (Oct), and ethylhexylmethoxycinnamate (Cin), with β-CD [31]. The inclusion complexation process is governed by a variety of forces, including bond angle bending, dihedral angle bending, electrostatic forces, and van der Waals interactions. van der Waals forces are the primary contributors to the binding energy, while the electrostatic contribution is modest. Other forces were considered negligible in comparison [31]. Hydrogen bonding is the primary polar intermolecular force in CDs. CDs have hydroxyl groups on their outer surface, which can interact with similar functional groups on guest molecules, acting as either hydrogen bond donors or acceptors [32]. The interior cavity of CDs is hydrophobic and has no ability to form hydrogen bonds with guest molecules, as it lacks hydrogen donors and acceptors. However, it contributes to the formation and stability of CD complexes, enhancing their stability in conjunction with van der Waals forces [30].

3.9 CDS FOR ENHANCING THE STABILITY OF UVR ABSORBER MOLECULES

3.9.1 UVR Absorbers-CDs Inclusion Complexes

CDs have been shown to protect UV absorbers from photodegradation. The photostability of a UV absorber within a CD inclusion complex is crucial, particularly in applications where protection against light-induced deterioration is essential. In CD complexes, each UV absorber is individually surrounded by the CD and micro-encapsulated, enhancing its chemical and physical properties by temporarily confining them within the host cavity, facilitating beneficial changes [20]. The characteristics of CD inclusion complexes offer several benefits, such as enhanced solubility for poorly soluble UV absorbers, stabilization of sensitive UV absorbers against degradation from environmental factors like oxidation, light, and heat, control of volatility and sublimation, physical separation of incompatible compounds (via chromatography), taste masking, odour elimination, and regulated release of drugs and flavours. These complexes have significant applications in cosmetics, food, agriculture, pharmaceuticals, textiles, and plastics [33]. The stability of CD inclusion complexes produced with UV absorbers is determined by multiple factors. Some of them are discussed below:

(i) *Hydrophobic cavity:* CDs feature a hydrophobic cavity and a toroidal structure capable of accommodating UV-absorbing molecules. Their hydrophobic or less polar sections protect these molecules from their environment, contributing to enhanced stability [17].

(ii) *Non-covalent interactions:* The formation of CD inclusion complexes relies on non-covalent interactions, including van der Waals forces, hydrogen bonds, and hydrophobic interactions. These interactions help stabilize the complex by confining the guest molecule within the CD cavity [30].

(iii) *Protection from external factors:* UV-absorbing molecules are prone to degradation when exposed to UV radiation, heat, or other environmental conditions. Encapsulating UV absorbers within CD cavities protects them from direct UV exposure. This protection prevents the photodegradation of UV-absorbing compounds, maintain their efficacy over time [1].

(iv) *Solubility enhancement:* CDs enhance the solubility of UV-absorbing compounds, which typically have poor water solubility. This increased solubility improves the stability and effectiveness of the inclusion complex [1].

3.9.2 The Effectiveness of UV Absorber-CD Complexes in Sunscreen Formulations

Sunscreen is a crucial component of skincare, providing protection against the harmful effects of UVR from the sun. The efficiency of sunscreens relies mainly on the presence of UV filters. However, these UV filters can experience changes in their chemical or physical characteristics owing to exposure to heat, UVR, or other environmental variables. This can result in photocatalytic and photolytic reactions, which degrade the product's structure and reduce its efficacy. Since UV filters can't entirely protect the skin from UVR, sunscreen formulations commonly incorporate antioxidants to provide an extra layer of resistance. Antioxidants help prevent DNA damage, skin aging, and disease by neutralizing UV-induced reactive oxygen species and free radicals [33].

Antioxidants are compounds that oxidize themselves before other substances, effectively averting oxidative damage. They enhance the shelf life of products by preventing deterioration caused by enzymatic and non-enzymatic oxidation. Antioxidants interact with free radicals, disrupting chain reactions before important components are affected [34]. They are extensively employed in food, cosmetics, beverages, and pharmaceuticals, performing as health supplements, active ingredients, or stabilizing agents. Natural antioxidants such as vitamins (A, B, C, E, and K), polyphenols (resveratrol, green tea polyphenol), rutin, and ferulic acid are incorporated into food, cosmetics, and pharmaceutical products [35].

In sunscreen formulations, antioxidants provide an extra line of defence against the detrimental effects of UVR. While they can't provide complete protection, sunscreens with UV filters effectively reflect or absorb a significant amount of UV rays. Antioxidants help mitigate the damage UVR causes to both the skin and the UV absorbers in sunscreens. Their key advantage lies in scavenging UV-induced free radicals, which delays the degradation of UV absorbers and reduces skin damage, premature aging, and disease [36]. However, the efficiency of natural antioxidants is limited due to poor absorption, low solubility, and the propensity to form aggregates in water. Additionally, natural antioxidants deteriorate when exposed to oxidative stimuli such as light, heat, air, moisture, and UVR [37].

The use of CD inclusion complexes is a novel technique for incorporating UV filters and antioxidants into formulations. Their ability to enhance the solubility and stability of lipophilic antioxidants, as well as protect UV filters from photodegradation, makes them a tempting alternative. Lipophilic antioxidants, such as flavonoids, polyphenols, and carotenoids, often encounter solubility challenges. Consequently, antioxidant-CD inclusion complexes comprising these antioxidants lend tremendous potential for cosmetic applications [33, 38].

3.10 RECENT ADVANCES IN CDS INCLUSION COMPLEXES FOR ENHANCING THE STABILITY OF UVR ABSORBERS

In the cosmetics industry, UV absorbers play a vital role in sunscreens and various cosmetic products. However, photodegradation of these filters leads to a gradual decrease in the product's effectiveness. To address this, CDs and their derivatives are used with different UV absorbers, such as avobenzone, benzophenone derivatives, methylene bis-benzotriazolyl tetramethylbutylphenol, phenylbenzimidazole sulfonic acid, para-aminobenzoic acid, octocrylene, oxybenzene, and ethylhexyl-methoxycinnamate. These form inclusion complexes that enhance the solubility, thermal stability, and photostability of the filters. For instance, β-CD is commonly incorporated into sunscreen formulations as a carrier or encapsulation agent for certain active ingredients, including UV absorbers.

Avobenzone, an organic UVA absorber used in sunscreen, has a diketo-form structure. Upon irradiation, this diketo form produces radicals such as arylglyoxals and benzils, leading to reduced UVA protection and potential skin damage. The use of β-CD significantly enhances the photostability of avobenzone. Yuan et al. studied the synthesis of photostability of an avobenzone and (2-hydroxy) propyl-β-CD inclusion complex. UV-vis, phase-solubility, IR, SEM, NMR, and XRD data revealed that 2-HP-β-CD can effectively encapsulate avobenzone. Under controlled conditions, the photolysis percentages of avobenzone and its complex were compared in solution after 200 minutes of irradiation. UV-vis spectrophotometric analysis showed 49% photolysis for avobenzene, but only 8% for the avobenzone-2-HP-β-CD inclusion complex. This study supports the development of an inclusion complex with greater photostability than non-complexed avobenzone [1, 39]. Another notable example of β-CD application in sunscreen chemicals, such as ethylhexyl-methoxycinnamate (Cin), oxybenzone (Oxy), and octocrylene (Oct), was demonstrated by Rawashdeh et al. The production of inclusion complexes was proven experimentally using DSC and ^{13}C NMR, as well as theoretically by molecular mechanics (MM) calculations and modelling. Spectrophotometric analysis showed that incorporating β-CD can enhance the photostability of these selected sunscreen ingredients in lotion formulations [33].

In the food industry, the use of UV absorbers with CD is not as common as in sunscreen or cosmetics. Nonetheless, some food packaging materials incorporate UV absorbers containing CD to enhance their resistance to UVR. A notable example is polyethylene terephthalate (PET), a widely used polymer in packaging. PET is known for its excellent transparency, making it suitable for clear containers, packaging sheets, and beverage bottles for water, soft drinks, and juices. However, PET is vulnerable to UVR, which can cause deterioration and a reduction in its mechanical properties over time [40]. Huang and coworkers developed a PET-EG-β-CD packaging material and evaluated its effectiveness in preserving chilled pork. Their results showed that the PET-EG-β-CD material extended the shelf life of the chilled pork by 2 to 4 days. Consequently, the PET-EG-β-CD film effectively maintained the quality of chilled pork and demonstrated its potential as an active packaging solution in the food industry [38].

In the pharmaceutical industry, the use of CD inclusion complexes with UV absorbers is of great interest as it improves the stability, solubility, and bioavailability of some medicines. While the direct mixing of CDs with UV absorbers is not a common practice in pharmaceuticals, CDs are often employed to enhance the properties of drugs, some of which may have intrinsic UV absorption or photosensitivity [41].

3.11 CONCLUSION

Cyclodextrins are one of the most fascinating macrocyclic compounds with prominent theoretical as well as practical significance particularly in the field of pharmacology, chemistry, and biology owing to their unique structural features, biocompatible and biodegradable nature and well-known safety profile in clinical applications for humans. By tailoring the host–guest chemistry of CD-inclusion complexes, properties like solubility, stability, bioavailability, photodegradation of guest molecules particularly UVR absorbers can be significantly altered. Physical and chemical UV absorbers or filters are the major active elements of sunscreens that protect skin from deteriorating UVR. CDs provide a protecting encapsulation that promotes the stability of antioxidants and UV filters upon prolonged exposure to sunlight as well as oxygen. In spite of significant exploration of CD inclusion complexes for wide applications, however, studies regarding novel and sustainable synthetic strategies to improve the stability of UVR absorbers using CDs along with their mechanism of action are still extremely important.

REFERENCES

1. Yuan, L., Li, S., Huo, D., Zhou, W., Wang, X., Bai, D. and Hu, J. (2019) 'Studies on the preparation and photostability of avobenzone and (2-hydroxy) propyl-β-cyclodextrin inclusion complex' *J. Photochem. Photobiol. A*, Vol 369, pp 174–180.
2. Agostino, S., Azzali, A., Casali, L., Taddei, P. and Grepioni, F. (2020) 'Environmentally friendly sunscreens: Mechanochemical synthesis and characterization of β-CD inclusion complexes of Avobenzone and Octinoxate with improved photostability' *ACS Sustainable Chem. Eng.*, Vol 8, pp 13215–13225.
3. Tian, L., Wu, Y., Hou, Y., Dong, Y., Ni, K. and Guo, M. (2024) 'Environmentally friendly UV absorbers: Synthetic characterization and biosecurity studies of the Host–Guest supramolecular complex' *Int. J. Mol. Sci.*, Vol 25, p 8476.
4. Nitulescu, G., Lupuliasa, D., Adam-Dima, I. and Nitulescu, G. M. (2023) 'Ultraviolet filters for cosmetic applications' *Cosmetics*, Vol 10, p 101.
5. Chisvert, A. and Salvador, A. (2007) 'UV filters in sunscreens and other cosmetics. Regulatory aspects and analytical methods' *Anal. Cosmet. Prod.*, pp 83–120.
6. Aguilera, J., Gracia-Cazaña, T. and Gilaberte, Y. (2023) 'New developments in sunscreens' *Photochem. Photobiol. Sci.*, Vol 22, pp 2473–2482.

7. Geisler, A. N., Austin, E., Nguyen, J., Hamzavi, I., Jagdeo, J. and Lim, H. W. (2021) 'Visible light. Part II: Photoprotection against visible and ultraviolet light' *J. Am. Acad. Dermatol.*, Vol 84, pp 1233–1244.
8. Mancuso, J. B., Maruthi, R., Wang, S. Q. and Lim, H. W. (2017) 'Sunscreens: An update' *Am. J. Clin. Dermatol.*, Vol 18, pp 643–650.
9. Rai, R., Shanmuga, S. C. and Srinivas, C. (2012) 'Update on photoprotection' *Indian J. Dermatol.*, Vol 57, pp 335–342.
10. Braga, S. S., Lysenko, K., El-Saleh, F. and Paz, F. A. A. (2020) 'Cyclodextrin-Efavirenz complexes investigated by solid state and solubility studies' *Proceedings*, Vol 78, p 15.
11. Crumling, M. A., King, K. A. and Duncan, R. K. (2017) 'Cyclodextrins and Iatrogenic hearing loss: New drugs with significant risk' *Front. Cell. Neurosci.*, Vol 11, p 355.
12. Ho, S., Thoo, Y. Y., Young, D. J. and Siow, L. F. (2017) 'Cyclodextrin encapsulated catechin: Effect of pH, relative humidity and various food models on antioxidant stability' *LWT – Food Sci. Technol.*, Vol 85, pp 232–239.
13. Zhou, J., Jia, J., He, J., Li, J. and Cai, J. (2022) 'Cyclodextrin inclusion complexes and their application in food safety analysis: Recent developments and future prospects' *Foods*, Vol 11, p 3871.
14. Li, Z., Chen, S., Gu, Z., Chen, J. and Wu, J. (2014) 'Alpha-cyclodextrin: Enzymatic production and food applications' *Trends Food Sci. Technol.*, Vol 35, pp 151–160.
15. Valle, E. M. M. D. (2004) 'Cyclodextrins and their uses: A review' *Process Biochem.*, Vol 39, pp 1033–1046.
16. Astray, G., González-Barreiro, C., Mejuto, J. C., Rial-Otero, R. and Simal-Gándara, J. (2009) 'A review on the use of cyclodextrins in foods' *Food Hydrocoll.*, Vol 23, pp 1631–1640.
17. Christaki, S., Spanidi, E., Panagiotidou, E., Athanasopoulou, S., Kyriakoudi, A., Mourtzinos, I. and Gardikis, K. (2023) 'Cyclodextrins for the delivery of bioactive compounds from natural sources: Medicinal, food and cCosmetics applications' *Pharm.*, Vol 16, p 1274.
18. Shelley, H. and Babu, R. J. (2018) 'Role of cyclodextrins in nanoparticle-based drug delivery systems' J. Pharm. Sci., Vol 107, pp 1741–1753.
19. Cid-Samamed, A., Rakmai, J., Mejuto, J. C., Simal-Gándara, J. and Astray, G. (2022) 'Cyclodextrins inclusion complex: Preparation methods, analytical techniques and food industry applications' *Food Chem.*, Vol 384, p 132467.
20. Cheirsilp, B. and Rakmai, J. (2016) 'Inclusion complex formation of cyclodextrin with its guest and their applications' *Biol. Eng. Med.*, Vol 2, p 1–6.
21. Pandey, D., Panwar, V. S., Mishra, H., Adhikari, L., Pandey, M. and Semalty, M. (2021) 'Cyclodextrin based nanoparticles for improved solubility and drug delivery' *J. Mountain Res.*, Vol 16, pp 187–199.
22. Zia, V., Rajewski, R. A. and Stella, V. J. (2001) 'Effect of cyclodextrin charge on complexation of neutral and charged substrates: comparison of (SBE)7M-Beta-CD to HP-Beta-CD' *Pharm. Res.*, Vol 18, pp 667–673.
23. Jug, M. and Mura, P. A. (2018) 'Grinding as Solvent-Free green chemistry approach for Cyclodextrin inclusion complex preparation in the solid state' *Pharmaceutics*, Vol 10, pp 189.
24. Hernández-Sánchez, P., López-Miranda, S., Guardiola, L., Serrano-Martínez, A., Gabaldón, J. A. and Núñez-Delicado, E. (2016) 'Optimization of a method for preparing solid complexes of essential clove oil with β-cyclodextrins' *J. Sci. Food Agric.*, Vol 97, pp 420–426.

25. Kaur, K., Jindal, R. and Jindal, D. (2019) 'Synthesis, characterization and studies on host-guest interactions of inclusion complexes of metformin hydrochloride with β–cyclodextrin' *J. Mol. Liq.*, Vol 282, pp 162–168.
26. Liu, Y., Chen, Y., Gao, X., Fu, J. and Hu, L. (2020) 'Application of cyclodextrin in food industry' Crit. Rev. Food Sci. Nutr., Vol 62, pp 2627–2640.
27. Tian, Y., Zhu, Y., Bashari, M., Hu, X., Xu, X. and Jin, Z. (2013) 'Identification and releasing characteristics of high-amylose corn starch–cinnamaldehyde inclusion complex prepared using ultrasound treatment' *Carbohydr. Polym.*, Vol 91, pp 586–589.
28. Mura, P. (2014) 'Analytical techniques for characterization of cyclodextrin complexes in aqueous solution: A review' *J. Pharm. Biomed. Anal.*, Vol 101, pp 238–250.
29. Liu, L. and Guo, Q. (2002) 'The driving forces in the inclusion complexation of Cyclodextrins' *J. Incl. Phenom. Macrocycl. Chem.*, Vol 42, pp 1–14.
30. Poulson, B. G., Alsulami, Q. A., Sharfalddin, A., Agammy, E. F. El, Mouffouk, F., Emwas, A. H., Jaremko, L. and Jaremko, M. (2022) 'Cyclodextrins: Structural, chemical, and physical properties, and applications' *Polysaccharides*, Vol 3, pp 1–31.
31. Al-Rawashdeh, N. A. F., Al-Sadeh, K. S. and Al-Bitar, M. B. (2013) 'Inclusion complexes of sunscreen agents with β-Cyclodextrin: Spectroscopic and molecular modeling studies' *J. Spectrosc.*, Vol 2013, pp 1–11.
32. Song, L. X., Bai, L., Xu, X. M., He, J. and Pan, S. Z. (2009) 'Inclusion complexation, encapsulation interaction and inclusion number in cyclodextrin chemistry' *Coord. Chem. Rev.*, Vol 253, pp 1276–1284.
33. Dahabra, L., Broadberry, G., Gresley, A. L., Najlah, M. and Khoder, M. (2021) 'Sunscreens containing cyclodextrin inclusion complexes for enhanced efficiency: A strategy for skin cancer prevention' *Molecules*, Vol 26, p 1698.
34. Cherubim, D. J. D. L., Martins, C. V. B., Fariña, L. O., Lucca, R. A. D. S. (2019) 'Polyphenols as natural antioxidants in cosmetics applications' *J. Cosmet. Dermatol.*, Vol 19, pp 33–37.
35. Hoang, H. T., Moon, J. Y. and Lee, Y. C. (2021) 'Natural antioxidants from plant extracts in skincare cosmetics: Recent applications, challenges and perspectives' *Cosmetics*, Vol 8, p 106.
36. Souza, C., Campos, P. M., Schanzer, S., Albrecht, S., Lohan, S. B., Lademann, J., Darvin, M. E. and Meinke, M. C. (2017) 'Radical-Scavenging activity of a sunscreen enriched by antioxidants providing protection in the whole solar spectral range' *Skin Pharmacol. Physiol.*, Vol 30, pp 81–89.
37. Wang, H., Wang, S., Zhu, H., Wang, S. and Xing, J. (2019) 'Inclusion complexes of Lycopene and B-Cyclodextrin: Preparation, characterization, stability and antioxidant activity' *Antioxidants*, Vol 8, p 314.
38. Huang, H., Fan, L., Wu, L., Du, J. and Sun, W. (2023) 'Preparation of PET-EG-β-CD mat for the preservation of chilled pork' *Int. J. Food Sci. Technol.*, Vol 58, pp 4572–4581.
39. Yadav, H. K. S., Kasina, S. and Raizaday, A. (2016) 'Sunscreens' *Nanobiomaterials in Galenic Formulations and Cosmetics*, pp 201–230, William Andrew Publishing.
40. Benyathiar, P., Kumar, P., Carpenter, G., Brace, J. and Mishra, D. K. (2022) 'Polyethylene Terephthalate (PET) bottle-to-bottle recycling for the beverage industry: A review' *Polymers*, Vol 14, p 2366.
41. Eid, E. E. M., Almaiman, A. A., Alshehade, S. A., Alsalemi, W., Kamran, S., Suliman, F. E. O. and Alshawsh, M. A. (2023) 'Characterization of Thymoquinone-Sulfobutylether-B-Cyclodextrin inclusion complex for anticancer applications' *Molecules*, Vol 28, p 4096.

4 Cyclodextrins Unveiled

Revolutionizing UV Protection Across Diverse Sectors

Santosh Kumar Badampudi, K.V. Nagalakshmi, and Santosh Kumar Nadikatla

4.1 INTRODUCTION

The growing interest in UV radiation highlights its benefits across various sectors. UVB plays a key role in human skin functions like vitamin D synthesis [1], melanogenesis [2], wound healing [3], and dermatological treatments. In agriculture, UV affects plant growth and stress responses, while pharmaceuticals use it for sterilization, phototherapy, and compound synthesis. The beverage industry employs UV for microbial disinfection and product preservation.

However, UV radiation also causes degradation, discoloration, and loss of efficacy in products, and affects the quality and safety of crops and food. UVB and UVA rays can harm skin, leading to sunburn, aging, hyperpigmentation, and increased skin cancer risk.

To address these issues, cyclodextrins, known for their molecular architecture, have been used to enhance the stability and photoprotection of products. They form inclusion complexes with UV absorbers, improving solubility, controlled release, and reducing toxicity. This chapter explores cyclodextrins as UV filters in cosmetics, pharmaceuticals, food, and beverages, detailing their mechanisms and providing examples.

4.1.1 Cyclodextrins

Cyclodextrins are crystalline, water-soluble, cyclic oligosaccharides that are structurally related natural products. They're created through bacterial breakdown of cellulose, comprising glucose monomers arranged in a ring structure resembling a donut [4]. The most common types of cyclodextrins include α, β, and γ, which have six, seven, or eight anhydro-glucose units in their ring structure, respectively (Figure 4.1) [5].

Cyclodextrins serve as "molecular cages" and find widespread applications in the cosmetic [6], pharmaceutical [7], agrochemical [8], food industries [9], etc. These

DOI: 10.1201/9781003492757-4

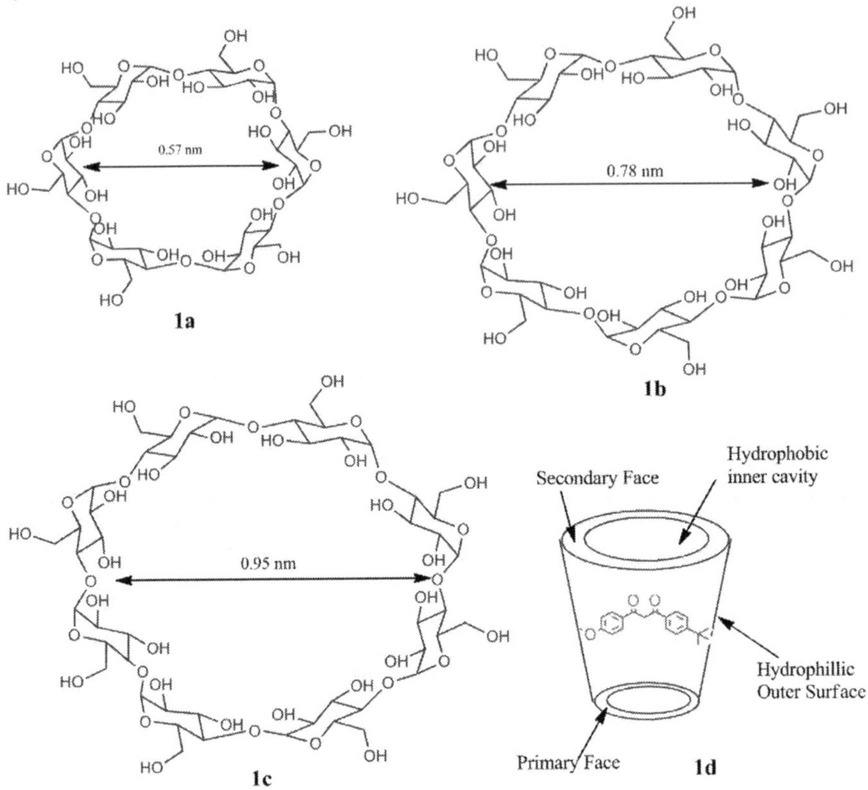

FIGURE 4.1 Structures of (a) α-cyclodextrin, (b) β-cyclodextrin, and (c) γ-cyclodextrin, (d) inclusion complex formation.

valuable tools enhance the solubility, stability, and delivery of active ingredients in various formulations because of the following properties.

4.2 ROLE OF CYCLODEXTRINS AS UVR FILTERS IN THE COSMETIC SECTOR

Scientific research has highlighted the harmful effects of UV radiation on skin, including erythema, premature aging, depigmentation, and increased skin cancer risk. This has led to a greater emphasis on sun protection and the development of effective cosmetic formulations. In response, the cosmetics industry is focusing on creating antioxidants and sunscreens that offer broad-spectrum protection, are water-resistant, non-comedogenic, and suitable for various skin types. Consumers also prefer lightweight, easy-to-apply, and aesthetically appealing products, prompting further innovation in sunscreen formulations. Figure 4.2 illustrates the effects of UV radiation across the layers of the skin.

The cosmetics industry's focus on antioxidants and sunscreens aims to meet the growing demand for UV protection, safety, efficacy, and user experience. Research

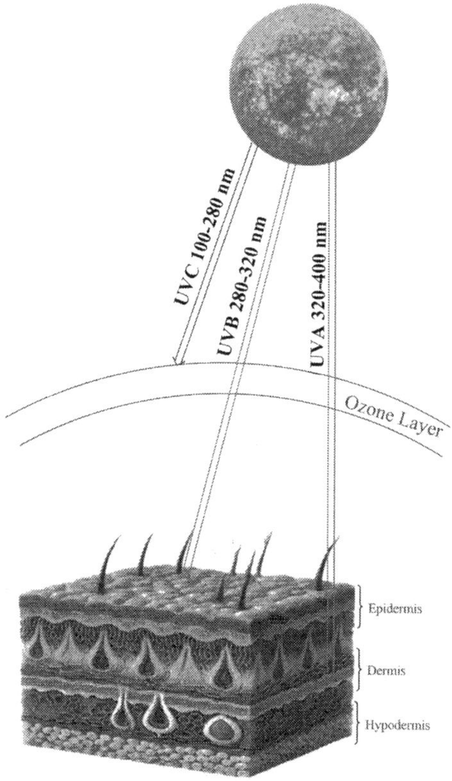

FIGURE 4.2 Effect of UV rays on skin. UVC (100–280 nm) absorbed by the ozone layer and does not reach the Earth's surface. UVB (280–320 nm) absorbed by the epidermis, responsible for sunburn and contributes to skin cancer. UVA (320–400 nm) penetrates deeper into the dermis, associated with long-term skin damage and photoaging.

has led to the use of cyclodextrins in sunscreen formulations, thanks to their ability to encapsulate UV-sensitive compounds and improve solubility. Cyclodextrins offer benefits such as enhanced photoprotection, stability, skin compatibility, reinforced antioxidant activity, moisture retention, fragrance encapsulation, and improved delivery of bioactive compounds.

4.2.1 The Key Functions of Cyclodextrins in Cosmetics

- Enhancing the apparent water solubility of hydrophobic molecules.
- Safeguarding labile molecules from degradation.
- Preserving the stability of emulsions and suspensions.
- Transforming fluids into dry powders, reducing skin sensitivity.
- Preventing undesirable interactions between formulation components.
- Masking unpleasant odours.
- Modulating skin permeability while controlling the release of active ingredients.

Recent data has highlighted the utilization of cyclodextrins in cosmetic formulations, particularly in perfumes, where they form host–guest inclusion complexes to enable controlled fragrance release [10]. Furthermore, cyclodextrins offer numerous advantages in cosmetics, including applications in anti-aging skincare, sunscreens, deodorants, antiperspirants, and shampoos. Commercialized cosmetics and cosmeceuticals containing free cyclodextrins or active ingredient-cyclodextrin inclusion complexes are summarized in Table 4.1.

TABLE 4.1
Some Commercialized Cosmetics Containing Cyclodextrins

Type of Cosmetic	Commercial Name	Active Ingredient with Cyclodextrin
Skincare	Olay Regenerist Micro-Sculpting Cream	CD-peptide complexes for anti-aging
	Neutrogena Hydro Boost Water Gel	CD stabilize hyaluronic acid for hydration
Sunscreens	La Roche-Posay Anthelios Melt-In Sunscreen SPF 60	CD-encapsulated UV filter for sun protection
	Coppertone Sport Sunscreen SPF 50	Cyclodextrins improve UV filter stability
Hair Care	Pantene Pro-V Daily Moisture Shampoo	Cyclodextrin complexes stabilize Pro-Vitamin B5
	Redken Extreme Anti-Snap Treatment	Cyclodextrins encapsulate ceramides for hair repair
Makeup	MAC Studio Fix Fluid Foundation	Cyclodextrin-encapsulated pigments for long-lasting coverage
	Maybelline Super Stay Matte Ink Lipstick	Cyclodextrins enhance pigment stability and vibrancy
Oral Care	Colgate Total Whitening Toothpaste	Cyclodextrin-hydrogen peroxide complex for stain removal
	Listerine Zero Mouthwash	Cyclodextrins encapsulate essential oils for fresh breath
Shampoos	Head & Shoulders Classic Clean Shampoo	Cyclodextrin-encapsulated Pyrithione zinc
	Herbal Essences Bio: Renew Argan Oil Shampoo	Cyclodextrin-encapsulated Argan oil extract
Deodorants	Secret Clinical Strength Antiperspirant	Cyclodextrin-encapsulated aluminum compounds
Skincare	Olay Regenerist Micro-Sculpting Cream	Cyclodextrin-peptide complexes for anti-aging
	Neutrogena Hydro Boost Water Gel	Cyclodextrins stabilize hyaluronic acid for hydration
Sunscreens	La Roche-Posay Anthelios Melt-In Sunscreen SPF 60	CD-encapsulated UV filter for sun protection
	Coppertone Sport Sunscreen SPF 50	Cyclodextrins improve UV filter stability

In addition to cyclodextrins' roles in cosmetics, ideal sunscreens should provide broad-spectrum UVA and UVB protection, demonstrate photostability, and prevent skin penetration to avoid adverse effects. They should be easy to apply, non-irritating, transparent, and regulatory-approved. Environmental safety is also crucial, ensuring sunscreen ingredients do not harm ecosystems when washed off into natural waters.

4.2.2 Essential Features of an Effective Sunscreen: Leveraging the Power of Cyclodextrin

4.2.2.1 Protection Against UVA and UVB Radiation

Effective sunscreens must offer broad-spectrum protection against UVA and UVB radiation. Due to the adverse effects of chemical UV filters, researchers have explored cyclodextrins (CDs) to encapsulate UV filters. CDs, particularly hydroxypropyl-β-CD (HP-β-CD), reduce UV filter penetration into the skin, forming inclusion complexes that shield UV-sensitive compounds. These complexes absorb and scatter UV rays, enhancing stability and prolonged UV protection.

4.2.2.2 Photostability

Cyclodextrin-based sunscreens offer superior photostability by encapsulating UV filters, preventing degradation from UV radiation, and maintaining sunscreen efficacy over time.

4.2.2.3 Surface Adherence

Cyclodextrins improve surface adherence by forming stable films on the skin, preventing deep penetration of UV filters and ensuring continuous protection.

4.2.2.4 Sustained Efficacy

By encapsulating active ingredients, cyclodextrins ensure sustained sunscreen efficacy, preventing degradation and maintaining consistent UV protection.

4.2.2.5 Ease of Use and Safety

Cyclodextrin-based sunscreens are easy to apply, non-irritating, and compatible with various skin types, offering user-friendly application with minimal adverse reactions.

4.2.2.6 Aesthetic Enhancement

Cyclodextrins enhance cosmetics by masking unfavourable odours and flavours.

4.2.2.7 Environmentally Friendly

Cyclodextrin-based sunscreens are biodegradable and low in toxicity, reducing their environmental footprint and protecting ecosystems.

4.2.2.8 Enhanced SPF Value

Cyclodextrins in sunscreen formulations can increase SPF value, providing equivalent protection to higher SPF commercial sunscreens. See Table 4.2 for examples of sunscreens with enhanced SPF after cyclodextrin inclusion.

TABLE 4.2
List of Various Sunscreens with Enhanced SPF Due to Inclusion of Cyclodextrins

Sunscreen	Cyclodextrin Used	SPF		Ref.
		Before Inclusion	After Inclusion	
Avobenzone	Hydroxypropyl-β-cyclodextrin	15	30	[11]
Oxybenzone	Methyl-β-cyclodextrin	20	40	[12]
Octocrylene	Random methyl-β-cyclodextrin	25	35	[13]
Ethylhexyl methoxycinnamate (EHMC)	Hydroxypropyl-γ-cyclodextrin	30	45	[14]
Homosalate	Hydroxypropyl-β-cyclodextrin	15	25	[15]
Octinoxate	Methyl-β-cyclodextrin	20	35	[16]
Octisalate	Methyl-β-cyclodextrin	15	30	[17]
Padimate O	Hydroxypropyl-γ-cyclodextrin	10	20	[18]
Ensulizole	Hydroxypropyl-β-cyclodextrin	25	40	[19]
Mexoryl SX	Hydroxypropyl-β-cyclodextrin	30	50	[20]

4.2.3 Working Mechanism of Cyclodextrins as UV Filters in Cosmetics

Cyclodextrins (CDs) act as protective shields for encapsulated antioxidants and UV filters, enhancing their stability against external factors and oxidative stress [21, 22]. These compounds are essential components of sunscreens, reflecting, absorbing, or dispersing UV radiation to prevent its penetration into the skin's stratum corneum (SC). While effective in reducing UV radiation exposure, UV filters can potentially penetrate the skin, leading to systemic effects. Chemical sunscreens absorb UV radiation, while physical sunscreens containing mineral particles like zinc oxide or titanium dioxide reflect and scatter UV rays, offering protection against various skin diseases.

The chemical stability of CDs stems from the absence of reducing glycosidic units in their structures, along with hydrogen bonds formed between adjacent hydroxyl groups on carbon atoms of the glucopyranose units [21]. Additionally, CDs resist enzymatic hydrolysis by burying bridge oxygens inside the non-polar ring center [22]. These CDs bind to encapsulated antioxidants or UV filters, limiting their interaction with extracellular molecules [23]. Moreover, the hydrophilic outer surface of CD-antioxidant inclusion complexes enhances their solubility and bioavailability, facilitating controlled release from topically applied sunscreens [24]. CDs in sunscreen formulations can also form a protective film and act as UV filter adjuvants by absorbing UV radiation [25, 26].

Cyclodextrins (CDs) are effective in protecting UV filters from photodegradation, oxidation, and preventing deeper skin penetration [27, 28]. For example, β-CD stabilizes UV filters like oxybenzone, octocrylene, and ethylhexyl-methoxycinnamate, improving their photostability [29]. Research also shows that CDs enhance the solubility and photostability of 4-methylbenzylidene camphor (4-MBC), a common UVB filter [30].

CDs not only stabilize antioxidants but also improve their solubility, potentially reducing costs and enhancing therapeutic outcomes [31, 32]. The formation of antioxidant-CD inclusion complexes depends on factors like solvent, pH, temperature, and mixing time, with CDs protecting active ingredients from degradation [33, 34]. Their unique structure allows them to encapsulate hydrophobic molecules, enhancing stability and facilitating targeted delivery [35, 36].

In conclusion, cyclodextrins are essential in sunscreen formulations, offering UV protection, photostability, surface adherence, and sustained efficacy. As demand for effective sun protection grows, CDs will continue to play a key role in advancing sunscreen technology, promoting skin health, and preventing UV-induced damage.

4.3 ROLE OF CYCLODEXTRINS AS UV FILTERS IN THE PHARMACEUTICAL SECTOR

In the pharmaceutical industry, protecting sensitive drug compounds from the detrimental effects of ultraviolet (UV) radiation is of paramount importance to ensure the stability, efficacy, and safety of pharmaceutical formulations. UV radiation, particularly in the UVB and UVA spectrums, can induce photodegradation of drugs, leading to reduced potency, altered pharmacokinetics, and potential toxicity. To address this challenge, cyclodextrins have emerged as indispensable tools in pharmaceutical formulation due to their unique molecular structure and properties [37].

4.3.1 Various Types of Drugs are Affected by UV Radiation and Are Protected by Cyclodextrin

4.3.1.1 Photolabile Drugs

Certain drugs are known to be sensitive to light and can undergo photodegradation upon exposure to UV radiation. For example, retinoids, which are used in the treatment of acne and other skin conditions, are sensitive to light. By encapsulating retinoids within cyclodextrins, their stability can be improved, allowing for the development of more effective topical formulations [38].

A few examples of photolabile drugs that can undergo degradation upon exposure to UV radiation are shown below in Table 4.3. Protecting these drugs from light exposure, such as through the use of opaque packaging or UV filters, is important to ensure their stability and effectiveness. Cyclodextrins can be used as UV filters to protect these photolabile drugs from degradation caused by exposure to UV radiation, thereby preserving their stability and efficacy.

4.3.1.2 Peptide and Protein Drugs

Peptide and protein drugs, such as insulin and growth hormones, are also sensitive to UV radiation and can undergo degradation when exposed to light. Encapsulating these drugs within cyclodextrins can protect them from UV-induced degradation, thereby improving their stability and bioavailability. In Table 4.4 some examples for peptide and protein drugs are shown [39].

TABLE 4.3
Effect of UV Radiation on Some Photolabile Drugs

Drug Type / Drug Name	Activity	Effect of UV Radiation
Furosemide	A diuretic medication used to treat fluid retention (edema) and high blood pressure	It is sensitive to light and can degrade upon exposure to UV radiation, potentially reducing its effectiveness.
Ciprofloxacin	A broad-spectrum antibiotic used to treat various bacterial infections.	Photolabile and can degrade when exposed to UV radiation.
Nitrofurantoin	Antibiotic used to treat urinary tract infections.	It is sensitive to light and can undergo photodegradation, which may result in decreased antimicrobial activity.
Chlorpromazine	An antipsychotic medication used to treat schizophrenia and other psychiatric disorders.	It is known to be photolabile and can degrade when exposed to light, potentially affecting its therapeutic efficacy.

TABLE 4.4
Effect of UV Radiation on Some Peptide and Protein Drugs

Drug Type/Drug Name	Activity	Effect of UV Radiation
Insulin	Regulates blood sugar in diabetes.	Sensitive to UV; exposure causes degradation and potency loss.
Growth Hormone	Treats growth hormone deficiency.	UV exposure leads to degradation and reduced activity.
Calcitonin	Treats osteoporosis and Paget's disease.	UV exposure degrades the drug, reducing efficacy.
Enzymes	Used in replacement therapy and digestion.	UV causes denaturation and loss of activity.
Vaccines	Contain peptide/protein antigens.	UV degrades antigens, reducing immune response efficacy.

Protecting peptide and protein drugs from light exposure, such as through the use of opaque packaging or UV filters, is essential to ensure their stability and effectiveness. Cyclodextrins can be used as UV filters to protect these sensitive drugs from degradation caused by exposure to UV radiation, thereby preserving their stability and therapeutic efficacy.

4.3.1.3 Photosensitive Chemotherapeutic Agents

Some chemotherapeutic agents used in cancer treatment are photosensitive and can degrade upon exposure to UV radiation. By encapsulating these drugs within cyclodextrins, their stability can be enhanced, leading to improved therapeutic outcomes and reduced side effects [40]. Some examples for these types of drugs are shown below in Table 4.5.

Protecting photosensitive chemotherapeutic agents from light exposure, such as through the use of opaque packaging or UV filters, is important to ensure their stability and effectiveness in cancer treatment. Cyclodextrins can be used as UV filters to protect these sensitive drugs from degradation caused by exposure to UV radiation, thereby preserving their stability and therapeutic efficacy.

4.3.1.4 Miscellaneous

UV Protection in Pharmaceuticals Using Cyclodextrins

- **Diclofenac Sodium**: Diclofenac sodium, a non-steroidal anti-inflammatory drug (NSAID), is sensitive to UV radiation, which can degrade its efficacy. Cyclodextrins, like hydroxypropyl-beta-cyclodextrin (HPβCD), protect diclofenac by encapsulating it within their hydrophobic cavity, forming inclusion complexes that shield the drug from UV exposure, preserving its stability and effectiveness.
- **Retinoids (e.g., Tretinoin)**: Retinoids, used for acne and photoaging treatment, are also UV-sensitive. Cyclodextrins, such as HPβCD and methyl-beta-cyclodextrin (MβCD), protect retinoids from photodegradation

TABLE 4.5
Effect of UV Radiation on Some Photosensitive Chemotherapeutic Agents

Drug Type/Drug Name	Activity	Effect of UV Radiation
Vinca Alkaloids (e.g., vinblastine, vincristine)	Used for leukemia, lymphoma, and solid tumours	UV exposure leads to degradation and reduced potency.
Anthracyclines (e.g., doxorubicin, daunorubicin)	Treats leukemia, lymphoma, breast cancer	Photosensitive; degradation reduces effectiveness.
Taxanes (e.g.,paclitaxel, docetaxel)	Used for breast, lung, ovarian cancers	Sensitive to UV; degradation causes potency loss.
Platinum-Based Drugs (e.g., cisplatin, carboplatin)	Treats ovarian, lung, testicular cancers	UV exposure causes degradation and loss of efficacy.
Methotrexate	Used for leukemia, lymphoma, autoimmune diseases	Photosensitive; degradation reduces effectiveness.

by encapsulating them, ensuring stability and efficacy in light-exposed formulations.

- **Cyclodextrins' Mechanism**: Cyclodextrins, cyclic oligosaccharides with a hydrophobic cavity and hydrophilic exterior, can encapsulate poorly soluble or UV-sensitive drugs. This encapsulation enhances drug solubility and bioavailability while protecting them from UV degradation, making cyclodextrins valuable UV filters in pharmaceutical formulations.

Cyclodextrins play a crucial role in the pharmaceutical industry, safeguarding drug stability and enhancing efficacy in light-sensitive formulations.

4.3.2 Significance Applications of Cyclodextrin in Pharmaceuticals

UV Protection: Cyclodextrins can encapsulate UV-sensitive pharmaceutical compounds within their cavity, effectively shielding them from harmful UV radiation. This protection helps prevent photodegradation of drugs and active ingredients, ensuring their stability and efficacy.

The photodegradation of an antibiotic drug **tetracycline** is protected by encapsulation in hydrophobic cavity of cyclodextrins. This encapsulation shields tetracycline from direct exposure to light, thereby preventing the initiation of photodegradation reactions. Moreover, the stable and rigid structure of cyclodextrins provides a protective environment that stabilizes the molecular structure of tetracycline, minimizing the formation of reactive intermediates that could lead to degradation upon light exposure. Additionally, the encapsulation of tetracycline within cyclodextrins enhances its solubility and bioavailability, facilitating better absorption and distribution within the body. Consequently, this protective mechanism extends the shelf life of tetracycline formulations, ensuring sustained efficacy even under light-exposed storage conditions. [41]

Enhanced Solubility: Many UV-sensitive compounds have poor solubility in aqueous solutions, which can limit their effectiveness in pharmaceutical formulations. Cyclodextrins can improve the solubility of these compounds by forming inclusion complexes, making them more suitable for use in various dosage forms such as tablets, creams, and solutions.

Curcumin, a drug with low solubility and susceptibility to degradation by UV radiation, limiting its bioavailability and therapeutic efficacy. hydroxypropyl-β-cyclodextrin (HP-β-CD), can encapsulate curcumin molecules within their hydrophobic cavities, shielding them from direct exposure to UV light. This encapsulation not only enhances the solubility of curcumin but also protects it from photodegradation, thereby improving its stability and bioavailability. As a result, cyclodextrin-complexed curcumin formulations offer a promising strategy to overcome the challenges associated with its poor solubility and UV sensitivity, ensuring optimal therapeutic outcomes.

Improved Bioavailability: By increasing the solubility and dissolution rate of UV-sensitive drugs, cyclodextrins can enhance their bioavailability. This means that more of the active ingredient is absorbed into the bloodstream, leading to better therapeutic outcomes for patients.

Retinol, a form of vitamin A, is commonly used in skincare products. Retinol is sensitive to UV light, which can lead to its degradation and reduced effectiveness in promoting skin health. By encapsulating retinol within cyclodextrin molecules, such as β-cyclodextrin, its stability against UV radiation is increased. Additionally, cyclodextrin enhances the solubility of retinol, facilitating its absorption into the skin and improving its bioavailability. This allows for more efficient delivery of retinol to target cells, resulting in enhanced skin rejuvenation and anti-aging effects.

Versatility in Formulation: Cyclodextrins are compatible with a wide range of pharmaceutical excipients and formulation techniques, allowing for flexibility in product design. They can be incorporated into different dosage forms and formulations to provide UV protection tailored to specific patient needs.

Octinoxate, also known as octyl methoxycinnamate, a common chemical UV filter used in sunscreens. Octinoxate can experience photodegradation when exposed to UV radiation, leading to a reduction in its effectiveness in providing sun protection. Cyclodextrins, such as hydroxypropyl-β-cyclodextrin (HP-β-CD), can encapsulate octinoxate molecules, protecting them from direct UV exposure and preventing photodegradation. This encapsulation enhances the stability of octinoxate in sunscreen formulations, ensuring prolonged protection against harmful UV rays.

Stability Enhancement: UV radiation can induce degradation of pharmaceutical compounds, leading to loss of potency and efficacy. Cyclodextrins act as stabilizing agents, protecting UV-sensitive molecules from degradation and ensuring the stability of pharmaceutical formulations over time.

Ibuprofen, a commonly used nonsteroidal anti-inflammatory drug (NSAID). Ibuprofen can be susceptible to degradation when exposed to environmental factors such as light, moisture, and temperature variations. Cyclodextrins, such as hydroxypropyl-β-cyclodextrin (HP-β-CD), can encapsulate Ibuprofen molecules, providing protection against these destabilizing factors. This encapsulation enhances the stability of Ibuprofen, ensuring its effectiveness and shelf life in pharmaceutical formulations.

4.3.3 Drug Delivery Systems

Cyclodextrins play a crucial role as UV filters in drug delivery systems, primarily responsible for protecting UV-sensitive pharmaceutical compounds from degradation caused by exposure to ultraviolet (UV) radiation. This function is critical as it directly contributes to maintaining the stability and effectiveness of medications that are prone to degradation when exposed to light [9].

Cyclodextrins act as protective agents in drug delivery systems by encapsulating UV-sensitive pharmaceutical compounds within inclusion complexes. This encapsulation shields active ingredients from UV exposure, reducing degradation reactions and preserving drug stability and potency. By preventing UV-induced degradation, cyclodextrins help maintain the therapeutic efficacy and safety of medications throughout their shelf life. Their application highlights their importance in developing stable and effective drug delivery systems.

4.3.3.1 Mechanism of Action

The mechanism involves the encapsulation of the UV-sensitive drug molecules within the hydrophobic cavity of cyclodextrins, shielding them from direct UV exposure and preventing photodegradation. Here's a detailed step by step explanation along with examples:

Encapsulation: Cyclodextrins have a unique structure consisting of a hydrophobic cavity and a hydrophilic exterior. UV-sensitive drug molecules can be encapsulated within the hydrophobic cavity of cyclodextrins through non-covalent interactions such as van der Waals forces, hydrogen bonding, and hydrophobic interactions.

Protection from UV Radiation: Once encapsulated, the drug molecules are shielded from direct exposure to UV radiation. UV radiation can cause photodegradation of the drug molecules, leading to a decrease in their potency and efficacy. However, by being encapsulated within cyclodextrins, the drug molecules are protected from UV-induced degradation.

Enhanced Stability: The encapsulation of drug molecules by cyclodextrins enhances their stability by preventing interactions with UV radiation. Cyclodextrins act as a physical barrier, absorbing and scattering UV radiation, thereby preventing it from reaching the encapsulated drug molecules and inducing degradation reactions.

Controlled Release: Cyclodextrins can release the encapsulated drug molecules in a controlled manner. When the drug delivery system is administered, the cyclodextrin molecules gradually release the encapsulated drug molecules, ensuring a sustained and controlled release profile.

Table 4.6 presents a compilation of several commercially available drugs utilized for various purposes, all of which contain cyclodextrin along with its respective role. In the majority of these drugs, the commonly employed cyclodextrin ishydroxypropyl-beta-cyclodextrin (HPβCD) or sulfobutylether-beta-cyclodextrin (SBEβCD).

Overall, cyclodextrins offer pharmaceutical developers a versatile and effective solution for incorporating UV protection into drug products. Their ability to enhance solubility, bioavailability, and stability makes them invaluable tools in the formulation of UV-sensitive pharmaceuticals, contributing to the development of safer and more effective medications.

TABLE 4.6
List of Some Commercially Available Drugs Containing Cyclodextrins and Its Role

Name of the Drug	Activity of the Drug	Role of Cyclodextrin
Ciclodan (Ciclopirox Olamine)	Antifungal	Improves solubility of ciclopirox olamine.
Fungirox (Ciclopirox Olamine)	Antifungal	Enhances solubility and bioavailability.
Sporanox (Itraconazole)	Antifungal	Enhances solubility, absorption, and bioavailability.
Tracleer (Bosentan)	PAH treatment	Improves solubility and bioavailability of bosentan.
Aldurazyme (Laronidase)	Treats Hurler syndrome	Stabilizer to improve stability and shelf life.
Voriconazole	Antifungal	Enhances solubility and bioavailability for fungal infections.
Saxenda (Liraglutide)	Weight management	Improves stability and solubility for effective delivery.
Rapamune (Sirolimus)	Immunosuppressive	Enhances solubility and bioavailability for transplant rejection prevention.
Firazyr (Icatibant)	Treats hereditary angioedema	Improves solubility and stability for rapid absorption.
Victoza (Liraglutide)	Type 2 diabetes treatment	Enhances solubility and stability for effective blood glucose control.

4.4 ROLE OF CYCLODEXTRINS AS UV FILTERS IN THE FOOD SECTOR

In the food industry, where innovation and quality are crucial, cyclodextrins emerge as unsung heroes [42, 43]. These ring-shaped molecules, composed of glucose units, excel at encapsulating UV-sensitive compounds, shielding them from harmful UV radiation.

Beyond protection, cyclodextrins enhance preservation and quality. They prevent UV-induced degradation, extend shelf life, and maintain food's colour, flavour, and texture. By safeguarding nutritional integrity, they ensure food remains fresh and enjoyable.

For instance, cyclodextrins stabilize light-sensitive additives like beta-carotene, curcumin, and vitamins (D, A, and E), preventing degradation and preserving potency in fortified foods. These applications highlight their vital role as UV filters in ensuring food quality and safety.

4.4.1 Mechanism of Action

Cyclodextrins act as UV filters by forming inclusion complexes with UV-sensitive molecules, creating a protective barrier that prevents UV-induced degradation.

Additionally, they function as antioxidants, scavenging free radicals generated by UV radiation, thus preventing oxidative damage and enhancing food protection.

4.4.2 Use of Cyclodextrin on Various Food Items Affected by UV Radiation

Here are some examples of food items and how they can be affected by UV radiation, along with how cyclodextrins can protect them. Table 4.7 shows examples of food items affected by UV radiation and their protection through cyclodextrins.

TABLE 4.7
Use of Cyclodextrin on Various Food Items Affected by UV Radiation

Type of Food Item	Affect UV Radiation	Influence of Cyclodextrin
Beta-carotene in Orange Juice	Degrades beta-carotene, causing color and nutrient loss.	Encapsulates beta-carotene, preserving colour and nutrients.
Curcumin in Curry Paste	Fades colour and reduces flavour via photochemical reactions.	Protects curcumin, maintaining colour and flavour.
Vitamin D in Fortified Milk	Degrades vitamin D, reducing nutritional quality.	Shields vitamin D, preserving nutritional value.
Vitamin A in Margarine	Degrades vitamin A, reducing stability and nutrition.	Protects vitamin A, ensuring stability and nutrition.
Vitamin E in Cooking Oil	Degrades, reducing antioxidant activity and shelf life.	Preserves antioxidant properties and oil quality.
Omega-3 in Fish Oil	Oxidizes, causing rancid odour and taste.	Prevents oxidation, maintaining freshness and efficacy.
Lycopene in Tomato Sauce	Degrades, causing colour and nutrient loss.	Shields lycopene, preserving colour and health benefits.
Polyphenols in Green Tea	Degrades, reducing potency and antioxidant activity.	Protects polyphenols, ensuring activity and bioavailability.
Anthocyanins in Berry Juices	Fades colour, reducing nutritional value.	Preserves colour and health benefits.
Probiotics in Yogurt	Reduces viability, lowering efficacy.	Protects probiotics, maintaining viability and effectiveness.

4.5 ROLE OF CYCLODEXTRINS AS UV FILTERS IN AGRICULTURE

4.5.1 UV Protection in Agriculture

UV radiation significantly impacts crop growth, yield, and quality by reducing photosynthetic activity and causing DNA damage. Both UVB and UVA rays can have beneficial and detrimental effects on plants, necessitating effective UV management [44].

- **Plant Growth and Development**: Moderate UV exposure can stimulate photosynthesis, while excessive exposure inhibits it and disrupts cellular processes, hindering plant development.
- **Crop Yield and Quality**: UV radiation damages plant tissues, reduces chlorophyll, and alters metabolic pathways, affecting yield and quality. UV protection measures like shading and UV-blocking films can mitigate these effects [45].
- **Disease Management**: UV radiation helps control pathogens but excessive exposure weakens plant immunity. UV protection maintains a balance, supporting disease management.
- **Stress Response**: UV exposure generates reactive oxygen species (ROS), causing oxidative stress. UV protection reduces ROS and enhances antioxidant defenses, improving plant resilience.
- **Economic Impact**: UV-induced crop damage can lead to significant economic losses. UV protection mitigates damage, preserving crop value and farmers' livelihoods [46].
- **Sustainable Agriculture**: UV protection promotes resource-efficient practices, reduces chemical inputs, and minimizes environmental impact. Sustainable methods like natural UV filters and biological protectants contribute to eco-friendly farming.
- **Future Food Security**: With fluctuating UV radiation due to climate change and ozone depletion, UV protection is essential for safeguarding crop production and ensuring global food security.

4.5.2 Cyclodextrins in Agriculture

Cyclodextrins, though less common than synthetic chemical UV filters, are used in agriculture for UV protection. They are included in formulations to shield plants from UV radiation, offering a natural alternative to chemical protectants.

4.5.3 Applications of Cyclodextrin-Based UV Filters in Agriculture

- **Crop Protection Formulations**: Cyclodextrins protect UV-sensitive active ingredients in pesticides, herbicides, and fungicides, preventing degradation during storage and application, ensuring efficacy [47].
- **Foliar Sprays**: Cyclodextrins in foliar sprays shield nutrients or growth regulators from UV degradation, enhancing their absorption and promoting plant health.

- **Greenhouse Films**: Cyclodextrin-infused greenhouse films reduce UV infiltration, safeguarding plants and creating optimal growth conditions [48].
- **Seed Coatings**: Cyclodextrins in seed coatings protect seeds, microorganisms, and nutrients from UV damage, improving germination and crop yield [49, 50].
- **Soil Amendments**: Cyclodextrins in soil amendments protect microorganisms and nutrients from UV degradation, supporting soil health and fertility [51].
- **Plant Growth Regulators**: Cyclodextrins shield UV-sensitive plant hormones and regulators, maintaining their effectiveness in promoting crop growth [52, 53].

The use of cyclodextrins as UV filters in agriculture offers a versatile solution to counter UV-induced crop damage, including sunburn, reduced photosynthesis, and oxidative stress. Figure 4.3 illustrates the protective role of cyclodextrins against UV-induced stress in plants. By enhancing the solubility, stability, and bioavailability of UV filters, cyclodextrins improve their effectiveness. This approach protects crop yields and quality while supporting sustainable practices with minimal environmental

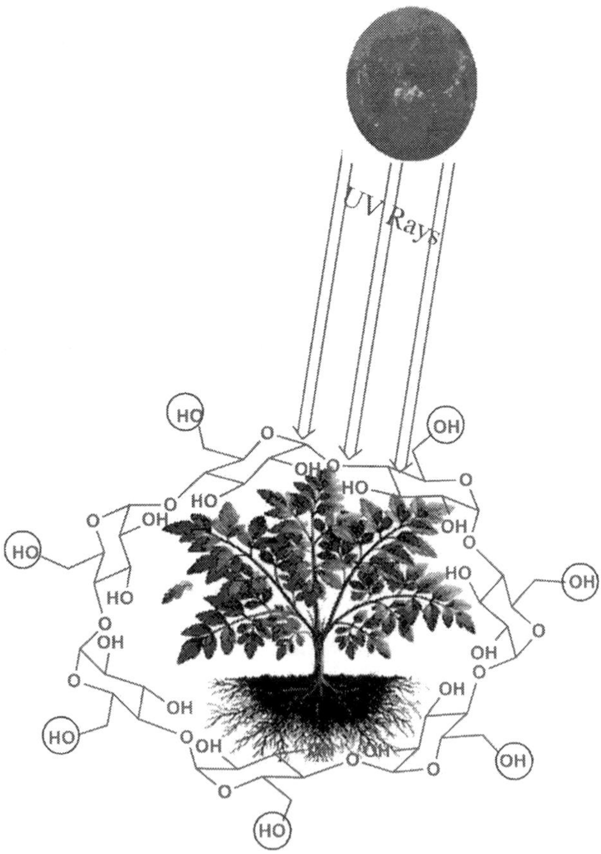

FIGURE 4.3 Cyclodextrin protected plant by formation of inclusion complex

impact. As research progresses, cyclodextrins hold promise for resilient crops and global food security under rising UV radiation levels.

4.6 CONCLUSIONS

The versatile application of cyclodextrins as UV filters across diverse industries including cosmetics, pharmaceuticals, food, and agriculture underscores their pivotal role in safeguarding human health, enhancing product stability, and mitigating environmental impacts. Through encapsulation, cyclodextrins offer an elegant solution to the challenges posed by UV radiation, providing effective photoprotection while addressing issues of solubility, stability, and bioavailability of UV filters. Their adaptability and compatibility with various formulations highlight their potential as invaluable tools for UV protection in a range of consumer products. As research and innovation continues to advance, the integration of cyclodextrins into these industries not only ensures product efficacy and safety but also contributes to the promotion of sustainable practices and the preservation of both human and environmental well-being.

REFERENCES

1. Young, A. R., Morgan, K. A., Harrison, G. I., Lawrence, K. P., Petersen, B., Wulf, H. C., & Philipsen, P. A. (2021). A revised action spectrum for vitamin D synthesis by suberythemal UV radiation exposure in humans in vivo. *Proceedings of the National Academy of Sciences*, Vol *118*(40). https://doi.org/10.1073/pnas.2015867118
2. Gilchrest, B. A., Park, H., Eller, M. S., & Yaar, M. (1996). Mechanisms of ultraviolet light-induced pigmentation. *Photochemistry and Photobiology*, Vol *63*(1), pp 1–10. https://doi.org/10.1111/j.1751-1097.1996.tb02988.x
3. Gupta, A., Avci, P., Dai, T., Huang, Y.-Y., & Hamblin, M. R. (2013). Ultraviolet radiation in wound care: Sterilization and stimulation. *Advances in Wound Care*, Vol *2*(8), pp 422–437. https://doi.org/10.1089/wound.2012.0366
4. Poulson, B. G., Alsulami, Q. A., Sharfalddin, A., El Agammy, E. F., Mouffouk, F., Emwas, A.-H., Jaremko, L., & Jaremko, M. (2021). Cyclodextrins: Structural, chemical, and physical properties, and applications. *Polysaccharides*, Vol *3*(1), pp 1–31. https://doi.org/10.3390/polysaccharides3010001
5. Rajkumar, T., Kukkar, D., Kim, K.-H., Sohn, J. R., & Deep, A. (2019). Cyclodextrin-metal–organic framework (CD-MOF): From synthesis to applications. *Journal of Industrial and Engineering Chemistry*, Vol *72*, pp 50–66. https://doi.org/10.1016/j.jiec.2018.12.048
6. Feng, T., Zhuang, H., & Yang, N. (2018). Applications in cosmetics. *Cyclodextrins*, pp 143–207. https://doi.org/10.1142/9789813229662_0006
7. Ferreira, L., Mascarenhas-Melo, F., Rabaça, S., Mathur, A., Sharma, A., Giram, P. S., Pawar, K. D., Rahdar, A., Raza, F., Veiga, F., Mazzola, P. G., & Paiva-Santos, A. C. (2023). Cyclodextrin-based dermatological formulations: Dermopharmaceutical and cosmetic applications. *Colloids and Surfaces B: Biointerfaces*, Vol *221*, p 113012. https://doi.org/10.1016/j.colsurfb.2022.113012
8. Morillo E. (2006). Application of cyclodextrins in agrochemistry. In: *Cyclodextrins and their Complexes. Chemistry, Analytical Methods, Applications*. Dodziuk H, ed. Wiley-VCH, Verlag GmbH & Co. KGaA, Weinheim, Germany, Chapter 16 pp. 459–467.

9. Liu, Y., Chen, Y., Gao, X., Fu, J., & Hu, L. (2020). Application of Cyclodextrin in food industry. *Critical Reviews in Food Science and Nutrition*, Vol *62*(10), pp 2627–2640. https://doi.org/10.1080/10408398.2020.1856035
10. Sharma, N., & Baldi, A. (2014). Exploring versatile applications of Cyclodextrins: An overview. *Drug Delivery*, Vol *23*(3), pp 729–747. https://doi.org/10.3109/10717544.2014.938839
11. Riviere, J. E., Monteiro-Riviere, N. A., Baynes, R. E., Xia, X.-R., & Smith, C. (2004). Quantitating the Percutaneous Absorption of Mechanistically-Defined Chemical Mixtures. https://doi.org/10.21236/ada422081
12. Tønnesen, H. H., Másson, M., & Loftsson, T. (2002). Studies of curcumin and curcuminoids. XXVII. cyclodextrin complexation: Solubility, chemical and photochemical stability. *International Journal of Pharmaceutics*, Vol *244*(1–2), pp 127–135. https://doi.org/10.1016/s0378-5173(02)00323-x
13. Velasco, M. V. (2004). Cosmetic dermatology: Pincipies and practice. *Revista Brasileira de Ciências Farmacêuticas*, Vol 40(1), pp 112–112. https://doi.org/10.1590/s1516-93322004000100018
14. Puglia, C., Frasca, G., Musumeci, T., Rizza, L., Puglisi, G., Bonina, F., & Chiechio, S. (2012). Curcumin loaded NLC induces histone hypoacetylation in the CNS after intraperitoneal administration in mice. *European Journal of Pharmaceutics and Biopharmaceutics*, Vol 81(2), pp 288–293. https://doi.org/10.1016/j.ejpb.2012.03.015.
15. Manca, M. L., Castangia, I., Zaru, M., Nácher, A., Valenti, D., Fernàndez-Busquets, X., Fadda, A. M., & Manconi, M. (2015). Development of curcumin loaded sodium hyaluronate immobilized vesicles (hyalurosomes) and their potential on skin inflammation and wound restoring. *Biomaterials*, Vol *71*, pp 100–109. https://doi.org/10.1016/j.biomaterials.2015.08.034
16. Alvarez, A. I., Pérez, M., Prieto, J. G., Molina, A. J., Real, R., & Merino, G. (2008). Fluoroquinolone efflux mediated by ABC Transporters. *Journal of Pharmaceutical Sciences*, Vol *97*(9), pp 3483–3493. https://doi.org/10.1002/jps.21233
17. Tønnesen, H. H., & Karlsen, J. (2002). Alginate in drug delivery systems. *Drug Development and Industrial Pharmacy*, Vol *28*(6), pp 621–630. https://doi.org/10.1081/ddc-120003853
18. Stalin, T., & Rajendiran, N. (2006). Intramolecular charge transfer associated with hydrogen bonding effects on 2-aminobenzoic acid. *Journal of Photochemistry and Photobiology A: Chemistry*, Vol *182*(2), pp 137–150. https://doi.org/10.1016/j.jphotochem.2006.02.002
19. Scognamiglio, J., Jones, L., Letizia, C. S., & Api, A. M. (2012). Fragrance material review on Anisyl alcohol. *Food and Chemical Toxicology*, Vol *50*. https://doi.org/10.1016/j.fct.2011.10.012
20. Li, H., Yang, J., & Porter, N. A. (2005). Preparation and photochemistry of O-aminocinnamates. *Journal of Photochemistry and Photobiology A: Chemistry*, Vol *169*(3), pp 289–297. https://doi.org/10.1016/j.jphotochem.2004.06.021
21. Rincón-López, J., Almanza-Arjona, Y. C., Riascos, A. P., & Rojas-Aguirre, Y. (2021). Technological evolution of cyclodextrins in the pharmaceutical field. *Journal of Drug Delivery Science and Technology*, Vol *61*, 102156. https://doi.org/10.1016/j.jddst.2020.102156
22. Pipkin J.D., Rajewski R., Mainous B. (2016). Compositions Containing Silymarin and Sulfoalkyl Ether Cyclodextrin and Methods of Using the Same Patent WO 2016/149685A1.

23. Vyas, A., Kumar Sonker, A., & Gidwani, B. (2014). Carrier-based drug delivery system for treatment of acne. *Scientific World Journal, 2014*, pp 1–14. https://doi.org/10.1155/2014/276260
24. Do Nascimento Pedrosa, T., De Vuyst, E., Mound, A., Lambert de Rouvroit, C., Maria-Engler, S. S., & Poumay, Y. (2016). Methyl-β-cyclodextrin treatment combined to incubation with interleukin-4 reproduces major features of atopic dermatitis in a 3D-culture model. *Archives of Dermatological Research*, Vol *309*(1), pp 63–69. https://doi.org/10.1007/s00403-016-1699-7
25. Mancuso, J. B., Lee, S. S., Paller, A. S., Ohya, Y., & Eichenfield, L. F. (2021). Management of severe atopic dermatitis in pediatric patients. *Journal of Allergy and Clinical Immunology: In Practice*, Vol *9*(4), pp 1462–1471. https://doi.org/10.1016/j.jaip.2021.02.017
26. Niu, X.-Q., Zhang, D.-P., Bian, Q., Feng, X.-F., Li, H., Rao, Y.-F., Shen, Y.-M., Geng, F.-N., Yuan, A.-R., Ying, X.-Y., & Gao, J.-Q. (2019). Mechanism investigation of ethosomes transdermal permeation. *International Journal of Pharmaceutics: X*, Vol *1*, p 100027. https://doi.org/10.1016/j.ijpx.2019.100027
27. Akhtar, N., Verma, A., & Pathak, K. (2016). Investigating the penetrating potential of nanocomposite β-cycloethosomes: Development using central composite design, *in vitro* and *ex vivo* characterization. *Journal of Liposome Research*, Vol *28*(1), pp 35–48. https://doi.org/10.1080/08982104.2016.1254241
28. Takada, H., Yonekawa, J., Matsumoto, M., Furuya, K., & Sokabe, M. (2017). Hyperforin/hp-*β*-cyclodextrin enhances mechanosensitive ca2+ signaling in HaCaT keratinocytes and in atopic skin ex vivo which accelerates wound healing. *BioMed Research International, 2017*, pp 1–9. https://doi.org/10.1155/2017/8701801
29. Varela, P., Amorim, I., Massa, A., Sanches, M., & Silva, E. (1998). Piroxicam-β-cyclodextrin and photosensitivity reactions. *Contact Dermatitis*, Vol *38*(4), pp 229–229. https://doi.org/10.1111/j.1600-0536.1998.tb05726.x
30. Kumar, S., Prasad, M., & Rao, R. (2021). Topical delivery of Clobetasol propionate loaded nanosponge hydrogel for effective treatment of psoriasis: Formulation, physicochemical characterization, antipsoriatic potential and biochemical estimation. *Materials Science and Engineering: C*, Vol *119*, p 111605. https://doi.org/10.1016/j.msec.2020.111605
31. Kettel, M. J., Heine, E., Schaefer, K., & Moeller, M. (2016). Chlorhexidine loaded cyclodextrin containing PMMA Nanogels as antimicrobial coating and delivery systems. *Macromolecular Bioscience*, Vol *17*(2), p 1600230. https://doi.org/10.1002/mabi.201600230
32. Teixeira, K. I., Denadai, A. M., Sinisterra, R. D., & Cortés, M. E. (2014). Cyclodextrin modulates the cytotoxic effects of chlorhexidine on microrganisms and cells*in vitro*. *Drug Delivery*, Vol *22*(3), pp 444–453. https://doi.org/10.3109/10717544.2013.879679
33. Rohani Shirvan, A., & Nouri, A. (2020). Medical textiles. In S, ul-Islam & B. S. Butola (eds.). *Advances in Functional and Protective Textiles*, Advances in Functional and Protective Textiles, Woodhead Publishing, pp 291–333. https://doi.org/10.1016/b978-0-12-820257-9.00013-8
34. Reddersen, K., Finger, S., Zieger, M., Wiegand, C., Buschmann, H.-J., Elsner, P., & Hipler, U.-C. (2016). Cytocompatibility testing of cyclodextrin-functionalized antimicrobial textiles—a comprehensive approach. *Journal of Materials Science: Materials in Medicine*, Vol *27*(12). https://doi.org/10.1007/s10856-016-5804-4
35. Mehenni, L., Lahiani-Skiba, M., Ladam, G., Hallouard, F., & Skiba, M. (2018). Preparation and characterization of spherical amorphous solid dispersion with

amphotericin B. *Pharmaceutics*, Vol *10*(4), pp 235. https://doi.org/10.3390/pharmaceutics10040235

36. Loftsson, T., & Brewster, M. E. (2010). Pharmaceutical applications of Cyclodextrins: Basic science and product development. *Journal of Pharmacy and Pharmacology*, Vol *62*(11), 1607–1621. https://doi.org/10.1111/j.2042-7158.2010.01030.x
37. Ioele, G., De Luca, M., Garofalo, A., & Ragno, G. (2017). Photosensitive drugs: A review on their photoprotection by liposomes and Cyclodextrins. *Drug Delivery*, Vol *24*(2), pp 33–44. https://doi.org/10.1080/10717544.2017.1386733
38. Irie, T. (1999). Cyclodextrins in peptide and protein delivery. *Advanced Drug Delivery Reviews*, Vol *36*(1), pp 101–123. https://doi.org/10.1016/s0169-409x(98)00057-x
39. Ben Mihoub, A., Larue, L., Moussaron, A., Youssef, Z., Colombeau, L., Baros, F., Frochot, C., Vanderesse, R., & Acherar, S. (2018). Use of cyclodextrins in anticancer photodynamic therapy treatment. *Molecules*, Vol *23*(8), pp 1936. https://doi.org/10.3390/molecules23081936
40. Pandey, M., Rani, P., Adhikari, L., Gupta, M., Semalty, A., & Semalty, M. (2021). Preparation and characterization of cyclodextrin complexes of doxycycline hyclate for improved photostability in aqueous solution. *Journal of Inclusion Phenomena and Macrocyclic Chemistry*, Vol *02*(3–4), pp 271–278. https://doi.org/10.1007/s10847-021-01116-z
41. Tiwari, G., Tiwari, R., & Rai, A. (2010). Cyclodextrins in delivery systems: Applications. *Journal of Pharmacy And Bioallied Sciences*, Vol *2*(2), pp 72. https://doi.org/10.4103/0975-7406.67003
42. Fenyvesi, Éva & Otta, K. & Kolbe, I. & Novák, Cs & Szejtli, József. (2004). Cyclodextrin complexes of UV filters. *Journal of Inclusion Phenomena,* Vol *48*, pp 117–123. 10.1023/B:JIPH.0000022518.38097.55
43. Kfoury, M., Landy, D., & Fourmentin, S. (2018). Characterization of cyclodextrin/volatile inclusion complexes: A Review. *Molecules*, Vol *23*(5), pp 1204. https://doi.org/10.3390/molecules23051204
44. Susanne, N.t, & Schreiner, M. (2018). UVB and UVA as eustressors in horticultural and agricultural crops. *Scientia Horticulturae*, Vol *234*, pp 370–381. https://doi.org/10.1016/j.scienta.2018.02.021.
45. Szejtli, J. (1998). Introduction and general overview of Cyclodextrin Chemistry. *Chemical Reviews*, Vol *98*(5), pp 1743–1754. https://doi.org/10.1021/cr970022c
46. Lagarce, F. (2020). Nanomedicines: Promises and reality. *Drug Discovery Today*, Vol *25*(3), pp 473–474. https://doi.org/10.1016/j.drudis.2020.01.001
47. Liu, H., Li, S., Pan, W., Wang, Y., Han, F., & Yao, H. (2006). Investigation into the potential of low-frequency ultrasound facilitated topical delivery of cyclosporin A. *International Journal of Pharmaceutics*, Vol *326*(1–2), pp 32–38. https://doi.org/10.1016/j.ijpharm.2006.07.022
48. Pyra, A., Filik, K., Szermer-Olearnik, B., Czarny, A., & Brzozowska, E. (2020). New insights on the feature and function of tail tubular protein B and tail fiber protein of the lytic bacteriophage φyeo3-12 specific for yersinia enterocolitica serotype O:3. *Molecules*, Vol *25*(19), pp 4392. https://doi.org/10.3390/molecules25194392
49. Laun, K., Mebs, S., Duan, J., Wittkamp, F., Apfel, U.-P., Happe, T., Winkler, M., Haumann, M., & Stripp, S. T. (2018). Spectroscopical investigations on the redox chemistry of [Fefe]-hydrogenases in the presence of carbon monoxide. *Molecules*, Vol pp *23*(7), 1669. https://doi.org/10.3390/molecules23071669

50. Rubilar, J. F., Zúñiga, R. N., Osorio, F., & Pedreschi, F. (2015). Physical properties of emulsion-based hydroxypropyl methylcellulose/whey protein isolate (HPMC/WPI) edible films. *Carbohydrate Polymers*, Vol *123*, pp 27–38. https://doi.org/10.1016/j.carbpol.2015.01.010
51. Szejtli, J. (2004). Cyclodextrins: Applications. In Marcel Dekker (ed.), *Encyclopedia of Supramolecular Chemistry*, pp 405–413. https://dokumen.pub/the-history-of-cyclodextrins-1st-ed-9783030493073-9783030493080.html
52. Semerad, J., Pacheco, N. I., Grasserova, A., Prochazkova, P., Pivokonsky, M., Pivokonska, L., & Cajthaml, T. (2020). In vitro study of the toxicity mechanisms of nanoscale zero-valent iron (NZVI) and released iron ions using earthworm cells. *Nanomaterials*, Vol *10*(11), p 2189. https://doi.org/10.3390/nano10112189
53. Wickstroem, K., Karlsson, J., Ellingsen, C., Cruciani, V., Kristian, A., Hagemann, U. B., Bjerke, R. M., Ryan, O. B., Linden, L., Mumberg, D., Brands, M., & Cuthbertson, A. (2019). Synergistic effect of a HER2 targeted thorium-227 conjugate in combination with Olaparib in a BRCA2 deficient xenograft model. *Pharmaceuticals*, Vol *12*(4), p 155. https://doi.org/10.3390/ph12040155.

5 Cyclodextrin-Modified Antioxidants as UV-Ray Absorbers

Rima Das

5.1 INTRODUCTION

The term "ultraviolet" denotes "beyond violet," with violet representing the highest frequencies within the visible light spectrum. Ultraviolet light possesses a frequency greater than that of violet light. The discovery of UV radiation dates back to 1801, credited to the German physicist *Johann Wilhelm Ritter*, who noted the presence of invisible rays extending past the violet end of the visible spectrum [1, 2].

Medical organizations advise patients to safeguard themselves from UV radiation by employing sunscreen. Within sunscreen formulations, ingredients such as avobenzone, oxybenzone, and octyl methoxycinnamate act as organic chemical absorbers or "blockers," capable of absorbing UVA/UVB rays [3]. Sunscreen offers UV protection either by preventing the generation of free radicals through UV filters or by scavenging existing free radicals using antioxidants [4]. The harmful UV ray is either absorbed or reflect partially by the applied sunscreen but still our skin remains susceptible to low levels of UV radiation, which triggers the production of free radicals in the skin (Figure 5.1) [5].

The demand for radical-scavenging antioxidants arises in sunscreen formulations. Antioxidants are essential for neutralizing free radicals, often needing a combination of both lipid- and water-soluble types to ensure complete protection [6].

Antioxidants are substances that hinder oxidation, a chemical process that can generate free radicals. This process, known as autoxidation, contributes to the degradation of organic compounds, including living matter. Free radicals are byproducts of normal metabolic processes in the body and can inflict harm on cells, proteins, and DNA if not properly addressed. Acting as scavengers, antioxidants effectively neutralize free radicals, thereby reducing their damaging effects. Their primary role is to counteract free radicals and prevent oxidative damage to cells, proteins, and DNA [7].

Cells produce reactive oxygen species (ROS), including hydrogen peroxide (H_2O_2), hypochlorous acid (HClO), and free radicals such as the hydroxyl radical (•OH) and superoxide anion (O_2^-). These reactive oxygen species play essential roles in cellular functions, including redox signalling. Ideally, antioxidant systems aim not to completely eliminate oxidants but rather to maintain them at an optimal concentration [8].

Natural antioxidants (e.g., quercetin, rutin, ferulic acid, resveratrol, etc.) play a crucial role in stabilizing sunscreen formulations and enhancing their biocompatibility

DOI: 10.1201/9781003492757-5

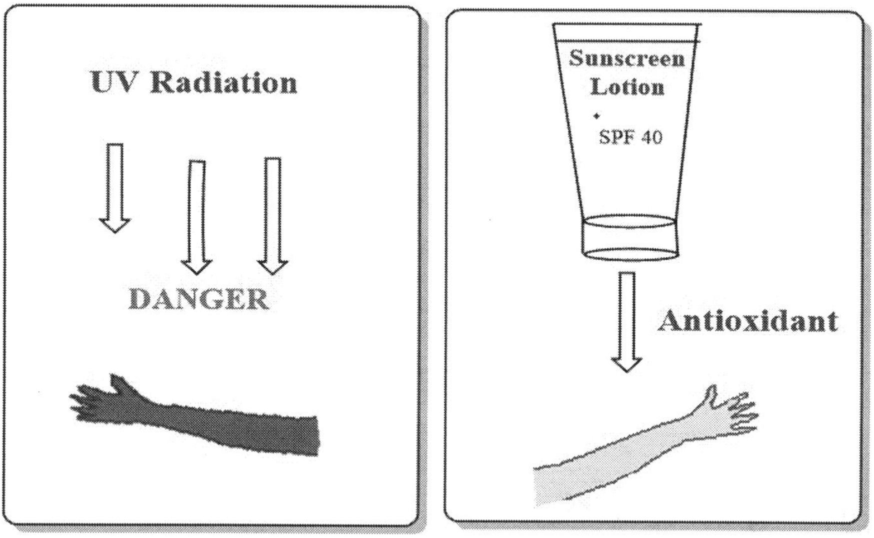

FIGURE 5.1 Penetration of UV rays into skin.

while minimizing toxicity. Additionally, these antioxidants offer the advantage of being more accessible and cost-effective compared to synthetic alternatives [9]. Some commonly utilized natural antioxidants, such as quercetin (QCT), rutin, trans-resveratrol, epigallocatechin gallate, and ferulic acid, are known to form complexes with CD.

Cyclodextrins (CDs) emerge as a promising solution, with the ability to create inclusion complexes with poorly soluble substances such as antioxidants. Dissolution challenges are particularly pronounced with lipophilic antioxidants like flavonoids, polyphenols, and carotenoids. Consequently, inclusion complexes of antioxidants with CD within these categories hold higher promise for the sunscreen industry.

Moreover, CDs are recognized for their resistance to enzymatic hydrolysis, achieved by concealing all bridging oxygens within the non-polar ring center. These bridges form bonds with captured antioxidants or UV filters, reducing their accessibility to interact with extracellular molecules [10]. Additionally, the hydrophilic exterior of CD-antioxidant inclusion complexes enhances their solubility, thereby improving bioavailability. Additionally, they can be used to control the release of antioxidants from topically applied sunscreens [11].

5.2 NATURAL ANTIOXIDANT-CD INCLUSION COMPLEX

5.2.1 Quercetin-CD Inclusion Complex

Quercetin is a type of flavonoid compound naturally present in plants, foods, and beverages. It belongs to a subset of vividly coloured polyphenolic phytochemicals. Researchers in pharmaceutical and medicinal chemistry are particularly interested in

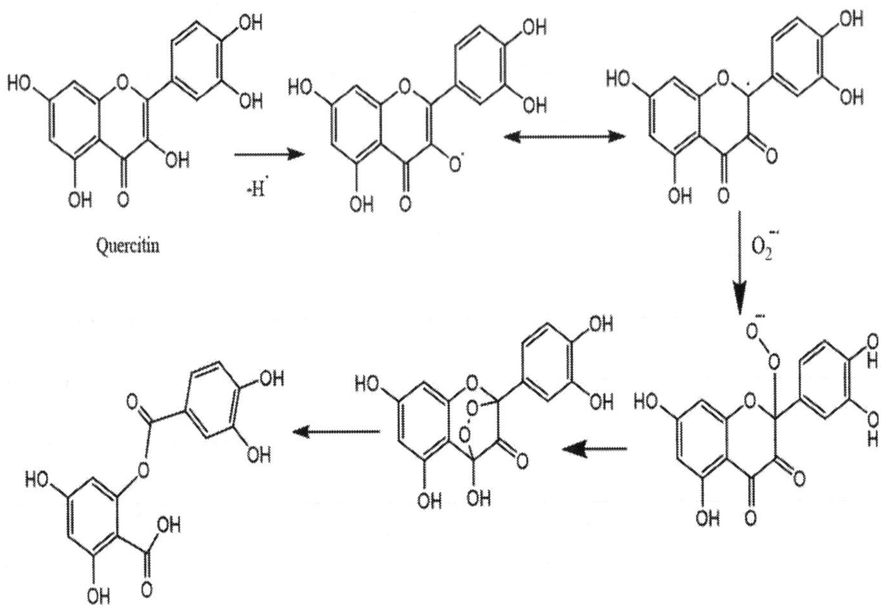

FIGURE 5.2 Quercetin mediated radical scavenging of superoxide anion [15].

the antioxidant, antitumour, and antibacterial properties exhibited by flavonoids like quercetin [12].

Extensive research has been conducted on the potential health benefits associated with the antioxidant properties of quercetin. However, its application via topical administration is hindered by its rapid photodegradation. In an effort to address this challenge, Jullian *et al.* investigated the utilization of slightly water-soluble flavonoid quercetin (QCT) in combination with various cyclodextrins in 2006 [13]. Figure 5.2 demonstrates the scavenging ability of QCT with superoxide anion. All CD-QCT complexes exhibited superior scavenging capabilities against both radicals when compared to quercetin alone. Additionally, these findings suggest that the formation of complexes proved beneficial in increasing the solubility of QCT without substantially diminishing its antioxidant efficacy [14].

5.2.2 Rutin-CD Inclusion Complex

Rutin, also known as quercetin 3-rhamnosylglucoside, is a safe and stable derivative of quercetin. Its chemical structure, defined by phenolic rings and free hydroxyl groups, allows it to form internal hydrogen bonds with free radicals. Presenting as a yellow-green or light-yellow powder with no discernible taste, rutin demonstrates poor solubility in water but readily dissolves in ethanol, isopropyl alcohol, and sodium hydroxide [16]. Classified as photosensitive, rutin showcases notable properties including anti-inflammatory, antidiabetic, antibacterial, and antimutagenic effects.

Moreover, its potent antioxidant capabilities primarily stem from the appearance of phenolic rings and free hydroxyl groups within its shape [17].

Savic *et al.* observed an interaction between rutin and cyclodextrins at a 1:1 molar ratio. These inclusion complexes effectively slowed down the light induced decomposition of rutin under UVB stream and notably boosted its antioxidant ability and phase solubility [18].

Paczkowska *et al.* examined the antioxidant and biological activity, as well as the penetrability of rutin complexes with beta-cyclodextrin compared to its free form. The results indicated that the complex exhibited greater stability and solubility than the free form. The study demonstrated that rutin forms inclusion complexes with β-cyclodextrins using a grinding technique. Importantly, it was confirmed that incorporating rutin into β-cyclodextrin cavities was particularly beneficial in improving antibacterial activity, chemical stability as well as solubility and dissolvability [19].

Taira *et al.* explored the scavenging capabilities of rutin and discovered its ability to inhibit tyrosinase activity, resulting in a reduction of melanin content by approximately 40% in murine melanoma cell lines. Importantly, the extract exhibited no cytotoxic effects, indicating its safety for incorporation into cosmetic and cosmeceutical formulations. Additionally, Corina *et al.* illustrated that rutin's antiproliferative effects were correlated with enhanced cellular assimilation, leading to expand apoptotic activity making antagonistic towards cancer cells [20].

5.2.3 Ferulic Acid-CD Inclusion Complex

Ferulic acid (FA) is a phenolic compound renowned for its potent antioxidant properties, serving as a photoprotective agent and exhibiting radical scavenging capabilities [21].

When exposed to light and air, ferulic acid (FA) undergoes functional and organoleptic changes, making it less suitable for cosmetic applications. Given the continuous quest in the cosmetic industry for novel and fruitful products with desirable biochemical activity and coherent skin delivery, various technological advancements have been pursued in cosmetic formulation development [22]. Centini *et al.* embarked on preparing a series of FA/cyclodextrin (CD) complexes to enhance its physicochemical stability. Their attention was primarily directed towards the FA/α-CD inclusion complex, which had previously shown significant stability in a thermodynamic study [23].

Anselmi showed that encapsulating ferulic acid (FA) with α-cyclodextrin improves its chemical stability under UVB exposure, effectively preventing the formation of degradation products when irradiated. Notably, this encapsulation prevented the emergence of the less active cis-isomer of FA, as well as other degeneration byproducts. Furthermore, it facilitated the improved bioavailability of FA on the skin, thereby slowing its release. Additionally, FA contributed to preserving the integrity of physiological cells when revealed to air and ultraviolet radiation (UVR), thus mitigating the risk of carcinogenic cell formation [24].

Due to its water-soluble nature, ferulic acid (FA) encounters challenges in reaching the brain, with a reported solubility of 780 μg/mL. To address this limitation, Inoue *et al.* engineered a molecular derivative of FA known as FAD012. FAD012 exhibits increased fat solubility and reduced water solubility compared to FA, enhancing its ability to penetrate the central nervous system. Consequently, FAD012 is presumed to mitigate the shortcomings associated with FA. To further enhance water solubility, a gamma cyclodextrin (γ-CD)-FAD012 inclusion complex was prepared [24]. The research focused on evaluating the physicochemical properties, including the solubility, of this complex.

5.2.4 Resveratrol-CD Inclusion Complex

Resveratrol (Res) is a naturally occurring polyphenol compound that belongs to the non-flavonoid group, characterized by its stilbene chemical structure. As a stilbene phytoalexin, resveratrol possesses a wide range of biological activities, prompting its incorporation into products developed by both the food and pharmaceutical industries [25]. Notably, resveratrol is recognized for its antioxidant, anti-inflammatory, and antimicrobial properties [26]. Additionally, it has demonstrated protective effects on the cardiovascular system and exhibits anticancer properties across various types of cancers [27].

Silva *et al.* endeavoured to enhance resveratrol's aqueous solubility by utilizing randomly methylated-β-cyclodextrin [28]. Their study encompassed the characterization of the inclusion complex (IC), dissolution assessments, and evaluation of the IC's antioxidant, cytotoxicity, and anti-Campylobacter activities. Yen and colleagues aimed to assess the potential of resveratrol nanofibres in enhancing the aqueous solubility of resveratrol [29].

Moreover, the antioxidant capabilities of resveratrol nanofibres have been shown to be comparable to those of vitamin C, a renowned natural antioxidant. Consequently, formulations of resveratrol nanofibres not only present significantly enhanced water solubility but also demonstrate commendable antioxidant activity. Our results indicate that resveratrol-loaded PVP/HPBCD nanofibres show potential for clinical applications as an antioxidant formulation [30]. Consequently, these nanofibres may find applications as formulations for topical skin treatment with antioxidant and anti-inflammatory effects.

5.2.5 Astaxanthin-CD Inclusion Complex

Astaxanthin (ASX) (Figure 5.3), also known as 3,30-dihydroxy-b,b0-carotene-4,40-dione, is a valuable carotenoid pigment renowned for its potent antioxidant properties. Its versatility extends across various industries, including nutraceuticals, cosmetics, food, and feed [31].

Astaxanthin (ASX) is not produced by animals and must be acquired through their diet. It is present in various aquatic organisms and plays several essential biological roles. These include protection against oxidation and UV light damage, bolstering immune responses, contributing to pigmentation, influencing communication, and affecting reproductive behaviour [32].

Because of its high degree of unsaturation, ASX is prone to degradation when exposed to light and oxygen, resulting in a reduction of its antioxidant properties.

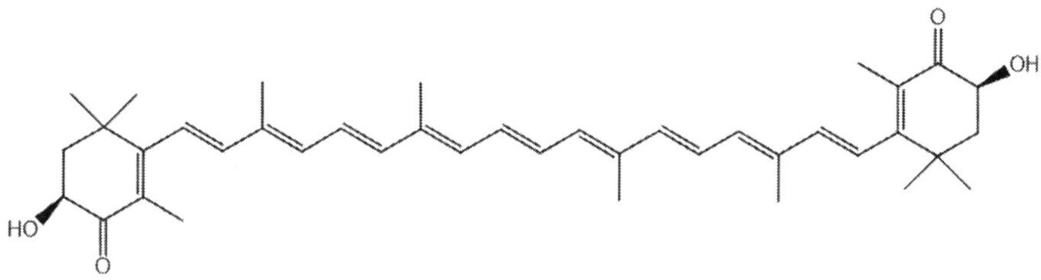

FIGURE 5.3 Chemical structure of astaxanthin.

Additionally, ASX's poor solubility in water restricts its utility as a water-phase antioxidant. Li *et al.* addressed these challenges by forming an inclusion complex between β-cyclodextrin (β-CD) and astaxanthin. This complex significantly improved the heat stability of astaxanthin after formation [33].

Jin and colleagues created an inclusion complex involving astaxanthin (ASX) and hydroxypropyl-β-cyclodextrin (HP-β-CD). The resulting inclusion complex exhibited a water solubility exceeding 1.0 mg/mL, significantly surpassing that of ASX alone. Moreover, the starting degradation temperature of ASX was increased to approximately 290 °C after complexation. Longevity tests conducted on the inclusion complex in solution revealed a notable enhancement in ASX stability against light and oxygen compared to its standalone form [34].

5.2.6 Chrysin-CD Inclusion Complex

Chrysin, a naturally occurring flavone, showcases a diverse array of significant biological activities. This flavonoid, also known as 5,7-dihydroxyflavone, serves as a primary active component in plants. The anticancer properties of chrysin have been extensively researched. Additionally, it also exhibits moderate inhibitory activity against tyrosinase and aromatase [35]. However, due to its insolubility in water, the application of chrysin is often restricted, prompting considerable interest in the development of suitable drug delivery systems.

In a recent study, Kim and colleagues explored the effect of β-cyclodextrin, heptakis (2,6-di-O-methyl)-β-cyclodextrin, and 2-hydroxypropyl-β-cyclodextrin on the aqueous solubility of three flavones—chrysin, apigenin, and luteolin—using phase-solubility analyses [36]. Furthermore, Basu and colleagues investigated the inclusion of chrysin in β-CD within an aqueous solution using both spectroscopic and theoretical approaches. This inclusion process not only elevated the solubility of chrysin but also enhanced its antioxidant potential, as evidenced by the ABTS radical scavenging assay [37].

Understanding the inclusion mechanism of chrysin could enable the design of cyclodextrin derivatives that are more specific and possess improved affinity and functionality toward chrysin. This could lead to enhanced bioavailability and antioxidant activity of chrysin, thereby increasing its potential as a viable drug candidate.

5.2.7 Kaempferol-CD and Myricetin-CD Inclusion Complex

Flavonoids represent a class of low molecular weight polyphenolic compounds found abundantly in vegetables and fruits. They share a common basic structure known as the flavan nucleus, comprised of 15 carbon atoms organized into three rings (C6–C3–C6), denoted as A, B, and C (refer to Figure 5.4). These structural disparities, particularly in substitution, significantly impact the stability of phenoxyl radicals and consequently influence the antioxidant properties of flavonoids [38].

The antioxidant qualities mentioned above predominantly stem from their ability to scavenge radicals. Numerous studies consistently highlight the remarkable antioxidant efficacy of flavonoids abundant in hydroxyl groups. Notably, flavonols rich in hydroxyl groups, such as myricetin and kaempferol, have been repeatedly proven as potent antioxidants across various research studies [39]. Furthermore, cyclodextrins (CDs) serve as effective agents for complexing flavonols, serving as reservoirs for controlled dosing.

Núñez-Delicado and colleagues studied the effect of complexing flavonols using the oxygen radical absorbance capacity-fluorescein (ORAC-FL) assay. They observed that this complexation process significantly heightened the antioxidant activity of all three flavonols, with maximum enhancement achieved when each flavonol was encapsulated within the hydrophobic cavity of CDs [40]. This increase in antioxidant ability was assigned to the protection of flavonols against fast oxidation by free radicals. Generally, the antioxidant effectiveness of flavonoids is contingent upon the structure and arrangement of their hydroxyl groups.

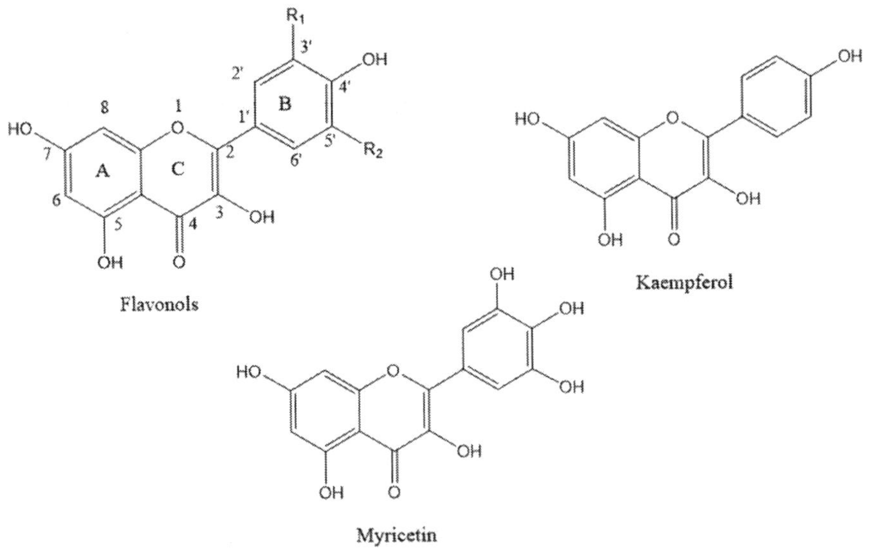

FIGURE 5.4 Structure of flavonol, kaempferol, and myricetin.

The radical scavenging activity was induced due to electron delocalization from ring B as initiated by the presence of double bond at C2-C3 position along with a 4-keto arrangement. The flavonols examined in this study possess varying numbers of hydroxyl groups in ring B. Their observed antioxidant activity followed this order: kaempferol > quercetin > myricetin, indicating an inverse relationship with the number of hydroxyl groups in the B ring. Both flavonols demonstrated superior antioxidant properties relative to myricetin, characterized by a galloyl structure in Ring B [40].

5.2.8 Garlic Oil-CD Inclusion Complex

Steam distillation is a common method utilized for extracting and concentrating volatile organosulfur compounds found in garlic, resulting in the production of garlic oil (GO). The primary constituents of GO include diallyl disulfide (DADS), diallyl trisulfide (DATS), allyl propyl disulfide, a small quantity of disulfide, and potentially diallyl polysulfide [41]. Despite its richness in organosulfur compounds, GO possesses antimicrobial and antioxidant properties. However, challenges such as volatility, low physicochemical stability, strong odour, and insolubility in water restrict its utility as a functional food ingredient [42]. Complexation of guest compounds with CDs can enhance stability, improve aqueous solubility, and offer protection against oxidation, light-induced decomposition, and heat-induced changes [43].

Wang and colleagues investigated the physicochemical properties and release characteristics of garlic oil (GO) inclusion complexes with β-cyclodextrin (β-CD). Using various analytical techniques, including Fourier transform infrared spectroscopy, they confirmed the formation of the GO/β-CD inclusion complex, which showed a 1:1 stoichiometric ratio. The phase solubility study demonstrated a notable enhancement in the water solubility of GO. Moreover, the liberation rate of GO from the inclusion complex was found to be manageable [44].

5.2.9 Vitamin E-CD Inclusion Complex

Vitamin E, particularly α-tocopherol, is extensively used in drug delivery and wound dressings due to its antioxidant properties. However, its application faces challenges because of its low water solubility and vulnerability to oxygen, light, and alkaline pH [45]. Cyclodextrins (CDs) offer a unique solution to these limitations through their hydrophobic cavities, enabling non-covalent host–guest inclusion complexation (IC) with various active compounds. The formation of CD-ICs with vitamin E can significantly improve its solubility and oxidative stability. Studies conducted by the Uyar group have demonstrated enhanced release, effective antioxidant properties, improved water solubility, and UV-light stability for vitamin E through CD inclusion complexes [46].

Furthermore, the vitamin E/HPβCD-IC NF web demonstrated enhanced photostability for the sensitive vitamin E, maintaining its antioxidant activity even after exposure to UV light [46].

5.2.10 Phloretin-CD Inclusion Complex

Phloretin, a natural active compound classified as a dihydrochalcone flavonoid. It is primarily found in the husk and root skin of apples, pears, and various other fruits and vegetables [47]. This compound has garnered significant attention due to its advantageous properties such as antioxidation, antitumour activity, antidiabetic effects, and antibacterial properties [48]. Furthermore, its ability to reduce melanin and enhance skin fairness has led to its incorporation into cosmetic formulations. However, the direct use of phloretin is hindered by its instability and poor water solubility, limiting its bioavailability and practical application [49].

Using a co-evaporation technique, Liang *et al.* devised an inclusion complex of phloretin and HP-β-CD [50]. Despite this alteration, the inclusion complex maintained a similar DPPH radical-scavenging capacity to free phloretin at equivalent concentrations. Moreover, the stability of phloretin was notably enhanced upon the formation of the inclusion complex with HP-β-CD. In a recent study, inclusion complexes of phloretin with methyl-β-cyclodextrin were synthesized using a freeze-drying technique [51]. The researchers concluded that phloretin was effectively encapsulated within the cyclodextrin cavity. Phase solubility studies revealed a 1:1 stoichiometry for the formed complexes.

Moreover, antioxidant activity assays demonstrated a significant improvement in the reducing ability and DPPH radical scavenging capability of the inclusion complexes. The phloretin-β-CD inclusion complexes offers a promising avenue for utilizing phloretin in water-based formulations as antioxidants [51].

5.2.11 Curcumin-CD Inclusion Complex

Curcumin finds wide application in food as a stabilizer in jellies and a natural colorant in cheeses [51]. Additionally, it is utilized in traditional medicine. Despite its effectiveness in inhibiting the growth of various tumor cells, its limited bioavailability and quick metabolism hinder its widespread use. However, complexation with cyclodextrins and their derivatives has proved to be a sure shot method to improve the solubility [52]. Liu *et al.* developed a curcumin/β-cyclodextrin polymer inclusion complex, which displayed novel antioxidant and anticancer properties [53].

5.2.12 Lycopene-CD Inclusion Complex

Lycopene, classified as a carotenoid, lend a red hue to various vegetables and fruits, such as tomatoes, watermelons, pink guavas, and pink grapefruits. Its biological significance lies in its antioxidant, anti-inflammatory, cancer-preventive, and cardiovascular protective properties [54]. Natural lycopene has found application in cosmetic formulations like age-defying therapies, facial creams, and eye creams. Nonetheless, its molecular structure, characterized by multiple unsaturated bonds, renders it highly unstable, prone to degradation when exposed to light, heat, and certain chemical environments [55].

Inclusion complexes of lycopene with β-CD were prepared and characterized by Xing *et al.* using a precipitation method. Lycopene was encapsulated within the cavity of β-CD in a 1:1 stoichiometric ratio. Additionally, the thermal and irradiant stability of lycopene was significantly enhanced through the formation of lycopene-β-CD complexes. The research outcomes indicated a noticeable increase in DPPH radical scavenging activity upon generation of the inclusion complex with β-cyclodextrin at concentrations ranging from 5 to 30 µg/mL [56].

5.2.13 Baicalein-CD Inclusion Complex

Baicalein (BAI) has garnered significant interest due to its multifaceted pharmacological benefits, including anti-inflammatory, antitumour, and antioxidative properties [57]. However, BAI is hindered by its poor water solubility and susceptibility to light and high temperatures, factors that can diminish its bioavailability and limit its clinical utility [58]. To address these challenges, supercritical fluid (SCF) technology has emerged as a promising alternative for preparing inclusion complexes [59].

Wang *et al.* utilized the PCA process to prepare an inclusion complex of BAI and HP-β-CD. Through PCA, BAI transitioned from a crystalline to an amorphous state, as confirmed by analyses including differential scanning calorimetry (DSC), X-ray diffraction (XRD), and scanning electron microscopy (SEM). The solubility and bioavailability of BAI were notably enhanced in the BAI/HP-β-CD inclusion complex compared to the pure bulk drug. Moreover, the coherent complex formation of BAI with HP-β-CD (Figure 5.5) resulted in improved antioxidant and antibacterial

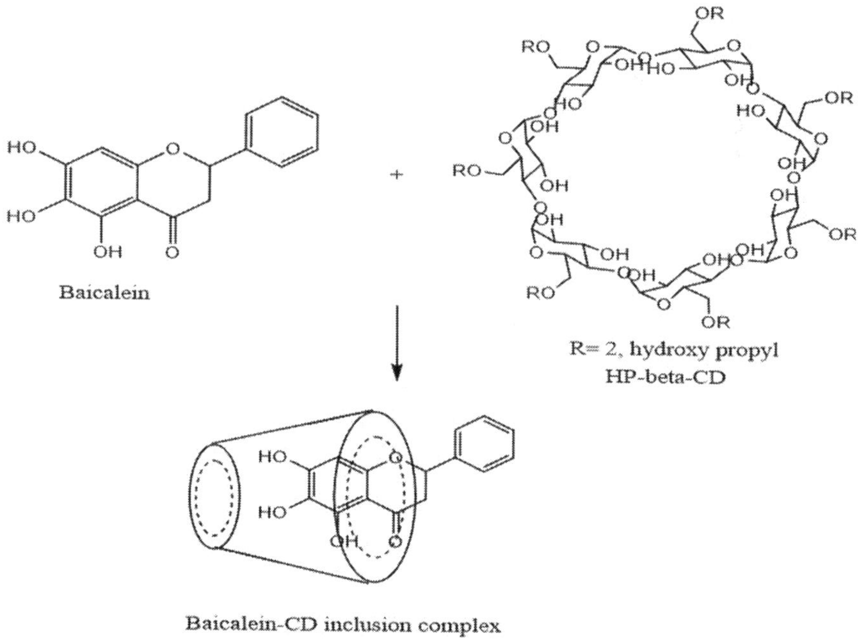

FIGURE 5.5 Interaction of BAI with HP-β-CD.

activities, further emphasizing the positive impact of enhanced solubility on BAI's therapeutic efficacy [60].

5.2.14 Saikosaponin-CD Inclusion Complex

Saikosaponin-d (SSD), (Figure 5.6) a triterpene saponin derived from Bupleurum plants, possesses a spectrum of biological properties, notably antioxidant, anti-inflammatory, and anticancer activities across various cancer types. However, despite its promising attributes, the biomedical utility of SSD is constrained by its poor aqueous solubility [61].

In an effort to overcome this limitation, Yan *et al.* developed SSD-HPBCD inclusion complexes by complexing SSD with hydroxypropyl-β-cyclodextrin (HPBCD) in different ratios [62]. The results revealed a significant enhancement in water solubility upon formation of the SSD-HPBCD inclusion complexes. Furthermore, the inclusion complex demonstrated notable anticancer effects against cutaneous squamous cell carcinoma cells, suggesting its potential as a viable drug formulation for the treatment of skin cancer.

5.2.15 Coenzyme Q10-CD Inclusion Complex

Coenzyme Q10 (CoQ10), alternatively familiar as ubiquinone, plays a vital role in various cellular processes and serves as a lipophilic substance. In its reduced form, CoQ10 functions as a radical scavenger, effectively shielding DNA, lipids, proteins, and other essential molecules from oxidative damage [63].

Lutka and Pawlaczyk endeavoured to form inclusion complexes of CoQ10 with various cyclodextrins (CDs). Through methods such as "kneading" and "heating," they successfully prepared inclusion complexes of CoQ10 with γ-CD and substituted

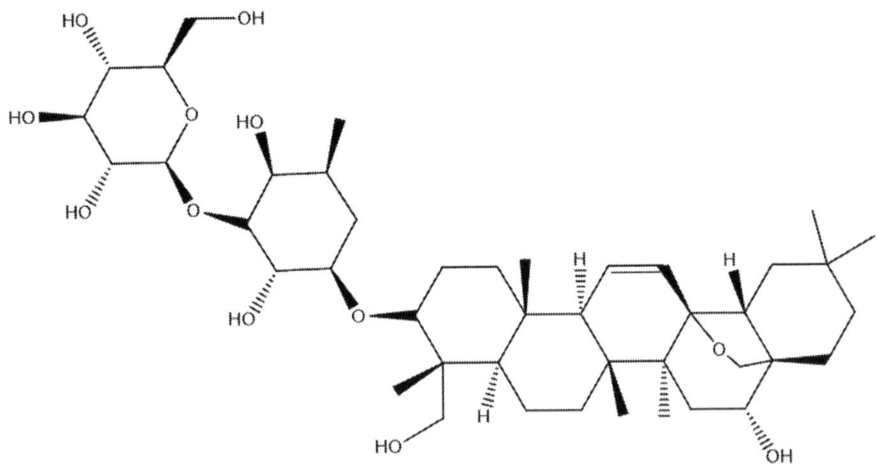

FIGURE 5.6 Chemical structures of (A) saikosaponin-d (SSD).

β-CDs [64]. Similarly, Firet *et al.* explored the microencapsulation of CoQ10 with beta-CD and gamma-CD using the "coprecipitation" method in aqueous solution. Their aim was to enhance CoQ10's water solubility and consequently its bioavailability [65].

Their studies demonstrated that the formation of these complexes led to increased water solubility, as well as improved thermo- and photostability of CoQ10. Notably, the complex of CoQ10 with *β*-CD exhibited promise as a pharmaceutical ingredient and food additive due to its enhanced properties [66].

5.3 CONCLUSIONS

Antioxidants are commonly incorporated into sunscreen formulations to mitigate the adverse effects of ultraviolet radiation (UVR) that may not be fully obstructed by UV filters. However, degradation of both antioxidants and UV filters is likely when they are exposed to UV radiation and other environmental factors.

Cyclodextrins (CDs) offer a protective barrier that enhances the stability of both antioxidants when they experience sunlight and oxygen. This chapter highlights that the formation of inclusion complexes of antioxidants within the cavities of CDs demonstrates a challenging approach to improve the activity and photostability of sunscreens. Moreover, it has been discovered that the water solubility of poorly soluble antioxidants can be remarkably increased by CDs, thereby enhancing their biological activities, particularly as anticancer agents. Additionally, CDs have the ability to regulate the liberation of encapsulated antioxidants and facilitate their permeability through various layers of the skin.

REFERENCES

1. Draper, J.W. **1842**, On a new imponderable substance and on a class of chemical rays analogous to the rays of dark heat., *London, Edinburgh, Dublin Philosophical Magazine J. Sci.*, LXXX, 453–461
2. Draper, J.W. **1844**, Description of the Tithonometer, *The Practical Mechanic and Engineer's Magazine*, January, 122–127.
3. Burnett, M.E.; Wang, S.Q. **2011**, Current sunscreen controversies: A critical review. *Photodermatol. Photoimmunol. Photomed.*, 27(2), 58–67. doi:10.1111/j.1600-0781.2011.00557.x.
4. Jung, K.; Seifert, M. **2007**, The role of antioxidants in cosmetic products to prevent free radical damage in Skin-an efficacy study. *Euro Cosmetics*, 15, 16.
5. Morabito, K.; Shapley, N.C.; Steeley, K.G.; Tripathi, A. **2011**, Review of sunscreen and the emergence of non-conventional absorbers and their applications in ultraviolet protection. *Int. J. Cosmet. Sci.*, 33, 385–390.
6. Osterwalder, U.; Herzog, B. **2009**, Sun protection factors: World wide confusion. *Br. J. Dermatol.*, 161 (Suppl. 3), 13–24.
7. Klemchuk, P.P. **2000**, "Antioxidants". In Peter P. Klemchuk (ed.) Ullmann's Encyclopedia of Industrial Chemistry. WILEY-VCH, doi:10.1002/14356007.a03_091. ISBN 3527306730
8. Lobo, J.M. **2014**, Photodegradation of avobenzone: Stabilization effect of antioxidants. *J. Photochem. Photobiol. B*, 140, 36–40.

9. Ebrahimzadeh, M.A.; Enayatifard, R.; Khalili, M.; Ghaffarloo, M.; Saeedi, M.; Yazdani Charati, J. **2014**, Correlation between sun protection factor and antioxidant activity, Phenol and Flavonoid contents of some medicinal plants. *Iran. J. Pharm. Res.*, 13, 1041–1047.
10. Buedenbender, S.; Schulz, G.E. **2009**, Structural base for enzymatic Cyclodextrin hydrolysis. *J. Mol. Biol.*, 385, 606–617.
11. Costa, R.; Santos, L. **2017**, Delivery systems for cosmetics—from manufacturing to the skin of natural antioxidants. *Powder Technol.*, 322, 402–416.
12. Formica, J.V.; Regelson, W. **1995**, Review of the biology of quercetin and related bioflavonoids. *Food Chem. Toxicol.*, 33(12), 1061–80. doi:10.1016/0278-6915(95)00077-1
13. Jullian, C.; Moyano, L.; Yanez, C.; Olea-Azar, C. **2007**, Complexation of quercetin with three kinds of cyclodextrins: An antioxidant study. *Spectrochimica Acta Part A,* 67, 230–234.
14. Carlotti, M.E.; Sapino, S.; Ugazio, E.; Caron, G. **2011**, On the complexation of quercetin with methyl-β-cyclodextrin: photostability and antioxidant studies. *J. Incl. Phenom. Macrocycl. Chem.*, 70(1–2), 81–90. doi:10.1007/s10847-010-9864-7
15. Nimse, S.B.; Pal, D. **2015**, Free radicals, natural antioxidants, and their reaction mechanisms. *RSC Adv.*, 5, 27986–28006.
16. Almeida, J.S.; Lima, F.; Da Ros, S.; Bulhoes, L.O.; de Carvalho, L.M.; Beck, R.C. **2010**, Nanostructured systems containing rutin: In vitro antioxidant activity and photostability studies. *Nanoscale Res. Lett.,* 5(10), 1603–1610.
17. Yoo, H.; Ku, S.K.; Baek, Y.D.; Bae, J.S. **2014**, Anti-inflammatory effects of rutin on HMGB1-induced inflammatory responses in vitro and in vivo. *Inflamm. Res.*, 63(3), 197–206.
18. Savic, I.M.; Savic-Gajic, I.M.; Nikolic, V.D.; Nikolic, L.B.; Radovanovic, B.C.; Milenkovic-Andjelkovic, A. **2016**, Enhencemnet of solubility and photostability of rutin by complexation with b-cyclodextrin and (2-hydroxypropyl)-bcyclodextrin. *J. Incl. Phenom. Macrocycl. Chem.,* 86, 33–43. doi:10.1007/s10847-016-0638-8.
19. Paczkowska, M.; Mizera, M.; Piotrowska, H.; Powałowska, D.S.; Lewandowska, K.; Goscianska, J.; Pietrzak, R.; Bednarski, W.; Majka, Z.; Cielecka-Piontek, J. **2015**, Complex of rutin with β-cyclodextrin as potential delivery system, *PLOS ONE,* 10, e0120858. DOI:10.1371/journal.pone.0120858.
20. Taira, J.; Tsuchida, E.; Uehara, M.; Ohhama, N.; Ohmine, W.; Ogi, T. **2015**, The leaf extract of Mallotus Japonicus and its major active constituent, rutin, suppressed on Melanin production in Murine B16F1 Melanoma. *Asian Pac. J. Trop. Biomed.*, 5, 819–823.
21. Anselmi, C.; Centini, M.; Maggiore, M.; Gaggelli, N.; Andreassi, M.; Buonocore, A.; Beretta, G.; Facino, R.M. **2008**, Non-covalent inclusion of ferulic acid with α-cyclodextrin improves photo-stability and delivery: NMR and modeling studies. *J. Pharm. Biomed. Anal.*, 46, 645–652.
22. Magdassi, S.; Touito E. **1999**, *Novel cosmetic delivery systems*, Marcel Dekker, New York.
23. Casolaro, M.; Anselmi, C.; Picciocchi, G. **2005**, The protonation thermodynamics of ferulic acid/c-cyclodextrin inclusion compounds. *Thermochim. Acta*, 425, 143–147.
24. Ikeda, N.; Inoue, Y.; Ogata, Y.; Murata, I.; Meiyan, X.; Takayama, J.; Sakamoto, T.; Okazaki, M.; Kanamoto, I. **2020**, Improvement of the solubility and evaluation of the physical properties of an inclusion complex formed by a new ferulic acid derivative and γ-Cyclodextrin. *ACS Omega*, 5, 12073–12080.

25. Gambini, J.; Ingles, M.; Olaso, G.; Lopez-Grueso, R.; Bonet-Costa, V.; Gimeno-Mallench, L.; Mas-Bargues, C.; Abdelaziz, K.M.; Gomez-Cabrera, M.C.; Vina, J. **2015**, Properties of resveratrol: In vitro and in vivo studies about metabolism, bioavailability, and biological effects in animal models and humans. *Oxidative Med.Cell. Longev.*, 2015, 837042.
26. Carrizzo, A.; Forte, M.; Damato, A.; Trimarco, V.; Salzano, F.; Bartolo, M.; Maciag, A.; Puca, A.A.; Vecchione, C. **2013**, Antioxidant effects of resveratrol in cardiovascular, cerebral and metabolic diseases. *Food Chem. Toxicol.*, 61, 215–226.
27. Aziz, S.W.; Aziz, M.H. **2017**, Protective molecular mechanisms of Resveratrol in UVR induced Skin carcinogenesis. *Photodermatol. Photoimmunol. Photomed.,* 34, 35–41.
28. Duarte, A.; Martinho, A.; Luís, Â.; Figueiras, A.; Oleastro, M.; Domingues, F.C.; Silva, F. **2015**, Resveratrol encapsulation with methyl-β-cyclodextrin for antibacterial and antioxidant delivery applications. *LWT – Food Sci. Technol.*, 63, 1254–1260. doi:10.1016/j.lwt.2015.04.004.
29. Lin, Y.-C.; Hu, S. C-S.; Huang, P-H.; Lin, T-C.; Yen, F-L. **2020**, Electrospun resveratrol-loaded polyvinylpyrrolidone/cyclodextrin nanofibers and their biomedical applications. *Pharmaceutics,* 12, 552. doi:10.3390/pharmaceutics12060552
30. Bilska, K.; Wojciechowska, N.; Alipour, S.; Kalemba, E.M. **2019**, Ascorbic acid-the little-known antioxidant in woody plants. *Antioxidants*, 8, 645.
31. Naguib, Y.M.A. **2000**, Antioxidant acitivities of astaxanthin and related carotenoids. *J. Agric. Food Chem.*, 48, 1150–1154.
32. Chen, X.; Chen, R.; Guo, Z.; Li, C.; Li, P. **2007**, The preparation and stability of the inclusion complex of astaxanthin with b-cyclodextrin, *Food Chem.,* 101, 1580–1584.
33. Yuan, C.; Jin, Z.; Xu, X.; Zhuang, H.; Shen, W. **2008**, Preparation and stability of the inclusion complex of astaxanthin with hydroxypropyl-b-cyclodextrin. *Food Chem.,* 109, 264–268.
34. Lee, S.J.; Yoon, J.H.; Song, K.S. **2007**, Chrysin inhibited stem cell factor (SCF)/c-Kit complex-induced cell proliferation in human myeloid leukemia cells. *Biochem. Pharmacol.*, 74, 215–225.
35. Kim, H.; Kim, H.W.; Jung, S. **2008**, Aqueous solubility enhancement of some flavones by complexation with cyclodextrins. *Bull. Kor. Chem. Soc.*, 29, 590–594.
36. Chakraborty, S.; Basu, S.; Lahiri, A.; Basak, S. **2010**, Inclusion of chrysin in b-cyclodextrin nanocavity and its effect on antioxidant potential of chrysin: A spectroscopic and molecular modeling approach. *J. Mol. Struct.,* 977, 180–188.
37. Huber, G.M.; Rupasinghe, V.; Shahidi, F. **2009**, Inhibition of oxidation of omega-3 polyunsaturated fatty acids and fish oil by quercetin glycosides. *Food Chem.*, 117, 290–295. doi:10.1016/j.foodchem.2009.04.007
38. Soobrattee, M.A.; Neergheen, V.S.; Luximon-Ramma, A.; Aruoma, O.I.; Bahorum, T. **2005**, Phenolics as potential antioxidant therapeutic agents: Mechanism and actions. *Mutat. Res.,* 579, 200–213.
39. Mercader-Ros, M.T.; Lucas-Abellán, C.; Fortea, M.I.; Gabaldón, J.A.; Núñez-Delicado, E. **2010**, Effect of HP-b-cyclodextrins complexation on the antioxidant activity of flavonols. *Food Chem.,* 118, 769–773.
40. Wu, C.-C.; Sheen, L.-Y.; Chen, H.-W.; Kuo, W.-W.; Tsai, S.J.; Lii, C.-K. **2002**, Differential effects of garlic oil and its three major organosulfur components on the hepatic detoxification system in rats. *J. Agric. Food Chem.*, 50, 378–383.
41. Corzo-Martínez, M.; Corzo, N.; Villamiel, M. **2007**, Biological properties of onions and garlic. *Trends Food Sci. Technol.*, 18, 609–625.
42. Hedges, A.R. **1998**. Industrial application of cyclodextrins. *Chem. Rev.*, 98, 2035–2044.

43. Wang, J.; Cao, Y.; Sun, B.; Wang, C. **2011**, Physicochemical and release characterisation of garlic oil-b-cyclodextrin inclusion complexes. *Food Chem.*, 127, 1680–1685.
44. Taepaiboon, P.; Rungsardthong, U.; Supaphol, P. **2007**, Vitamin-loaded electrospun cellulose acetate nanofiber mats as transdermal and dermal therapeutic agents of vitamin A acid and vitamin E. *Eur. J. Pharm. Biopharm.*, 67, 387–397.
45. Celebioglu, A.; Uya, T. **2017**, Antioxidant Vitamin E/Cyclodextrin inclusion complex electrospun nanofibers: Enhanced water-solubility, prolonged shelf-life and photostability of Vitamin E. *J. Agric. Food Chem.*, 65(26), 5404–5412. doi:10.1021/acs.jafc.7b01562
46. Crespy, V.; Aprikian, O.; Morand, C.; Besson, C.; Manach, C.; Demigne, C.; Remesy, C. **2001**, Bioavailability of phloretin and phloridzin in rats. *J. Nutr.,* 131, 3227–3230.
47. Rezk, B.M.; Haenen, G.R.M.M.; van der Vijgh, W.J.F.; Bast, A. **2002**, The antioxidant activity of phloretin: The disclosure of a new antioxidant pharmacophore in flavonoids. *Biochem. Biophys. Res. Commun.,* 295, 9–13.
48. Oresajo, C.; Stephens, T.; Hino, P.D.; Law, R.M.; Yatskayer, M.; Foltis, P.; Pillai, S.; Pinnell, S.R. **2009**, Protective effects of a topical antioxidant mixture containing vitamin C, ferulic acid, and phloretin against ultraviolet-induced photodamage in human skin. *J. Cosmet. Dermatol.*, 7, 290–297.
49. Wei, Y.; Zhang, J.; Memon, A.H.; Liang, H.; **2017**, Molecular model and in vitro antioxidant activity of a water-soluble and stable phloretin/hydroxypropyl-β-cyclodextrin inclusion complex. *J. Mol. Liq.,* 236, 68–75.
50. Hu, X.; Zhou, Z.; Han, L.; Li, S. and Zhou, W.*,* **2020**, Preparation and characterization of Phloretin by complexation with cyclodextrins. *New J. Chem.*, 44, 5218–5223. doi:10.1039/C9NJ0622 8A.
51. Chen, X.; Zou, L.Q.; Niu, J.; Liu, W.; Peng, S.F.; Liu, C.M. **2015**, The stability, sustained release and cellular antioxidant activity of curcumin nanoliposomes. *Molecules*, 20, 14293–14311.
52. Chen, J.; Qin, X.; Zhong, S.; Chen, S.; Su, W.; and Liu, Y. **2018**, Characterization of curcumin/cyclodextrin polymer inclusion complex and investigation on its antioxidant and antiproliferative activities. *Molecules*, 23, p 1179.
53. Kim, C.H.; Park, M.K.; Kim, S.K.; Cho, Y.H. **2014**, Antioxidant capacity and anti-inflammatory activity of lycopenein watermelon. *Int. J. Food Sci. Technol.*, 49, 2083–2091.
54. Ciriminna, R.; Fidalgo, A.; Meneguzzo, F.; Ilharco, L.M.; Pagliaro, M. **2016**, Lycopene: Emerging production methods and applications of a valued carotenoid. *ACS Sustain. Chem. Eng.,* 4, 643–650.
55. Wang, H.; Wang, S.; Zhu, H.; Wang, S.; Xing, J. **2019**, Inclusion complexes of Lycopene and β-Cyclodextrin: Preparation, characterization, stability and antioxidant activity. *Antioxidants*, 8, 314. doi:10.3390/antiox8080314.
56. Li, J.; Han, L.; Chao, J. **2012**, Preparation and characterization of the inclusion complex of baicalein with γ-cyclodextrin: an antioxidant ability study. *J. Incl. Phenom. Macrocycl. Chem.*, 73(1–4), 247–254.
57. Zhou, Q.; Wei, X.; Dou, W.; Chou, G.; Wang, Z. **2013**, Preparation and characterization of inclusion complexes formed between baicalein and cyclodextrins. *Carbohyd. Polym.*, 95, 733–739.
58. Jun, S.W., Kim, M.S., Kim, J.S., Park, H.J., Lee, S., Woo, J.S., Hwang, S.J. **2007**, Preparation and characterization of simvastatin/hydroxypropyl-β-cyclodextrin inclusion complex using supercritical antisolvent (SAS) process. *Eur. J. Pharm. Biopharm.*, 66, 413–421.

59. Hu, D., Lin, C., Liu, L., Li, S., Zhao, Y. **2012**, Preparation, characterization, and in vitro release investigation of lutein/zein nanoparticles via solution enhanced dispersion by supercritical fluids. *J. Food Eng.*, 109, 545–552.
60. Yan, T.; Ji, M.; Sun, Y.; Yan, T.; Zhao, J.; Zhang, H.; Wang, Z. **2020**, Preparation and characterization of baicalein/hydroxypropyl-β-cyclodextrin inclusion complex for enhancement of solubility, antioxidant activity and antibacterial activity using supercritical antisolvent technology. *J. Incl. Phenom. Macrocycl. Chem.,* 96, 285–295. https://doi.org/10.1007/s10847-019-00970-2
61. Yuan, B.; Yang, R.; Ma, Y.; Zhou, S.; Zhang, X.; Liu, Y. **2017**, A systematic review of the active saikosaponins and extracts isolated from Radix Bupleuri and their applications. *Pharma. Biol.*, 55, 620–635.
62. Hu, S. C-S.; Lai, Y-C.; Lin, C-L.; Tzeng, W-S.; Yen, F-L. **2019**, Inclusion complex of saikosaponin-d with hydroxypropyl-β-cyclodextrin:Improved physicochemical properties and anti-skin cancer activity *Phytomedicine*, 57, 174–182.
63. Bliznakov, E.G.; Wilkins, D.J. **1998**, Biochemical and clinical consequences of inhibiting coenzyme Q10 biosynthesis by lipid lowering HMG-CoA reductase inhibitors (statins): a critical overview. *Adv. Ther.*, 5, 218–228.
64. Lutka, A.; Pawlaczyk, J. **1995**, Inclusion complexation of coenzymeQ10 with cyclodextrins. *Acta Poloniae Pharm. Drug Res.*, 52, 379–386.
65. Fir, M.M.; Smidovnik, A.; Milivojevic, L.; Zmitek, J.; Prosek, M. **2009**, Studies of CoQ10 and cyclodextrin complexes: Solubility, thermo- and photo-stability. *J. Incl. Phenom. Macrocycl. Chem.,* 64, 225–232. doi:10.1007/s10847-009-9555-4
66. Prosek, M.; Smidovnik, A.; Fir, M.; Strazisar, M.; Andrensek, M.; Golc Wondra, A.; Zmitek, J. **2005**, New water-soluble form of coenzymeQ10 in the form of an inclusion complex with β-cyclodextrin, process of preparing, and use thereof. *SI Patent* (WO 2005/111224 A2, 24 Nov.).

6 Innovative Utilization of Cyclodextrins for Advanced UVR Filtration

Sarojini, Sr. S, A. Dinesh Karthik, and Mary Nancy Flora R.

6.1 INTRODUCTION

UV radiation (UVR) is a form of electromagnetic radiation that originates from the sun and is divided into three types based on wavelength: UVA, UVB, and UVC. UVA rays, with wavelengths between 320 and 400 nm, constitute about 95% of the UV radiation reaching the Earth's surface [1]. These rays penetrate deep into the skin, leading to premature ageing, wrinkles, and DNA damage that can ultimately result in skin cancer. Unlike UVB, UVA can pass through glass and clouds, meaning that exposure risks are present even on cloudy days or indoors. UVB rays, with wavelengths ranging from 290 to 320 nm, make up about 5% of the UV radiation that reaches us. These rays primarily affect the outer layer of the skin, causing sunburn and direct DNA damage, which significantly contributes to the development of skin cancers, including melanoma. UVB radiation is most intense between 10 a.m. and 4 p.m., and, unlike UVA, it does not penetrate through glass [2]. UVC rays, with the shortest wavelengths between 100 and 290 nm, are completely absorbed by the Earth's atmosphere and do not reach the surface. Despite being the most harmful type of UV radiation, their high energy is harnessed in artificial sources for sterilization purposes due to their germicidal properties.

The importance of effective UVR filtration arises from the considerable health and environmental risks associated with prolonged UV exposure. For humans, extended exposure to UVR can lead to both acute and chronic health issues. These include various forms of skin cancer, such as melanoma and non-melanoma, cataracts that impair vision, and suppression of the immune system [3]. UVA and UVB are particularly responsible for photoaging and carcinogenesis, highlighting the importance of adequate protection. From an environmental perspective, UVR affects ecosystems by damaging phytoplankton populations, which are crucial to aquatic food webs. UV radiation also contributes to the degradation of both natural and synthetic materials, leading to the deterioration of plastics, textiles, and other materials exposed to

DOI: 10.1201/9781003492757-6

sunlight. This degradation results in economic costs and increased environmental pollution as these materials break down [4].

Current UVR filtration technologies include chemical sunscreens, physical sunscreens, and UV-blocking films and fabrics. Chemical sunscreens use organic compounds like oxybenzone and avobenzone to absorb UV radiation [5]. However, these compounds can cause skin irritation, hormone disruption, and environmental pollution, particularly in aquatic ecosystems where they accumulate and affect marine life. Physical sunscreens use minerals such as zinc oxide and titanium dioxide to block or scatter UV radiation physically. While effective, these sunscreens often leave a white residue on the skin and may not provide uniform protection [6]. UV-blocking films and fabrics are applied to windows and used in clothing to shield against UV radiation. Although effective initially, their performance can degrade over time due to environmental exposure and repeated washing.

The limitations of these existing technologies are significant. Many suffer from photodegradation, reducing their effectiveness over time [7]. The need for frequent reapplication or replacement adds to their inconvenience and cost. Additionally, there are potential health and environmental concerns associated with their use, prompting the search for more sustainable and effective solutions. Innovative approaches, such as the use of CDs, offer promising solutions to these limitations. CDs, with their ability to form inclusion complexes with UVR absorbers, enhance the solubility, stability, and efficacy of these absorbers. This improvement can lead to more effective and durable UVR filtration technologies, addressing the current challenges and providing better protection for human health and the environment.

6.2 CYCLODEXTRINS: STRUCTURE AND PROPERTIES

6.2.1 Structure of Cyclodextrins

CDs are cyclic oligosaccharides composed of glucose units linked by α-1,4-glycosidic bonds. The most common types of cyclodextrins are α-, β-, and γ-cyclodextrins, which differ in the number of glucose units they contain (Figure 6.1).

6.2.2 α-Cyclodextrin (α-CD) Complexes

α-cyclodextrin (α-CD), composed of six glucose units, forms complexes with small to medium-sized UV absorbers, enhancing their solubility and stability. With an inner cavity diameter of 4.7–5.3 Å and an outer diameter of 14.6 Å, α-CD is well-suited for encapsulating small aromatic and hydrophobic molecules. For example, α-CD can form stable inclusion complexes with p-aminobenzoic acid (PABA), a common UV filter, and benzophenone-3 (oxybenzone), protecting them from photodegradation and increasing their solubility. These complexes improve the UV absorbers' stability, reduce their degradation, and enhance their bioavailability, making them more effective in sunscreen formulations. The hydrophobic interactions between the α-CD cavity and the UV absorber drive the complexation process, while van der Waals forces and potential hydrogen bonding at the rim of the cyclodextrin further stabilize

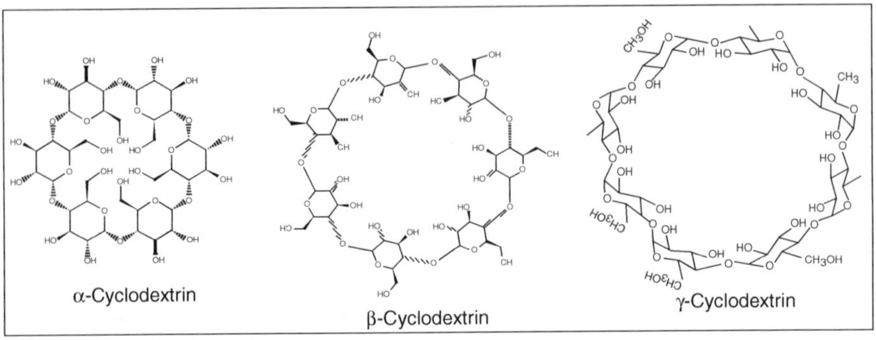

FIGURE 6.1 Illustrating the structure of various cyclodextrins.

the complex [8]. Overall, *α*-CD complexes offer significant benefits in enhancing the performance and longevity of UV absorbers in various applications.

6.2.3 β-Cyclodextrin (β-CD) Complexes

β-cyclodextrin (*β*-CD), comprising seven glucose units, is known for its versatility in forming inclusion complexes with a wide range of UV absorbers due to its moderate inner cavity diameter of 6.0–6.5 Å and an outer diameter of 15.4 Å. This makes *β*-CD suitable for encapsulating moderately sized aromatic and aliphatic UV-absorbing compounds. Notable examples include octyl methoxycinnamate (OMC), a prevalent UVB filter, and avobenzone, a broad-spectrum UVA filter. Encapsulation in *β*-CD significantly enhances these compounds' solubility, stability, and photostability, preventing photodegradation and increasing their effectiveness in sunscreen formulations. The inclusion complexes formed through hydrophobic interactions are stabilized further by van der Waals forces and potential hydrogen bonding, particularly between the hydroxyl groups on the cyclodextrin rim and the UV absorbers [9]. Consequently, *β*-CD complexes offer improved solubility in both aqueous and oil phases, enhanced stability, and extended protection, making them valuable in developing effective and durable sunscreen products.

6.2.4 γ-Cyclodextrin (γ-CD) Complexes

γ-cyclodextrin (*γ*-CD), composed of eight glucose units, features a larger inner cavity diameter of 7.5–8.3 Å and an outer diameter of 17.5 Å, making it ideal for encapsulating larger and bulkier UV absorbers. This capability allows *γ*-CD to form inclusion complexes with sizable UV-absorbing molecules such as homosalate, a UVB filter, and ethylhexyl triazone, a high-performance UV filter. The encapsulation within *γ*-CD enhances the solubility, stability, and photostability of these UV absorbers, protecting them from photodegradation and oxidative damage. The formation of these complexes is driven primarily by hydrophobic interactions, with additional stabilization from van der Waals forces and hydrogen bonding between the hydroxyl groups on the cyclodextrin rim and functional groups on the UV absorbers [10].

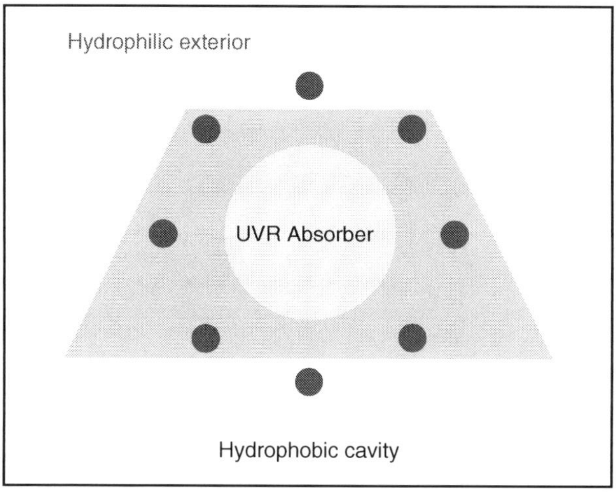

FIGURE 6.2 Illustrating the structure of cyclodextrins and their properties relevant to UVR filtration.

These γ-CD complexes improve the solubility of UV absorbers in various solvents, increase their resistance to photodegradation, and provide prolonged protection in sunscreen formulations, making them highly beneficial for enhancing the efficacy and durability of UV-protective products.

6.2.5 Unique Properties Relevant to UVR Filtration

Cyclodextrins have a truncated cone shape with a hydrophilic exterior due to the hydroxyl groups and a hydrophobic interior cavity (Figure 6.2). This structural feature allows them to form inclusion complexes with various guest molecules, including UVR absorbing molecules. The unique properties relevant to UVR filtration include (i) hydrophobic inner cavity: the hydrophobic cavity can encapsulate UVR-absorbing molecules, protecting them from photodegradation and enhancing their stability. (ii) hydrophilic exterior: the hydrophilic exterior ensures good solubility in aqueous environments, facilitating the incorporation of cyclodextrin-UVR absorber complexes in pharmaceutical formulations. (iii) Size and shape selectivity: different types of cyclodextrins can accommodate guest molecules of various sizes, providing versatility in the types of UVR absorbers that can be included [11].

6.3 MECHANISMS OF UVR FILTRATION BY CYCLODEXTRIN COMPLEXES

6.3.1 Inclusion Mechanism

CDs encapsulate UVR-absorbing molecules through an inclusion mechanism driven by non-covalent interactions (Figure 6.3). The hydrophobic inner cavity of cyclodextrins accommodates the UVR absorbers, while the hydrophilic exterior

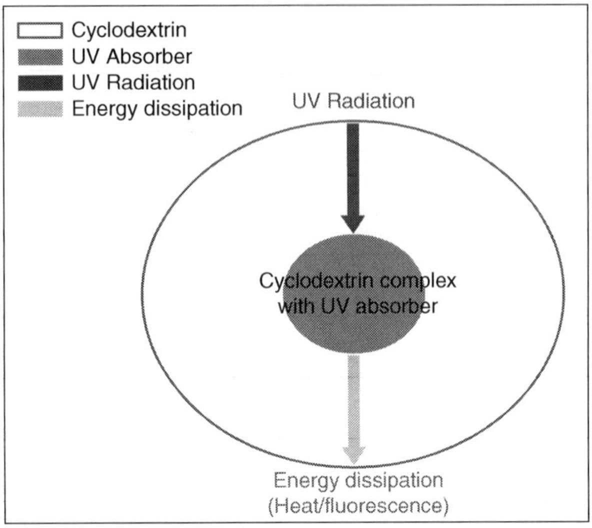

FIGURE 6.3 Mechanisms of UVR filtration by cyclodextrin complexes.

maintains solubility in aqueous environments [12]. This encapsulation is facilitated by hydrophobic interactions, van der Waals forces, and hydrogen bonding, which stabilize the inclusion complex. The complexation can alter the absorption spectrum of UVR absorbers, often enhancing their solubility and bioavailability. By embedding the UVR absorbers within the cyclodextrin cavity, the complexes protect the absorbers from environmental factors, thus maintaining their effectiveness and extending their functional lifespan.

6.3.2 Photostability Enhancement

The encapsulation of UVR absorbers by cyclodextrins significantly enhances their photostability, protecting them from photodegradation caused by UVR. This protection is due to the physical shielding provided by the cyclodextrin cavity, which isolates the UVR absorbers from direct exposure to light and reactive species. Comparative analysis shows that cyclodextrin-complexed UVR absorbers exhibit higher stability and longer durability than their free counterparts, which are more prone to degradation. The enhanced photostability not only improves the efficacy of the UVR absorbers but also ensures prolonged protection, making cyclodextrin complexes superior in maintaining the integrity and performance of UVR-filtering agents in various applications [13].

6.3.2 Synthesis Methods

6.3.2.1 Co-Precipitation

Co-precipitation is a widely used method for synthesizing cyclodextrin-UVR absorber complexes due to its simplicity and effectiveness. The procedure typically involves dissolving both the cyclodextrin and the UVR absorber in a suitable solvent, followed by the addition of a precipitating agent, such as an antisolvent, that induces the formation of a solid inclusion complex. The process starts with dissolving cyclodextrin and the UVR absorber in water or a compatible solvent. Then, an antisolvent like ethanol or acetone is added gradually under stirring, leading to the co-precipitation of the cyclodextrin-UVR absorber complex. The precipitate is then collected by filtration, washed to remove impurities, and dried. The advantages of the co-precipitation method include its ability to produce high-purity complexes with controlled stoichiometry and particle size [14]. This method is particularly effective for producing complexes with hydrophilic UVR absorbers. For example, the co-precipitation of β-cyclodextrin with octyl methoxycinnamate (OMC) enhances the solubility and photostability of OMC, making it more effective in sunscreen formulations. Additionally, this method can be scaled up easily, making it suitable for industrial applications.

6.3.2.2 Kneading Method

The kneading method is a straightforward and efficient technique for synthesizing cyclodextrin-UVR absorber complexes, particularly suited for solid-state processes. The step-by-step synthesis involves mixing cyclodextrin and the UVR absorber with a minimal amount of solvent to form a paste. Typically, cyclodextrin and the UVR absorber are combined in a mortar with a small quantity of water, ethanol, or another suitable solvent. The mixture is kneaded thoroughly to ensure intimate contact between the cyclodextrin and the UVR absorber, facilitating the formation of the inclusion complex. After kneading for a specified time, usually 30–60 minutes, the resulting paste is dried under reduced pressure or at room temperature. The kneading method is advantageous due to its simplicity, low solvent requirement, and ability to form complexes without the need for high temperatures or extensive equipment [15]. This method is particularly useful for UVR absorbers that are sensitive to heat or those with limited solubility in common solvents. An example is the kneading of γ-cyclodextrin with ethylhexyl triazone, a UV filter, which results in a stable complex with improved photostability and solubility. This method is also environment friendly, as it uses minimal solvents and generates little waste.

6.3.2.3 Solution Mixing

Solution mixing is a versatile and commonly used technique for synthesizing cyclodextrin-UVR absorber complexes. This method involves dissolving both the cyclodextrin and the UVR absorber in a common solvent, followed by thorough mixing to facilitate complex formation. The process starts with dissolving cyclodextrin in water, ethanol, or a mixture of solvents, depending on the solubility of the UVR absorber. The UVR absorber is then added to the solution, and the mixture is stirred continuously to ensure complete dissolution and interaction between the molecules. After sufficient mixing, the solvent is removed by evaporation, typically under reduced pressure, leaving behind the solid inclusion complex [16]. The main advantage of solution mixing is its ability to produce highly homogeneous complexes with precise control over the concentration and composition. This method is suitable for both hydrophilic and hydrophobic UVR absorbers, provided a suitable solvent system is chosen. For example, β-cyclodextrin and avobenzone can be dissolved in a water-ethanol mixture to form a stable complex with enhanced solubility and photostability. The choice of solvent and mixing conditions, such as temperature and stirring rate, are critical factors that influence the efficiency of complex formation and the quality of the final product.

6.3.2.4 Freeze-Drying

Freeze-drying, also known as lyophilization, is an advanced technique used to enhance the stability and storage of cyclodextrin-UVR absorber complexes. The process involves freezing a solution containing cyclodextrin and the UVR absorber, followed by sublimation of the solvent under reduced pressure. Initially, the cyclodextrin and UVR absorber are dissolved in a suitable solvent and mixed thoroughly to form the inclusion complex. The solution is then frozen rapidly, usually by immersing it in liquid nitrogen or placing it in a deep freezer. Once frozen, the solution is subjected to low pressure in a freeze-dryer, causing the solvent to sublimate directly from the solid phase to the gas phase, leaving behind a dry, porous complex [16]. The primary advantage of freeze-drying is the preservation of the structural integrity and stability of the inclusion complex. This method minimizes thermal and oxidative degradation, making it ideal for sensitive UVR absorbers. Freeze-dried complexes also exhibit excellent rehydration properties, allowing them to be easily reconstituted in aqueous solutions. For example, freeze-drying can be used to prepare γ-cyclodextrin complexes with homosalate, resulting in a stable, dry powder that retains its UV-absorbing properties over extended storage periods. Additionally, freeze-drying produces a lightweight, porous product that is convenient for handling, transportation, and formulation into various products.

6.4 CYCLODEXTRIN-UVR ABSORBER COMPLEXES

CDs have shown remarkable versatility in enhancing the efficacy of UVR absorbers through the formation of inclusion complexes. Each type of cyclodextrin—α, β, and γ—has a different cavity size, making them suitable for various UV absorbers based on their molecular dimensions. α-cyclodextrin complexes are ideal for encapsulating small UV absorbers such as p-aminobenzoic acid. These complexes significantly increase the solubility and stability of the UV absorbers, ensuring better dispersion and prolonged effectiveness. β-cyclodextrin complexes are suited for moderate-sized UV absorbers like octyl methoxycinnamate. The formation of these complexes enhances the photostability and bioavailability of the UV absorbers, protecting them from degradation due to sunlight exposure and improving their absorption into the skin. γ-cyclodextrin complexes, with their larger cavity, accommodate bigger UV absorbers such as homosalate. These complexes offer enhanced protection and longevity, ensuring that the UV absorbers maintain their effectiveness over extended periods. By forming inclusion complexes with CDs, these UV absorbers benefit from improved physical and chemical properties, leading to more effective UVR filtration and protection in various applications, from sunscreens to UV-protective coatings [16].

6.5 COMPREHENSIVE ASSESSMENT OF CYCLODEXTRIN-BASED UVR FILTERS

6.5.1 Solubility

Cyclodextrin-UVR absorber complexes significantly improve the solubility of UV absorbers in both aqueous and solvent-based systems. UV absorbers often exhibit poor solubility in water, limiting their application in formulations that require a homogeneous mixture, such as sunscreens and cosmetic products. Cyclodextrins, with their hydrophobic interior and hydrophilic exterior, can encapsulate these poorly soluble UV absorbers within their cavities, thereby enhancing their solubility. This inclusion complex formation allows the UV absorbers to be better dispersed in aqueous solutions, improving their effectiveness. Similarly, in solvent-based systems, cyclodextrin complexes can increase the compatibility of UV absorbers with various solvents, leading to more stable and effective formulations. This enhanced solubility not only improves the ease of formulation but also ensures a more uniform distribution of the UV absorbers, resulting in consistent UV protection [17].

6.5.2 Stability

The stability of UV absorbers is a critical factor in their effectiveness, as they can degrade under the influence of light (photodegradation), oxygen (oxidation), and heat (thermal degradation). Cyclodextrin-UVR absorber complexes provide a protective environment that shields the encapsulated UV absorbers from these degradative processes. By sequestering the UV absorbers within their hydrophobic cavities, cyclodextrins protect them from direct exposure to UV light, significantly reducing the rate of photodegradation. Additionally, the complexation with cyclodextrins can impede oxidative reactions by limiting the UV absorbers' exposure to oxygen [18]. Thermal degradation, which can compromise the stability and efficacy of UV absorbers, is also mitigated as the cyclodextrin complexes provide a stabilizing effect, maintaining the integrity of the UV absorbers even under elevated temperatures. This enhanced stability translates to longer-lasting and more reliable UV protection in various applications.

In a study by Martini *et al.*, the solubility and photostability of various UV absorbers were significantly improved by forming inclusion complexes with cyclodextrins [19]. The researchers focused on the UV absorber oxybenzone, which is commonly used in sunscreen formulations but has poor water solubility and can degrade under UV exposure. By forming an inclusion complex with β-cyclodextrin, oxybenzone's solubility in water increased dramatically. The study demonstrated that the complexation not only enhanced solubility but also improved the photostability of oxybenzone. The inclusion complex protected the UV absorber from photodegradation, resulting in a more stable and effective UV protection formulation.

6.5.3 Efficacy in UV Protection

The efficacy of UVR absorbers in providing UV protection is markedly enhanced through their complexation with cyclodextrins. Cyclodextrin-UVR absorber complexes

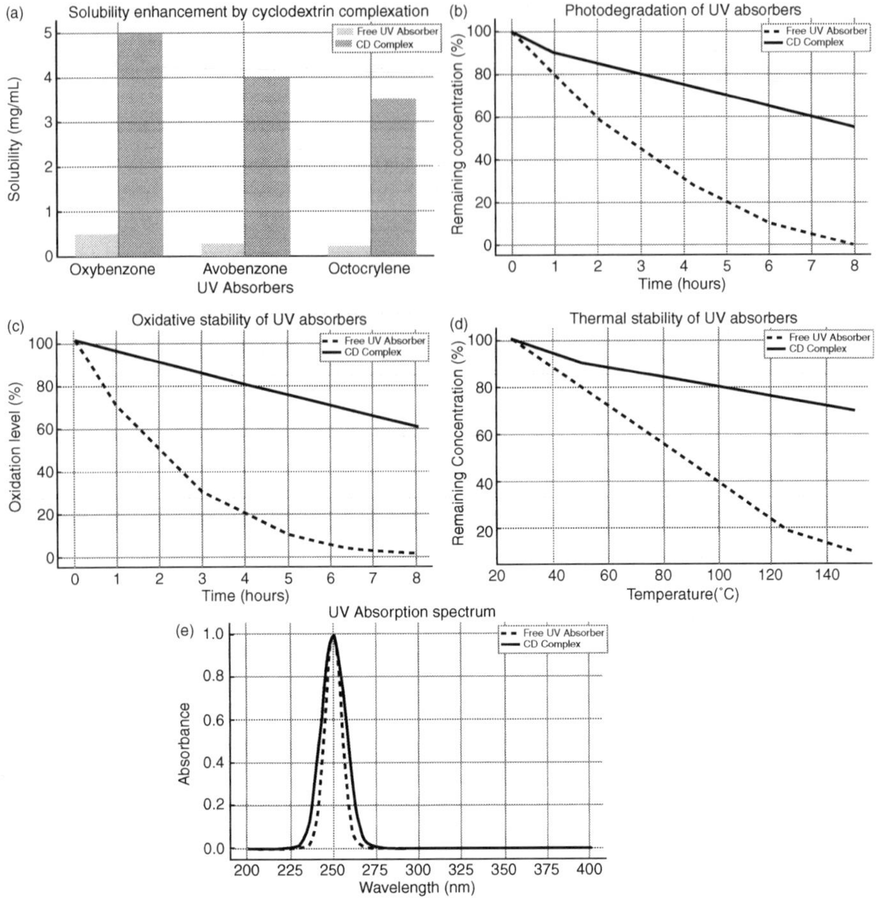

FIGURE 6.4 Comprehensive assessment of cyclodextrin-based UVR filters: (a) solubility enhancement of by cyclodextrin complexation; (b) photodegradation of UV absorbers; (c) oxidative stability analysis; (d) thermal stability; and (e) UV spectrum of the CD complex.

exhibit improved UV absorption profiles due to the stabilized and solubilized state of the UV absorbers. This stabilization ensures that the UV absorbers remain active for longer periods, providing sustained and effective UV protection [19]. Furthermore, the controlled release properties of cyclodextrin complexes allow for a gradual and consistent release of the UV absorbers over time. This sustained release is particularly beneficial in applications such as sunscreens, where prolonged UV protection is desirable. The enhanced UV absorption and sustained release profiles of cyclodextrin-UVR absorber complexes not only improve the overall performance of UV protection formulations but also ensure that the protection remains effective under various environmental conditions.

Figure 6.4(a)–(e) presents a series of graphs illustrating the benefits of cyclodextrin complexes in enhancing the performance of UV absorbers based on the literature survey on CDs for UVR filteration. Panel (a) shows a significant increase in solubility for UV absorbers when complexed with cyclodextrins. Panel (b) demonstrates improved photostability, as indicated by a slower degradation rate. Panel (c) highlights enhanced oxidative stability, with lower oxidation levels over time. Panel (d) depicts better thermal stability, maintaining higher concentrations at elevated temperatures. Panel (e) illustrates broadened and enhanced UV absorption efficiency across different wavelengths. These results collectively emphasize the advantages of cyclodextrin complexes in improving the solubility, stability, and effectiveness of UV absorbers in various applications.

6.6 APPLICATIONS IN PHARMACEUTICALS

6.6.1 Role of Cyclodextrins in Enhancing UV Filter Performance

Cyclodextrins play a crucial role in enhancing the performance of UV filters in sunscreen formulations. By forming inclusion complexes with UV filters, cyclodextrins can significantly improve their solubility and stability. This enhancement ensures a more uniform distribution of the UV filters within the sunscreen formulation, leading to more consistent and effective UV protection. Cyclodextrins also protect the UV filters from photodegradation, which is the breakdown of these molecules under UV radiation [20]. This protection extends the efficacy of the sunscreen, reducing the need for frequent reapplication. Moreover, cyclodextrins can enhance the bioavailability of the UV filters, ensuring that they are better absorbed by the skin, thus providing more effective protection against harmful UV rays.

6.6.2 Case Studies and Commercial Products

Several studies and commercial products have demonstrated the benefits of cyclodextrin-UV filter complexes in sunscreen formulations [21–23]. For instance, a study involving the inclusion complex of octyl methoxycinnamate (a common UVB filter) with β-cyclodextrin showed significant improvements in the stability and efficacy of the UV filter. This complex not only provided enhanced UV protection but also showed reduced skin irritation compared to the non-complexed UV filter. Commercially, products like UV Cut SPF 50+ from Kose Corporation incorporate cyclodextrin-UV filter complexes to offer superior UV protection with enhanced stability and reduced skin irritation. These examples highlight the practical benefits of cyclodextrins in developing more effective and user-friendly sunscreen products.

Figure 6.5 illustrates the global usage of cyclodextrins from 2014 to 2023. Over the past decade, there has been a significant and steady increase in cyclodextrin usage, starting from 10 metric tons in 2014 to 65 metric tons in 2023. This trend reflects the growing recognition and application of cyclodextrins in various fields, including pharmaceuticals, food industries, and UVR filtration, due to their unique ability to form inclusion complexes with various guest molecules.

6.6.3 Drug Delivery Systems – Protection of Photosensitive Drugs

Photosensitive drugs, which degrade upon exposure to light, pose a significant challenge in pharmaceutical formulations. Cyclodextrins can form inclusion complexes with these drugs, providing a protective environment that shields them from light-induced degradation. This protection not only preserves the efficacy of the

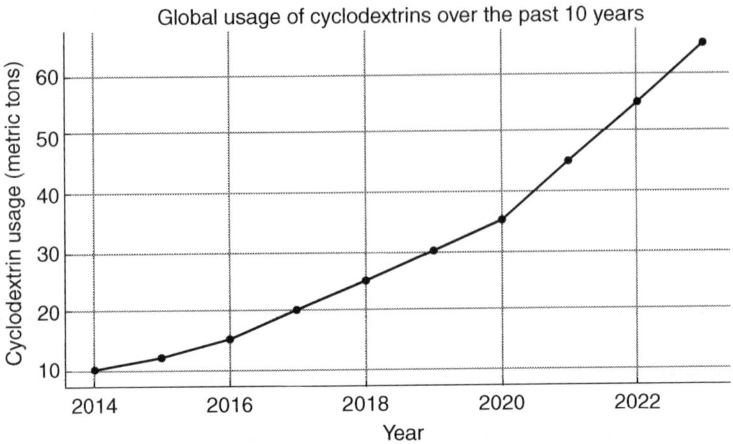

FIGURE 6.5 Graphical representation of the global usage of cyclodextrins over the past 10 years.

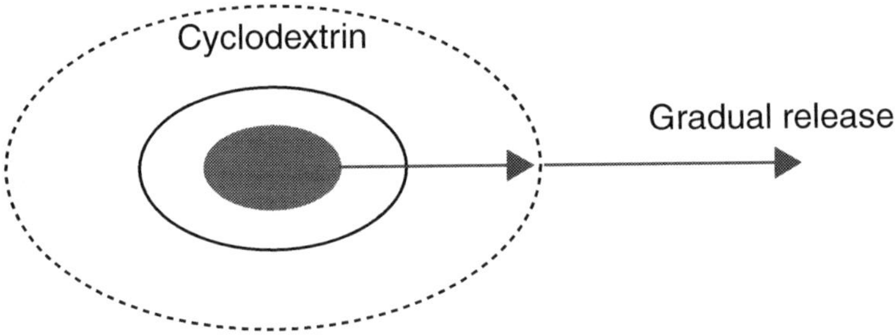

FIGURE 6.6 Mechanism of controlled release using cyclodextrins.

drugs but also extends their shelf-life. For example, cyclodextrin complexes of photosensitive drugs like nifedipine (a calcium channel blocker) and ketoprofen (an anti-inflammatory drug) have shown improved stability under light exposure, ensuring that the therapeutic benefits of these drugs are maintained until they are administered to patients [24].

6.6.4 Controlled Release Mechanisms

Cyclodextrins also play a pivotal role in the controlled release of drugs (Figure 6.6). By forming inclusion complexes with active pharmaceutical ingredients (APIs), cyclodextrins can modulate the release rate of these drugs. This controlled release mechanism ensures that the drug is delivered at a consistent rate over an extended period, enhancing therapeutic outcomes and reducing the frequency of dosing. For instance, the inclusion complex of diclofenac (an anti-inflammatory drug) with β-cyclodextrin has demonstrated a sustained release profile, providing prolonged pain relief with fewer doses [25].

6.6.5 Drug-Cyclodextrin Complexes

Cyclodextrin-drug complexes have been extensively studied for various therapeutic applications. One notable example is the inclusion complex of quercetin (a potent antioxidant) with γ-cyclodextrin. This complex not only improves the solubility and bioavailability of quercetin but also provides a controlled release, enhancing its antioxidant effects over time. Similarly, the inclusion complex of curcumin (an anti-inflammatory compound) with β-cyclodextrin has shown improved solubility, stability, and sustained release, making it more effective in managing inflammatory conditions [26]. These examples underscore the versatility of cyclodextrins in enhancing the performance of a wide range of pharmaceutical agents, from antioxidants to anti-inflammatory drugs, through improved stability, solubility, and controlled release mechanisms.

6.7 APPLICATIONS IN COSMETICS

6.7.1 Inclusion of Cyclodextrins in Creams and Lotions

Cyclodextrins are increasingly being incorporated into skin care products like creams and lotions to enhance their performance and stability. The unique structure of cyclodextrins, with their hydrophobic cavity and hydrophilic exterior, allows them to form inclusion complexes with various active ingredients. This encapsulation improves the solubility and stability of these ingredients, ensuring that they remain effective throughout the product's shelf-life. Cyclodextrins can also mask unpleasant odours and tastes, enhancing the overall sensory experience of the product [27].

6.7.2 Benefits for Skin Health and Product Stability

The inclusion of cyclodextrins in skin care products offers several benefits for skin health. By stabilizing and solubilizing active ingredients such as vitamins, antioxidants, and UV filters, cyclodextrins enhance their bioavailability and effectiveness. Cyclodextrin complexes of vitamin E and coenzyme Q10 provide better antioxidant protection, helping to combat the signs of ageing and improve skin texture [28]. Additionally, cyclodextrins can reduce the irritation potential of certain active ingredients, making skin care products more suitable for sensitive skin. The improved stability of these products also means that their beneficial effects are preserved over time, ensuring consistent performance and user satisfaction.

6.7.3 Literature Survey of Commercial Applications of CDs

Numerous commercial skin care products leverage the benefits of cyclodextrins. For instance, the Olay Regenerist Micro-Sculpting Cream incorporates cyclodextrin complexes to enhance the stability and efficacy of its anti-aging ingredients. Similarly, the Lancôme Visionnaire Advanced Skin Corrector uses cyclodextrin complexes to stabilize its active molecules, providing prolonged and effective skin correction [29]. These products exemplify how cyclodextrins can enhance the performance and

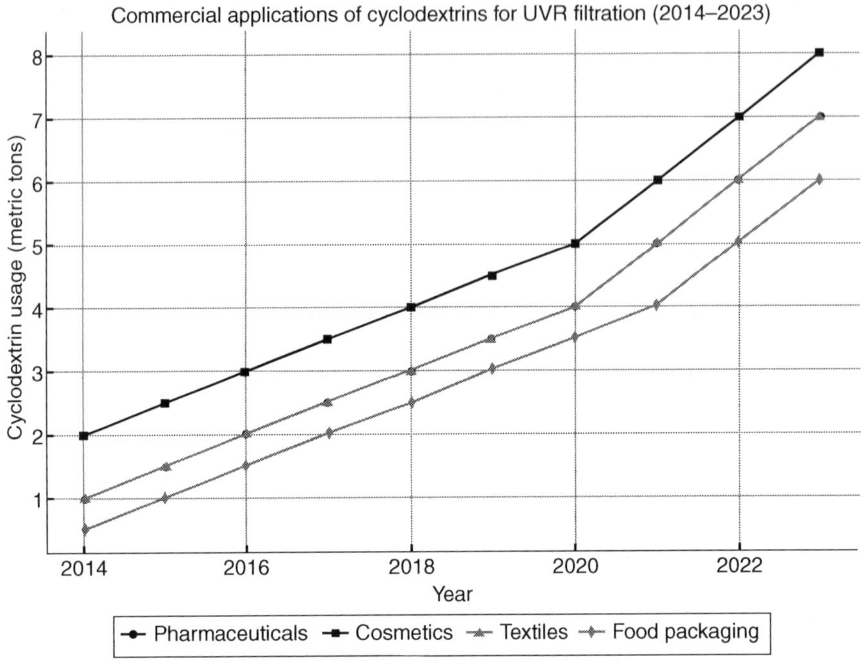

FIGURE 6.7 Statistical survey of the commercial applications of cyclodextrins for UVR filtration across four sectors: pharmaceuticals, cosmetics, textiles, and food packaging from 2014 to 2023.

user experience of skin care formulations. Figure 6.7 presents a statistical survey highlighting the commercial applications of cyclodextrins (CDs) in UVR filtration across four key sectors—pharmaceuticals, cosmetics, textiles, and food packaging—over the period from 2014 to 2023. The data reflect a steady increase in the use of CDs, with notable growth in the cosmetics and pharmaceutical industries, indicating their expanding role in enhancing UV protection and product stability.

6.7.4 Protection of Hair from UV Damage

Hair care products containing cyclodextrins can protect hair from the damaging effects of UV radiation. UV exposure can lead to hair damage, including colour fading, protein degradation, and loss of strength and elasticity. Cyclodextrin complexes with UV filters help to mitigate these effects by providing a protective barrier that absorbs and neutralizes UV rays [30]. This protection not only preserves the natural colour and strength of the hair but also maintains its overall health and appearance.

6.7.5 Enhanced Product Performance and Durability

In addition to UV protection, cyclodextrins enhance the overall performance and durability of hair care products. By stabilizing volatile and sensitive ingredients such as fragrances and active compounds, cyclodextrins ensure that these products remain effective for longer periods. This stabilization prevents the degradation of key ingredients, enhancing the longevity and effectiveness of shampoos, conditioners, and styling products [31]. Furthermore, cyclodextrins can improve the texture and feel of hair care formulations, providing a more pleasant and user-friendly experience.

6.7.6 Case Studies and Product Formulations

Several case studies highlight the effectiveness of cyclodextrin complexes in hair care products. For example, a study on the inclusion complex of hydrolyzed keratin with *β*-cyclodextrin demonstrated improved hair strength and resilience, protecting the hair from environmental stressors and mechanical damage [32]. Commercially, products like the L'Oréal Paris Elvive Total Repair 5 Protein Recharge Treatment use cyclodextrin complexes to enhance the stability and efficacy of their protein and nutrient components, providing comprehensive hair repair and protection. These formulations showcase the practical benefits of cyclodextrins in enhancing hair care product performance and durability.

6.8 FUTURE PERSPECTIVES

6.8.1 Innovations in Cyclodextrin Chemistry: Development of Modified Cyclodextrins

The future of cyclodextrin chemistry holds significant promise with the development of modified cyclodextrins tailored for specific applications. Researchers are actively working on synthesizing cyclodextrin derivatives with improved UV absorption properties and better compatibility with various UV filters. For instance, methylated and acetylated cyclodextrins have been shown to enhance the solubility and stability of a broader range of UV absorbers compared to their native forms. These modifications can lead to more effective sunscreen formulations with superior protective qualities. Moreover, these derivatives can be designed to possess specific functional groups that enhance their binding affinity and encapsulation efficiency, resulting in more robust and reliable UV protection [33].

6.8.2 Integration with Nanotechnology

Nanotechnology integration with cyclodextrin chemistry represents a cutting-edge approach to enhancing UV protection. Cyclodextrin-based nanocarriers are being developed to deliver UV absorbers more effectively to the skin or other surfaces. These nanocarriers can encapsulate UV filters within their core, protecting them from degradation and ensuring their controlled release over time. An example of this innovation is the use of cyclodextrin-coated nanoparticles, which can provide enhanced UV protection by maintaining a steady concentration of UV filters on the skin's surface [34]. This technology not only improves the efficacy of UV absorbers but also reduces the need for frequent reapplication, making sun protection more convenient and reliable.

6.8.3 Emerging Applications: Environmental Protection

The potential applications of cyclodextrin-UV absorber complexes extend beyond personal care products to environmental protection. For instance, incorporating these complexes into textiles and coatings can provide UV protection to fabrics and surfaces exposed to sunlight, such as outdoor furniture, automotive interiors, and building materials [35]. A practical example is the development of UV-protective coatings for outdoor fabrics, where cyclodextrin-UV absorber complexes enhance the durability and longevity of the material by protecting it from UV-induced degradation. These applications can significantly extend the life of products and reduce maintenance costs, offering both economic and environmental benefits.

6.8.4 Synergistic Effects with Other UV Protection Strategies

Cyclodextrin complexes can also synergize with other UV protection strategies, enhancing their overall efficacy. Combining cyclodextrins with antioxidants, for example, can provide dual protection by not only blocking UV radiation but also neutralizing free radicals generated by UV exposure. This combination can be particularly effective in antiageing formulations, where protecting the skin from UV damage and oxidative stress is crucial [36]. Another promising avenue is the incorporation of cyclodextrin-UV absorber complexes into formulations with other active ingredients, such as anti-inflammatory agents or skin-repairing compounds, creating multifunctional products that offer comprehensive skin care benefits. These synergistic formulations can provide enhanced protection and therapeutic effects, meeting the growing consumer demand for innovative and effective skincare solutions.

6.9 CONCLUSION

In summary, cyclodextrins offer significant advantages in UVR filtration due to their unique ability to form inclusion complexes with various UV absorbers. This capability enhances the solubility, stability, and efficacy of UV filters, providing more effective and long-lasting protection compared to traditional methods. Cyclodextrins improve the photostability of UV absorbers, reduce their photodegradation, and ensure controlled release, making them ideal for applications in personal care, pharmaceuticals, and environmental protection. The comparative benefits of cyclodextrin-based UV filters over conventional UV filters include better formulation compatibility, reduced skin irritation, and enhanced overall performance. Looking ahead, the potential for further research and development in cyclodextrin chemistry and its integration with nanotechnology holds promise for even more innovative UV protection solutions. Cyclodextrins represent a versatile and effective approach to UV protection, poised to revolutionize current strategies and offer comprehensive protection across various applications.

REFERENCES

1. Mohania, D., Chandel, S., Kumar, P., Verma, V., Digvijay, K., Tripathi, D., ... & Shah, D. (2017). Ultraviolet radiations: skin defense-damage mechanism. In Ahmad, S. I.

(Ed.). *Ultraviolet Light in Human Health, Diseases and Environment.* Cham: Springer International Publishing, (pp. 71–87).
2. Gies, P., Roy, C., & Udelhofen, P. (2004). Solar and ultraviolet radiation. In Hill, D., Elwood, J. M., & English, D. R. (Eds.). *Prevention of Skin Cancer.* Dordrecht: Springer Netherlands, (pp. 21–54).
3. Herman, J., McKenzie, R. L., Diaz, S. B., Kerr, J. B., Madronich, S., & Seckmeyer, G. (1998). Ultraviolet radiation at the Earth's surface. *UMBC Joint Center for Earth Systems Technology (JCET).*
4. Bernerd, F., Passeron, T., Castiel, I., & Marionnet, C. (2022). The damaging effects of long UVA (UVA1) rays: a major challenge to preserve skin health and integrity. *International Journal of Molecular Sciences*, 23(15), 8243.
5. Pathak, M. A. (1997). *Photoprotection against harmful effects of solar UVB and UVA radiation: an update* (pp. 59–79). New York, NY, USA: Marcel Dekker Inc.
6. DeBuys, H. V., Levy, S. B., Murray, J. C., Madey, D. L., & Pinnell, S. R. (2000). Modern approaches to photoprotection. *Dermatologic Clinics*, 18(4), 577–590.
7. Weihs, P., Schmalwieser, A. W., & Schauberger, G. (2012). UV effects on living organisms. In Laws, E. A. (Ed.), *Environmental Toxicology: Selected Entries from the Encyclopedia of Sustainability Science and Technology.* New York, NY: Springer New York, (pp. 609–688). Springer.
8. Majdi, M., Milani, B. Y., Movahedan, A., Wasielewski, L., & Djalilian, A. R. (2014, October). The role of ultraviolet radiation in the ocular system of mammals. *Photonics*, 1(4), 347–368.
9. Gao, W., Schmoldt, D., & Slusser, J. R. (2010). *UV radiation in global climate change* (p. 544). Tsinghua University Press.
10. Matsumura, Y., & Ananthaswamy, H. N. (2004). Toxic effects of ultraviolet radiation on the skin. *Toxicology and Applied Pharmacology*, 195(3), 298–308.
11. Khan, A., Nazir, A., Rehman, A., Naveed, M., Ashraf, M., Iqbal, K., ... & Maqsood, H. S. (2020). A review of UV radiation protection on humans by textiles and clothing. *International Journal of Clothing Science and Technology*, 32(6), 869–890.
12. Bais, A. F., Bernhard, G., McKenzie, R. L., Aucamp, P. J., Young, P. J., Ilyas, M., ... & Deushi, M. (2019). Ozone–climate interactions and effects on solar ultraviolet radiation. *Photochemical & Photobiological Sciences*, 18(3), 602–640.
13. Bais, A. F., Lucas, R. M., Bornman, J. F., Williamson, C. E., Sulzberger, B., Austin, A. T., ... & Heikkilä, A. M. (2018). Environmental effects of ozone depletion, UV radiation and interactions with climate change: UNEP Environmental Effects Assessment Panel, update 2017. *Photochemical & Photobiological Sciences*, 17(2), 127–179.
14. Kuthi, N. A., Basar, N., & Chandren, S. (2022). Nanonutrition-and nanoparticle-based ultraviolet rays protection of skin. In Das Talukdar, A., Sarker, S. D., & Patra, J. K. (Eds.), *Advances in Nanotechnology-Based Drug Delivery Systems* (pp. 227–280). Elsevier.
15. Svobodová, A., & Vostalova, J. (2010). Solar radiation induced skin damage: review of protective and preventive options. *International Journal of Radiation Biology*, 86(12), 999–1030.
16. Heisler, G. M. (2000). *Ultraviolet radiation, human health, and the urban forest* (Vol. 268). US Department of Agriculture, Forest Service, Northeastern Research Station.
17. Parisi, A. V., Igoe, D., Downs, N. J., Turner, J., Amar, A., & A Jebar, M. A. (2021). Satellite monitoring of environmental solar ultraviolet A (UVA) exposure and irradiance: a review of OMI and GOME-2. *Remote Sensing*, 13(4), 752.
18. Sahan, M. (2019, November). The measurements of the global solar radiation and solar ultraviolet radiation during 2018 year. In *AIP Conference Proceedings* (Vol. 2178, No. 1). AIP Publishing.

19. Araújo, R. G., Alcantar-Rivera, B., Meléndez-Sánchez, E. R., Martínez-Prado, M. A., Sosa-Hernández, J. E., Iqbal, H. M., ... & Martínez-Ruiz, M. (2022). Effects of UV and UV-vis irradiation on the production of microalgae and macroalgae: new alternatives to produce photobioprotectors and biomedical compounds. *Molecules*, 27(16), 5334.
20. De Gruijl, F. R. (1997). Health effects from solar UV radiation. *Radiation Protection Dosimetry*, 72(3–4), 177–196.
21. Zerefos, C., Fountoulakis, I., Eleftheratos, K., & Kazantzidis, A. (2023). Long-term variability of human health-related solar ultraviolet-B radiation doses from the 1980s to the end of the 21st century. *Physiological Reviews*, 103(3), 1789–1826.
22. Rojas, J. O. H. N., Londoño, C. E. S. A. R., & Ciro, Y. (2016). The health benefits of natural skin UVA photoprotective compounds found in botanical sources. *International Journal of Pharmacy and Pharmaceutical Science*, 8(3), 13–23.
23. Korać, R. R., & Khambholja, K. M. (2011). Potential of herbs in skin protection from ultraviolet radiation. *Pharmacognosy Reviews*, 5(10), 164.
24. Ghoneim, A. A., Kadad, I. M., & Altouq, M. S. (2013). Statistical analysis of solar UVB and global radiation in Kuwait. *Energy*, 60, 23–34.
25. Krishnamoorthy, S., Allabasha, N., Mani, M. K., & Sarkar, A. K. (2023). Effect of β-cyclodextrin on benzophenones and its impact on UVR filtration of cotton fabric through 'guest absorber—host enhancer' mechanism. *Journal of Inclusion Phenomena and Macrocyclic Chemistry*, 103(3), 131–145.
26. Biloti, D. N., Dos Reis, M. M., Ferreira, M. M. C., & Pessine, F. B. T. (1999). Photochemical behavior under UVA radiation of β-cyclodextrin included Parsol® 1789 with a chemometric approach. *Journal of Molecular Structure*, 480, 557–561.
27. Felton, L. A., Wiley, C. J., & Godwin, D. A. (2004). Influence of cyclodextrin complexation on the in vivo photoprotective effects of oxybenzone. *Drug Development and Industrial Pharmacy*, 30(1), 95–102.
28. Godwin, D. A., Wiley, C. J., & Felton, L. A. (2006). Using cyclodextrin complexation to enhance secondary photoprotection of topically applied ibuprofen. *European Journal of Pharmaceutics and Biopharmaceutics*, 62(1), 85–93.
29. El-Tahlawy, K., El-Nagar, K., & Elhendawy, A. G. (2007). Cyclodextrin-4 Hydroxy benzophenone inclusion complex for UV protective cotton fabric. *Journal of the Textile Institute*, 98(5), 453–462.
30. Simeoni, S., Scalia, S., & Benson, H. A. E. (2004). Influence of cyclodextrins on in vitro human skin absorption of the sunscreen, butyl-methoxydibenzoylmethane. *International Journal of Pharmaceutics*, 280(1–2), 163–171.
31. Scalia, S., Villani, S., & Casolari, A. (1999). Inclusion complexation of the sunscreen agent 2-Ethylhexyl-p-dimethylaminobenzoate with Hydroxypropyl-β-cyclodextrin: effect on photostability. *Journal of Pharmacy and Pharmacology*, 51(12), 1367–1374.
32. Scalia, S., Villani, S., Scatturin, A., Vandelli, M. A., & Forni, F. (1998). Complexation of the sunscreen agent, butyl-methoxydibenzoylmethane, with hydroxypropyl-β-cyclodextrin. *International Journal of Pharmaceutics*, 175(2), 205–213.
33. de Souza de Bustamante Monteiro, M. S., Ozzetti, R. A., Vergnanini, A. L., de Brito-Gitirana, L., Volpato, N. M., de Freitas, Z. M. F., ... & dos Santos, E. P. (2012). Evaluation of octyl p-methoxycinnamate included in liposomes and cyclodextrins in anti-solar preparations: preparations, characterizations and in vitro penetration studies. *International Journal of Nanomedicine*, vol 7, 3045–3058.
34. Liu, J., Ma, X., Shi, W., Xing, J., Ma, C., Li, S., & Huang, Y. (2020). Anti-ultraviolet properties of β-cyclodextrin-grafted cotton fabrics dyed by broadleaf holly leaf extract. *Textile Research Journal*, 90(21–22), 2441–2453.

35. Almagro, L., Sabater-Jara, A. B., Belchí-Navarro, S., Fernández-Pérez, F., Bru, R., & Pedreño, M. A. (2011). Effect of UV light on secondary metabolite biosynthesis in plant cell cultures elicited with cyclodextrins and methyljasmonate. *Plants and Environment*, 1, 115–136.
36. Berbicz, F., Nogueira, A. C., Neto, A. M., Natali, M. R. M., Baesso, M. L., & Matioli, G. (2011). Use of photoacoustic spectroscopy in the characterization of inclusion complexes of benzophenone-3-hydroxypropyl-β-cyclodextrin and ex vivo evaluation of the percutaneous penetration of sunscreen. *European Journal of Pharmaceutics and Biopharmaceutics*, 79(2), 449–457.

Bibliography

Brewster, M. E., & Loftsson, T. (2007). Cyclodextrins as pharmaceutical solubilizers. *Advanced Drug Delivery Reviews*, 59(7), 645–666. DOI: 10.1016/j.addr.2007.05.012

Jansook, P., Ogawa, N., & Loftsson, T. (2018). Cyclodextrins in pharmaceutical formulations. *Journal of Pharmaceutical Sciences*, 107(7), 1743–1756. DOI: 10.1016/j.xphs.2018.02.026

Loftsson, T., & Brewster, M. E. (2012). Pharmaceutical applications of cyclodextrins: basic science and product development. *Journal of Pharmacy and Pharmacology*, 64(6), 844–863. DOI: 10.1111/j.2042-7158.2012. 01461.x

7 Unlocking the Potential of Cyclodextrins

Advantages, Applications, and Cyclodextrin-UVR Absorber Complexes in Pharmaceutical Formulations

Kamakshi Gnanasekaran, Janice Marianne John, and Shinomol George K.

7.1 INTRODUCTION

A cyclic α-1,4-glucan made up of six or more than 100 glucose units is called a cyclodextrin [1]. By using an intramolecular transglycosylation process, cyclodextrin glucanotransferase (EC 2.4.1.19, CGTase) converts amylose into cyclodextrins, which are made up of six to several hundred glucose units [2]. A unique extracellular enzyme called cyclodextrin glycosyltransferase (1,4-α-D-glucan 4-α-D-(1,4-α-D-glucano)-transferase) aids in the cyclization reaction that forms non-reducing cyclodextrins from starch, amylose, and other polysaccharides [3–6]. In addition to cyclization, cyclodextrin glucanotransferase has modest hydrolytic activity and catalyzes coupling and disproportionation transglycosylation events [4–7]. In these processes, an α-glycosidic bond is usually cleaved. As the substrate in coupling, cyclodextrins move a linear malto-oligosaccharide to an acceptor substrate [8, 9]. A linear malto-oligosaccharide is cleaved during the disproportionation process, and the shorter product is then transferred to an acceptor substrate. Beginning with the cleavage of polysaccharide bonds to form a covalent intermediate, the γ-cyclization reaction proceeds to produce a cyclic oligosaccharide through cyclization [10]. Asp229 and Glu257 are essential catalytic amino acid residues that cause bond cleavage. An α-1,4-glycosidic bond is resealed during bond formation [4]. Currently, organic solvents are used for selective crystallization of α-, β-, and γ-cyclodextrins, though mutant cyclodextrin glycosyltransferase enzymes, despite their cost and safety concerns, are gaining popularity for producing specific cyclodextrin forms. The enzyme primarily

DOI: 10.1201/9781003492757-7

forms cyclodextrins through intramolecular transglycosylation, while also catalyze other transglycosylation reactions. Figure 7.1 depicts the formation of cyclodextrin. Figure 7.2 portrays the structure of cyclodextrins [9, 11].

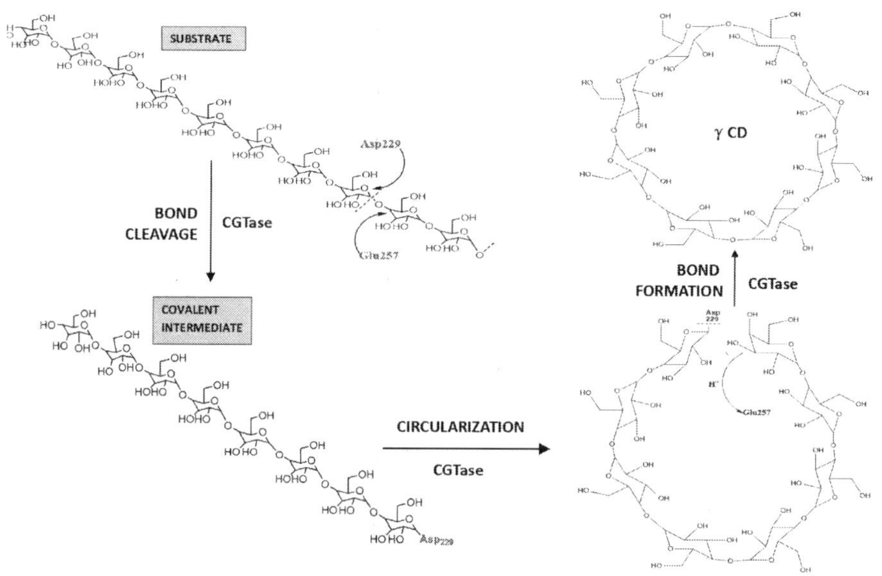

FIGURE 7.1 Cyclodextrin formation. Native cyclodextrins are modified by adding functional groups to the hydroxyl groups at positions 2, 3, and 6 on the glucose molecule to improve solubility and complex formation. Examples of such modifications include hydroxypropyl, methyl, carboxymethyl, and sulfobutylether *β*-cyclodextrins (HP*β*CD, M*β*CD, CM*β*CD, and SBE*β*CD [12–29].

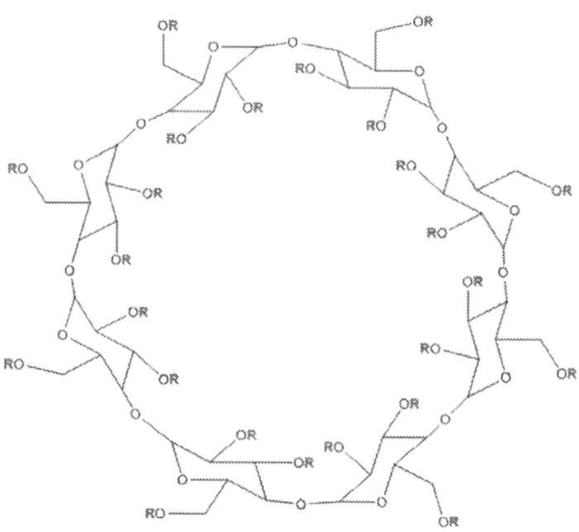

CYCLODEXTRIN	R-GROUP
Hydroxypropyl ß-Cyclodextrin	-CH3
Methyl ß-Cyclodextrin	-CH2CH(OH)CH3
Carboxymethyl ß-Cyclodextrin	-CH2COONa
Sulfobutylether ß-Cyclodextrin	-(CH2)4SO3Na

FIGURE 7.2 Cyclodextrin structure.

7.2 CONFORMATIONAL FLEXIBILITY OF CYCLODEXTRINS

There are six, seven, and eight d-glucose subunits in each of the natural cyclodextrins, α-cyclodextrins, β-cyclodextrins, and γ-cyclodextrins. Although they periodically switch to the $^1C_4$ chair conformation, these subunits predominantly adopt a $^4C_1$ chair conformation. The chair conformation increases the stability of the molecule by minimizing angular strain within the CD structure and placing the oxygen groups (hydroxyls, glycosidic groups) in the equatorial plane [18–22]. A few physiochemical characteristics of cyclodextrins are depicted in Table 7.1.

The hydrophobic cavities in α-, β-, and γ-cyclodextrins vary in size, with α-cyclodextrin having the smallest and γ-cyclodextrin the largest, corresponding to their respective glucose unit quantities [28, 29]. The inner cavity diameters of cyclodextrins are better described as a range rather than a single number, acknowledging the inherent flexibility of their structures [30]. Figure 7.3 represents the dimensions of α-CD, β-CD, and γ-CD.

Studies suggesting rigidity in glucose subunits have led to the idea that cyclodextrin molecules are rigid, a theory supported by X-ray crystallography, which reveals their

TABLE 7.1
Selected Physiochemical Characteristics of Cyclodextrins [3, 23–27]

Properties	α-CD	β-CD	γ-CD
Empirical formula	C36H60O30	C42H70O35	C48H80O40
Molecular weight (g/mol)	972	1135	1297
Number of glucopyranose units	6	7	8
Cavity diameter (nm)	0.47–0.57	0.60–0.78	0.83–0.95
Internal cavity volume (nm^3)	1740	2620	4720
Water molecules in the cavity	6	11	17
Aqueous solubility (g/L)	129.5	18.4	249.2
Temperature degradation (°C)	278	298	267

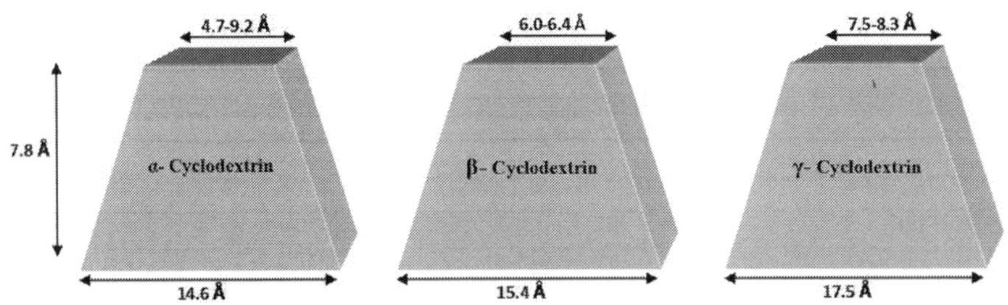

FIGURE 7.3 Dimensions of α-CD, β-CD, and γ-CD (height, inner cavity diameter, exterior diameter) [3, 31, 32].

truncated cone structure [33, 34]. X-ray crystallography, being limited to solid-state materials, cannot clearly demonstrate the rigidity of cyclodextrins or explain why they can form stable complexes with a variety of molecules [35–37]. Research shows that larger cyclodextrins, containing up to 31 glucose units, have more conformational flexibility than smaller ones like α-, β-, and γ-cyclodextrins, but this adaptability reduces their overall stability and limits their ability to form complexes like smaller cyclodextrins [32, 38–40].

7.3 COMPLEXATION PROPERTIES OF CYCLODEXTRIN

Cyclodextrins are versatile and can form stable complexes with a variety of molecules, including proteins, chiral substances, aromatic molecules, alkyl groups, and alkyl glycosides, resulting in large equilibrium constants [8, 37, 41, 42]. The cyclodextrin complexation mechanism is shown in Figure 7.4.

Molecular binding strength is highly dependent on the size of the cyclodextrin cavity [43, 44]. For example, β-cyclodextrin has a higher effect on cholesterol solubility and cell viability than derivatives of α-cyclodextrin and γ-cyclodextrin [45]. The binding strength of molecules to cyclodextrin cavities depends on the cyclodextrin size and the guest molecule's charge, intermolecular interactions, and size, with these factors varying in importance based on the type of guest molecule [46]. The ionic charge, intermolecular interactions, and size of each component influence the binding affinity of cyclodextrin complexes. Guest molecules' orientation within the cavity is affected by their charge, with positively charged derivatives pointing outward from β-cyclodextrin, while negatively charged derivatives show no preferential orientation [47].

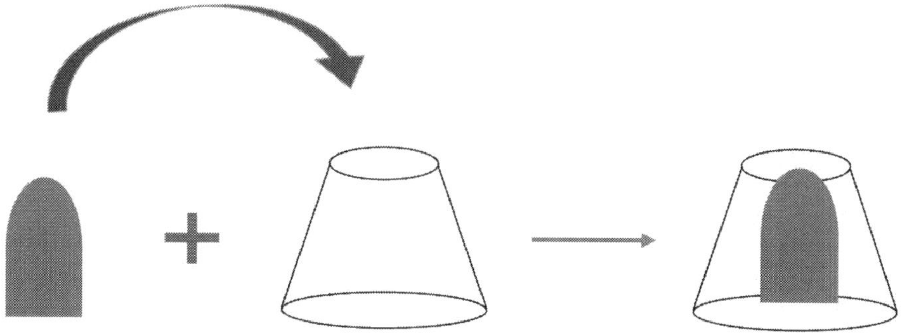

FIGURE 7.4 Cyclodextrin complexation mechanism. The orange-coloured moiety represents a guest molecule that approaches the empty cyclodextrin cavity, and complexation is completed when the guest molecule enters the cavity, either partially or fully.

7.4 CYCLODEXTRIN-UVR ABSORBER COMPLEXES IN PHARMACEUTICALS

7.4.1 UV Radiation and Photoprotection

The sun emits UVA, UVB, and UVC radiation, with the ozone layer protecting against UVC, while UVA and UVB can still harm the skin. UVA causes long-term DNA damage, and UVB contributes to skin cancer and short-term damage, posing significant health risks [48–51]. Ultraviolet radiation exposure is a major cause of skin allergies as well as malignancies such as malignant melanoma and basal and squamous cell carcinomas. The cosmetics industry has responded by introducing ultraviolet absorbing ingredients [52–54]. Sunscreens primarily use metal nanoparticles like ZnO and TiO_2 to absorb, reflect, and scatter ultraviolet light, but concerns about their negative effects on human health and the environment have prompted research into safer alternatives [55–57]. Some of these include encasing organic ultraviolet absorbers in organosilica nanoparticles or coating harmful nanomaterials with silica layers, which helps protect the UV filter molecules from external mechanical stress [58–62].

Sunscreens provide two different types of UV protection: they use antioxidants to scavenge free radicals and UV filters to stop free radicals from forming [63]. UV filters absorb, reflect, and scatter UV rays, while antioxidants eliminate free radicals. Despite limited UV exposure in the epidermis, free radicals still form, so antioxidants are added to cosmetic formulations to provide full protection against oxidative stress [61, 64]. These antioxidants are often combined with antioxidants that are soluble in lipids and water.

A sun protection formula should offer broad UVA and UVB protection, shield the skin from UV filters, and allow antioxidants to reach viable epidermal layers to work effectively [65]. Many sunscreens fail to meet these guidelines because UV filters can absorb UVR and undergo photoaddition, photofragmentation, and rearrangement, leading to the production of harmful free radicals and the degradation of other molecules [66, 67]. Antioxidants, being reactive with free radicals, are more prone to oxidation and degradation. Vitamin E, with its short half-life and limited UV stability, is not very effective at scavenging free radicals, while lipophilic antioxidants may aggregate, reducing their ability to penetrate and absorb [68, 69].

7.4.2 Cyclodextrin – UVR Absorber Complexes

Cyclodextrins form inclusion complexes with guest molecules, enhancing solubility, stability, and bioavailability, making them crucial in pharmaceutical formulations, particularly in complexing UVR absorbers [23, 70]. Cyclodextrin-UV absorber complexes protect ultraviolet absorbers, enhancing their stability by shielding them from external impacts and improving solubility for easier integration and dispersion [52, 71–73]. Cyclodextrin-UV absorber complexes provide controlled release for long-term UVR protection, enhance product durability, and reduce skin irritation by minimizing direct contact. They also improve photostability and increase the UPF factor of UV absorbers like benzophenones, boosting their effectiveness

[62, 74–76]. Cyclodextrin inclusion complexes enhance sunscreen formulations by increasing the solubility of lipophilic antioxidants and protecting UV filters from photodegradation. These complexes, including carotenoids, polyphenols, and flavonoids, also scavenge free radicals and reduce photoallergic/phototoxic effects [77].

Cyclodextrins play a critical role in promoting the stability and bioavailability of cyclodextrin encapsulated molecules, including antioxidants and UV filters [78]. Unique cyclodextrin structures based on reducing glycosidic units result in cyclodextrins with high resistance to chemical instability [79]. Hydrogen bond formation between the adjacent hydroxyl groups of glucopyranose units strengthen the chemical stability of the cyclodextrin structure and their water solubility [2, 80–82]. Cyclodextrins, resistant to enzymatic hydrolysis, encapsulate antioxidants and UV filters, minimizing interactions with surrounding molecules and improving their water solubility. Their hydrophilic outer surface enhances solubility, bioavailability, and skin penetration, improving sunscreen release profiles and dermal drug delivery for better skin health [83–86]. Hydroxypropyl-β-cyclodextrin enhances the solubility, stability, and efficacy of UV filters in sunscreens by forming inclusion complexes, improving water solubility and photostability. This addition increases the sunscreen's ability to protect skin from UV radiation, and using hydroxypropyl-β-cyclodextrin in high-performance liquid chromatography aids in the precise separation and identification of UV filters, ensuring the quality of commercial sunscreens [47, 87].

The ibuprofen–hydroxypropyl-β-cyclodextrin formulation enhances transdermal kinetics, improving secondary photoprotection, particularly at lower UV exposures. The complex significantly increases ibuprofen's aqueous solubility, aiding in the timely interruption of the inflammatory cascade triggered by UV exposure [88]. Cyclodextrins, including α-, β-, and γ-cyclodextrin, can efficiently protect against UV and IR radiation, potentially replacing traditional metal oxide UV absorbers like titanium and zinc. These cyclodextrin-based protection compositions can be formulated into various physical forms, such as aerosols, creams, gels, lotions, and powders [86, 89]. Cyclodextrins enhance skin permeability by interacting with phospholipids or cholesterol, facilitating the entry of active substances. In dermopharmaceutical formulations, cyclodextrins, such as silymarin-sulfobutylether-β-cyclodextrins, improve solubility, reduce oxidative stress, prevent skin ageing, and alleviate facial redness in rosacea, while also enhancing bioavailability and reducing toxicity of active compounds [90].

UVA rays (320–400 nm) can cause harmful biological effects, making chemical sunscreen filters essential for photoprotection [91, 92]. Photostability is crucial for these filters, as degradation not only reduces protection but may also produce toxic or allergenic byproducts. BMDBM, a widely used UVA filter, is commonly included in sunscreen formulations [93, 94]. However, some research has shown that UV-induced photodegradation of 4-tert-butyl-4′-methoxy dibenzoylmethane does place, which results in a reduction in its protective effectiveness. The breakdown products of 4-tert-butyl-4′-methoxy dibenzoylmethane may be dangerous and harm relevant biological molecules [17, 95].

The stabilizing effect of 2-hydroxypropyl-β-cyclodextrin on 4-tert-butyl-4′-methoxy dibenzoylmethane is reduced in oil-in-water emulsion carriers due to competition for space in the cyclodextrin cavity [77, 90]. Encapsulating the inclusion complex in lipid microparticles or lipospheres can minimize this issue [96, 97]. While β-cyclodextrin complexes showed minimal impact on photostability, they do reduce the risk of adverse reactions by containing photodegradation products within the cyclodextrin annulus [49, 98].

Various techniques have been used to minimize the photodecomposition of 4-tert-butyl-4′-methoxy dibenzoylmethane. β-cyclodextrin, being the most stable form of inclusion complex, enhances photostability. While α-cyclodextrin's smaller cavity may hinder stable complex formation, γ-cyclodextrin's larger cavity interferes with optimal guest-host interactions. Incorporating 4-tert-butyl-4′-methoxy dibenzoylmethane into randomly methylated β-cyclodextrin and 2-hydroxypropyl-β-cyclodextrin enhances its photostability [66, 98].

7.5 HYDROGEN BONDING AND AQUEOUS SOLUBILITY OF CYCLODEXTRIN

Cyclodextrins primarily rely on hydrogen bonding as their main polar intermolecular force in solution, although strong intramolecular hydrogen bonding limits the ability of outer hydroxyl groups to form bonds, especially in β-cyclodextrins [44]. The interior cavity of cyclodextrins does not facilitate hydrogen bonding with guest molecules. While hydrogen bonding is not crucial for stabilizing cyclodextrin complexes, it can contribute to the overall development and stability of complexes, often complementing van der Waals forces [99]. Research has indicated that hydrogen bonding plays a part in the encapsulation of some compounds by cyclodextrins, like methoprene and chloramphenicol, although its precise function is still unclear and needs more study [14, 100, 101]. Figure 7.5 displays the process by which water is released from the cyclodextrin cavity when a complex forms.

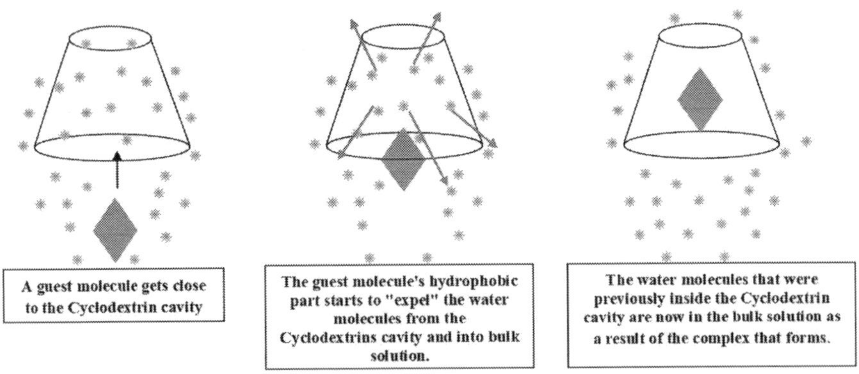

FIGURE 7.5 Process by which water is released from the cyclodextrin cavity when a complex is formed.

In aqueous solutions, the expulsion of water molecules from the cyclodextrin cavity is essential for complexation, while in the solid state, water molecules can form stabilizing interactions with cyclodextrin-drug complexes without being removed from the cavity [102].

7.6 PREPARATION OF CYCLODEXTRIN INCLUSION COMPLEXES

Common methods for forming cyclodextrin complexes include physical mixing, kneading with a small amount of solvent, solvent evaporation, lyophilization, co-precipitation, supercritical fluid technology, spray drying, and microwave irradiation. These methods vary in efficiency, with microwave irradiation offering faster results and fewer solvent residues compared to traditional techniques [85, 103, 104].

Cyclodextrins find applications in different drug delivery systems like oral drug delivery systems [94, 105, 106], rectal drug delivery systems, nasal drug delivery systems [85, 107], transdermal drug delivery systems, ocular drug delivery systems [62, 107], control and targeted drug delivery systems [108, 109], peptide and protein delivery, brain drug delivery and brain targeting [110], as excipients in drug formulations [55, 111], etc.

7.7 LIMITATIONS OF CYCLODEXTRIN

Cyclodextrins have limited water solubility, especially larger types like β-cyclodextrin, and their stability can be affected by environmental factors such as pH and temperature. They may also struggle to encapsulate large or highly hydrophobic molecules, with concerns over toxicity at high concentrations, cost, and complex production hindering widespread use [111, 112].

7.8 FUTURE PERSPECTIVES

Cyclodextrins and their derivatives are gaining attention in pharmaceutical drug delivery for enhancing solubility, dissolution, and bioavailability. Research into cyclodextrin-based nanocarriers and targeted delivery systems, including those for gene therapies and RNA therapeutics, is advancing solutions for complex diseases and improving drug delivery strategies [69, 113–117].

ACKNOWLEDGEMENTS

The authors thank the management of Kristu Jayanti college for their unwavering support.

REFERENCES

1. Wüpper S, Lüersen K and Rimbach G (2021) Cyclodextrins, natural compounds, and plant bioactives—A nutritional perspective, *Biomolecules*, 11(3):401, doi:10.3390/biom11030401.

2. Ross PD and Rekharsky MV (1996) Thermodynamics of hydrogen bond and hydrophobic interactions in Cyclodextrin complexes, *Biophysical Journal*, 71(4):2144–2154, doi:10.1016/S0006-3495(96)79415-8.
3. Del Valle EMM (2004) Cyclodextrins and their uses: A review, *Process Biochemistry*, 39(9):1033–1046, doi:10.1016/S0032-9592(03)00258-9.
4. Li Z, Wang M, Wang F, Gu Z, Du G, Wu J and Chen J (2007) γ-Cyclodextrin: A review on enzymatic production and applications, *Applied Microbiology and Biotechnology*, 77(2):245–255, doi:10.1007/s00253-007-1166-7.
5. Park J and Kim I (2013) Use of β-Cyclodextrin in an antimigration coating for polyester fabric, *Coloration Technology*, 129, doi:10.1111/cote.12026.
6. Liu J, Liu B, Tang Y, Zhang W, Wu W, Xie Y and Zhu W-H (2015) Highly efficient cosensitization of D–A–π–A benzotriazole organic dyes with porphyrin for panchromatic dye-sensitized solar cells, *Journal of Materials Chemistry C*, 3(42):11144–11150, doi:10.1039/C5TC02522B.
7. Perrone L, Prati L and Rossi M (1998) Removal of chlorinated organic compounds from water by catalytic dehydrohalogenation, *Applied Catalysis B: Environmental*, 15(3):241–246, doi:10.1016/S0926-3373(97)00051-9.
8. Biwer A, Antranikian G and Heinzle E (2002) Enzymatic production of cyclodextrins, *Applied Microbiology and Biotechnology*, 59(6):609–617, doi:10.1007/s00253-002-1057-x.
9. Wang H, Wang S, Zhu H, Wang S and Xing J (2019) Inclusion complexes of Lycopene and β-Cyclodextrin: Preparation, characterization, stability and antioxidant activity, *Antioxidants*, 8(8):314, doi:10.3390/antiox8080314.
10. Savarino P, Parlati S, Buscaino R, Piccinini P, Degani I and Barni E (2004) Effects of additives on the dyeing of polyamide fibres. Part I: β-Cyclodextrin, *Dyes and Pigments*, 60(3):223–232, doi:10.1016/S0143-7208(03)00142-6.
11. Jin Z-Y (2013) *Cyclodextrin Chemistry: Preparation and Application*, World Scientific.
12. Aiassa V, Garnero C, Zoppi A and Longhi MR (2023) Cyclodextrins and their derivatives as drug stability modifiers, *Pharmaceuticals*, 16(8):1074, doi:10.3390/ph16081074.
13. Akşit A and Yurdakul B (2006) Application of cyclodextrin to the textile dyeing and washing processes, *Journal of Applied Polymer Science*, 100:208–218, doi:10.1002/app.22863.
14. Audino PG, Masuh H and Zerba E (2005) Thermal behaviour, biological activity and conformational study of a [Methoprene/β-Cyclodextrin] complex in a smoke generating formulation, *Molecules: A Journal of Synthetic Chemistry and Natural Product Chemistry*, 10(3):534–544, doi:10.3390/10030534.
15. Bezerra FM, Lis MJ, Firmino HB, Dias da Silva JG, Curto Valle R de CS, Borges Valle JA, Scacchetti FAP and Tessaro AL (2020) The Role of β-cyclodextrin in the textile industry—review, *Molecules*, 25(16):3624, doi:10.3390/molecules25163624.
16. Burkinshaw S, Liu K and Salihu G (2017) The wash-off of dyeings using interstitial water Part 5: Residual dyebath and wash-off liquor generated during the application of disperse dyes and reactive dyes to polyester/cotton fabric, *Dyes and Pigments*, 171, doi:10.1016/j.dyepig.2017.11.021.
17. Deflandre A and Lang G (1988) Photostability assessment of sunscreens. Benzylidene camphor and dibenzoylmethane derivatives, *International Journal of Cosmetic Science*, 10(2):53–62, doi:10.1111/j.1467-2494.1988.tb00002.x.
18. Savarino P, Piccinini P, Montoneri E, Viscardi G, Quagliotto P and Barni E (2000) Effects of additives on the dyeing of nylon-6 with dyes containing hydrophobic

and hydrophilic moieties, *Dyes and Pigments*, 47(1):177–188, doi:10.1016/S0143-7208(00)00074-7.

19. Savolainen J, Järvinen K, Matilainen L and Järvinen T (1998) Improved dissolution and bioavailability of phenytoin by sulfobutylether-β-Cyclodextrin ((SBE)7m-β-CD) and hydroxypropyl-β-Cyclodextrin (HP-β-CD) complexation, *European Journal of Pharmaceutics and Biopharmaceutics*, 165(1), 69–78, doi:10.1016/s0378-5173(98)00004-0.
20. Schmidt BVKJ and Barner-Kowollik C (2017) Dynamic macromolecular material design-the versatility of cyclodextrin-based Host-Guest chemistry, *Angewandte Chemie (International Ed. in English)*, 56(29):8350–8369, doi:10.1002/anie.201612150.
21. Schönbeck C (2018) Charge determines guest orientation: A combined NMR and molecular dynamics study of β-Cyclodextrins and adamantane derivatives, *The Journal of Physical Chemistry B*, 122(18):4821–4827, doi:10.1021/acs.jpcb.8b02579.
22. Schurig V (2010) Use of derivatized Cyclodextrins as chiral selectors for the separation of enantiomers by gas chromatography, *Annales Pharmaceutiques Francaises*, 68(2):82–98, doi:10.1016/j.pharma.2009.11.004.
23. Street KW Jr (2006) Cyclodextrin cavity polarity and chromatographic implications, *Journal of Liquid Chromatography*, 10:655–662, doi:10.1080/01483918708069016.
24. Dodziuk H (2006) Molecules with holes – cyclodextrins, in Helena Dodziuk (ed.) *Cyclodextrins Complexes*, Wiley online library, United States, doi:10.1002/3527608982.ch1.
25. Saad HY and Higuchi WI (1965) Water solubility of cholesterol, *Journal of Pharmaceutical Sciences*, 54(8):1205–1206, doi:10.1002/jps.2600540826.
26. Szejtli J (1998) Introduction and general overview of Cyclodextrin chemistry, *Chemical Reviews*, 98(5):1743–1754, doi:10.1021/cr970022c.
27. Szente L, Singhal A, Domokos A and Song B (2018) Cyclodextrins: Assessing the impact of cavity size, occupancy, and substitutions on Cytotoxicity and cholesterol homeostasis, *Molecules (Basel, Switzerland)*, 23(5):1228, doi:10.3390/molecules23051228.
28. Challa R, Ahuja A, Ali J and Khar RK (2005) Cyclodextrins in drug delivery: An updated review, *AAPS PharmSciTech*, 6(2):43, doi:10.1208/pt060243.
29. Saokham P and Loftsson T (2017) γ-Cyclodextrin, *International Journal of Pharmaceutics*, 516(1):278–292, doi:10.1016/j.ijpharm.2016.10.062.
30. Steed JW, Johnson CP, Barnes CL, Juneja RK, Atwood JL, Reilly S, Hollis RL, Smith PH, Clark DL. Supramolecular chemistry of p- Sulfonatocalix[5] arene: A water-soluble, bowl- shaped host with a large molecular cavity, *Journal of the American Chemical Society*, 117(46): 11426–11433. doi: 10.1021/ ja00151a006
31. Cal K and Centkowska K (2008) Use of cyclodextrins in topical formulations: Practical aspects, *European Journal of Pharmaceutics and Biopharmaceutics: Official Journal of Arbeitsgemeinschaft Fur Pharmazeutische Verfahrenstechnik e.V*, 68(3):467–478, doi:10.1016/j.ejpb.2007.08.002.
32. Sherje AP, Dravyakar BR, Kadam D and Jadhav M (2017) Cyclodextrin-based nanosponges: A critical review, *Carbohydrate Polymers*, 173:37–49, doi:10.1016/j.carbpol.2017.05.086.
33. Chu H, Zhang R, Huang Q, Bai C and Wang Z (2017) Chemical conjugation with Cyclodextrins as a versatile tool for drug delivery, *Journal of Inclusion Phenomena and Macrocyclic Chemistry*, 89:1–10, doi:10.1007/s10847-017-0743-3.
34. Muankaew C and Loftsson T (2018) Cyclodextrin-based formulations: A non-invasive platform for targeted drug delivery, *Basic & Clinical Pharmacology & Toxicology*, 122(1):46–55, doi:10.1111/bcpt.12917.

35. Dodziuk H (2003) Rigidity versus flexibility: A review of experimental and theoretical studies pertaining to the cyclodextrin non-rigidity, *Cheminform*, 34, doi:10.1002/chin.200313276.
36. Heidel JD and Schluep T (2012) Cyclodextrin-containing polymers: Versatile platforms of drug delivery materials, *Journal of Drug Delivery*, 2012:262731, doi:10.1155/2012/262731.
37. Schwack W (1995) Photochemistry of dibenzoyl methane UVA filters Part 1, *Journal of Photochemistry and Photobiology B: Biology*, 28(3):229–234, doi:10.1016/1011-1344(95)07118-L.
38. Caira MR, Griffith VJ, Nassimbeni LR and Oudtshoorn B van (1994) Unusual 1C4 conformation of a methylglucose residue in crystalline permethyl-β-Cyclodextrin monohydrate, *Journal of the Chemical Society, Perkin Transactions,* 2(10):2071–2072, doi:10.1039/P29940002071.
39. Larsen KL (2002) Large cyclodextrins, *Journal of Inclusion Phenomena and Macrocyclic Chemistry*, 43(1):1–13, doi:10.1023/A:1020494503684.
40. Srinivasan K, Radhakrishnan S and Stalin T (2014) Inclusion complexes of β-cyclodextrin-dinitrocompounds as UV absorber for ballpoint pen ink. *Spectrochimica Acta Part A: Molecular and Biomolecular Spectroscopy*, 129: 551–564, doi:10.1016/j.saa.2014.03.021.
41. Cai W, Yu Y and Shao X (2005) Chiral recognition of aromatic compounds by beta-Cyclodextrin based on bimodal complexation, *Journal of Molecular Modeling*, 11(3):186–193, doi:10.1007/s00894-004-0233-6.
42. Liu Y, Li L, Zhang H-Y, Fan Z and Guan X-D (2003) Selective binding of chiral molecules of cinchona alkaloid by beta- and gamma-Cyclodextrins and organoselenium-bridged bis(beta-Cyclodextrin)s, *Bioorganic Chemistry*, 31(1):11–23, doi:10.1016/s0045-2068(02)00512-6.
43. Szente L and Szejtli J (2004) Cyclodextrins as food ingredients, *Trends in Food Science & Technology*, 15(3):137–142, doi:10.1016/j.tifs.2003.09.019.
44. Tarimci N (2011) Cyclodextrins in the Cosmetic Field, in Erem Bilensoy (ed.) *Cyclodextrins in Pharmaceutics, Cosmetics, and Biomedicine*, Wiley, doi:10.1002/9780470926819.ch7.
45. Saenger W, Jacob J, Gessler K, Steiner T, Hoffmann D, Sanbe H, Koizumi K, Smith SM and Takaha T (1998) Structures of the common cyclodextrins and their larger analogues-beyond the Doughnut, *Chemical Reviews*, 98(5):1787–1802, doi:10.1021/cr9700181.
46. Szejtli J (1988) Cyclodextrins in Pharmaceuticals, in *Cyclodextrin Technology. Topics in Inclusion Science*, vol 1. Springer, Dordrecht. https://doi.org/10.1007/978-94-015-7797-7_3.
47. Seidel RW and Koleva BB (2009) β-Cyclo-dextrin 10.41-hydrate, *Acta Crystallographica. Section E, Structure Reports Online*, 65(Pt 12):o3162–3163, doi:10.1107/S160053680904865X.
48. Dunaway S, Odin R, Zhou L, Ji L, Zhang Y and Kadekaro AL (2018) Natural antioxidants: Multiple mechanisms to protect skin from solar radiation, *Frontiers in Pharmacology*, 9:392, doi:10.3389/fphar.2018.00392.
49. Păduraru DN, Niculescu G, Bolocan A, Andronic O, Grumezescu AM and Bîrlă R (2022) An updated overview of cyclodextrin-based drug delivery systems for cancer therapy, *Pharmaceutics,* 14(8), https://doi.org/10.3390/pharmaceutics14081748.
50. Park S, Kim H, Jang S and Won J (2014) Effects of Cyclodextrin complexes acting as barriers on TiO2 nanoparticles in DSSCs, *Journal of Photochemistry and Photobiology A: Chemistry*, 283:17–21, doi:10.1016/j.jphotochem.2014.03.007.

51. Qi Q, Mokhtar MN and Zimmermann W (2007) Effect of ethanol on the synthesis of large-ring Cyclodextrins by Cyclodextrin glucanotransferases, *Journal of Inclusion Phenomena and Macrocyclic Chemistry*, 57(1):95–99, doi:10.1007/s10847-006-9213-z.
52. Brewster ME and Loftsson T (2007) Cyclodextrins as pharmaceutical solubilizers, *Advanced Drug Delivery Reviews*, 59(7):645–666, doi:10.1016/j.addr.2007.05.012.
53. Buschmann H-J and Schollmeyer E (2002) Applications of cyclodextrins in cosmetic products: A review, *Journal of Cosmetic Science*, 53:185–91.
54. Kockler J, Oelgemöller M, Robertson S and Glass BD (2012) Photostability of sunscreens, *Journal of Photochemistry and Photobiology C: Photochemistry Reviews*, 13:91–110, doi:10.1016/j.jphotochemrev.2011.12.001.
55. Dahabra L, Broadberry G, Le Gresley A, Najlah M and Khoder M (2021) Sunscreens containing cyclodextrin inclusion complexes for enhanced efficiency: A strategy for skin cancer prevention, *Molecules*, 26(6):1698, doi:10.3390/molecules26061698.
56. Fernandes C, Encarnação I, Fonseca A, Garrido J, Borges F and Garrido E (2015) Cyclodextrins: A valuable resource to attain environmental sustainability. In *Cyclodextrins: Synthesis, Chemical Applications and Role in Drug Delivery*, pp 85-108
57. Wang R, Wang X, Zhan Y, Xu Z, Xu Z, Feng X, Li S and Xu H (2019) A dual network hydrogel sunscreen based on Poly-γ-glutamic acid/tannic acid demonstrates excellent anti-UV, Self-recovery, and skin-integration capacities, *ACS Applied Materials & Interfac*es, 11(41):37502–37512, doi:10.1021/acsami.9b14538.
58. Knežević NŽ, Ilić N, D Okić V, Petrović R and Janaćković DOE (2018) Mesoporous Silica and Organosilica nanomaterials as UV-blocking agents, *ACS Applied Materials & Interfaces,* 10(24):20231–20236, doi:10.1021/acsami.8b04635.
59. Li D, Qian L, Feng Y, Feng J, Tang P and Yang L (2014) Co-intercalation of Acid Red 337 and a UV absorbent into layered double hydroxides: Enhancement of photostability, *ACS Applied Materials & Interfaces,* 6(23):20603–20611, doi:10.1021/am506696k.
60. Nazeeruddin MdK, Bessho T, Cevey L, Ito S, Klein C, De Angelis F, Fantacci S, Comte P, Liska P, Imai H and Graetzel M (2007) A high molar extinction coefficient charge transfer sensitizer and its application in dye-sensitized solar cell, *Journal of Photochemistry and Photobiology A: Chemistry*, 185(2):331–337, doi:10.1016/j.jphotochem.2006.06.028.
61. Osterwalder U and Herzog B (2009) Sun protection factors: World wide confusion, *The British Journal of Dermatology*, 161(Suppl 3):13–24, doi:10.1111/j.1365-2133.2009.09506.x.
62. Uekama K, Hirayama F and Irie T (1998) Cyclodextrin drug carrier systems. *Chemical Reviews,* 98(5), 2045–2076, doi:10.1021/cr970025p.
63. Abid AR, Marciniak B, Pędziński T and Shahid M (2017) Photo-stability and photo-sensitizing characterization of selected sunscreens' ingredients, J*ournal of Photochemistry and Photobiology* A: Chemistry, 332:241–250, doi:10.1016/j.jphotochem.2016.08.036.
64. Numanoğlu U, Şen T, Tarimci N, Kartal M, Koo OMY and Önyüksel H (2007) Use of cyclodextrins as a cosmetic delivery system for fragrance materials: Linalool and benzyl acetate, *AAPS PharmSciTech*, 8(4):34–42, doi:10.1208/pt0804085.
65. Freitas JV, Lopes NP and Gaspar LR (2015) Photostability evaluation of five UV-filters, trans-resveratrol and beta-carotene in sunscreens, *European Journal of Pharmaceutical Sciences: Official Journal of the European Federation for Pharmaceutical Sciences*, 78:79–89, doi:10.1016/j.ejps.2015.07.004.

66. Chatelain E and Gabard B (2001) Photostabilization of butyl methoxydibenzoylmethane (Avobenzone) and ethylhexyl methoxycinnamate by bis-ethylhexyloxyphenol methoxyphenyl triazine (Tinosorb S), a new UV broadband filter, *Photochemistry and Photobiology,* 74(3):401–406, doi:10.1562/0031-8655(2001)074<0401:pobmaa>2.0.co;2.
67. Heriot GS, Tong SYC, Cheng AC, Liew D (2020) A scenario-based survey of expert echocardiography recommendations for Patients with Staphylococcus aureus Bacteremia at varying risk for Endocarditis, *JAMA Network Open*, 3(4):e202401, doi: 10.1001/jamanetworkopen.2020.2401. PMID: 32271391; PMCID: PMC7146099.
68. Sonnendecker C, Thürmann S, Przybylski C, Zitzmann FD, Heinke N, Krauke Y, Monks K, Robitzki AA, Belder D and Zimmermann W (2019) Large-ring cyclodextrins as Chiral selectors for enantiomeric pharmaceuticals, *Angewandte Chemie (International Ed. in English)*, 58(19):6411–6414, doi:10.1002/anie.201900911.
69. Xu Y, Lu W, Xu G and Chou T-W (2021) Structural supercapacitor composites: A review, *Composites Science and Technology*, 204:108636, doi:10.1016/j.compscitech.2020.108636.
70. Stella VJ and Rajewski RA (1997) Cyclodextrins: Their future in drug formulation and delivery, *Pharmaceutical Research*, 14(5):556–567, doi:10.1023/a:1012136608249.
71. Loftsson T and Brewster ME (1996) Pharmaceutical applications of Cyclodextrins. 1. Drug solubilization and stabilization, *Journal of Pharmaceutical Sciences*, 85(10):1017–1025, doi:10.1021/js950534b.
72. Loftsson T and Brewster ME (2011) Pharmaceutical applications of cyclodextrins: Effects on drug permeation through biological membranes, *Journal of Pharmacy and Pharmacology,* 63(9):1119–1135, doi:10.1111/j.2042-7158.2011.01279.x.
73. Loftsson T and Duchêne D (2007) Cyclodextrins and their pharmaceutical applications, *International Journal of Pharmaceutics*, 329(1–2):1–11, doi:10.1016/j.ijpharm.2006.10.044.
74. Krishnamoorthy S, Allabasha N, Mani MK, and Sarkar AK (2023) Effect of β-cyclodextrin on benzophenones and its impact on UVR filtration of cotton fabric through 'guest absorber—Host enhancer' mechanism, *Journal of Inclusion Phenomena and Macrocyclic Chemistry*, 103: 131–145, https://doi.org/10.1007/s10847-023-01184-3.
75. Uekama K (2004) Design and evaluation of cyclodextrin-based drug formulation, *Chemical & Pharmaceutical Bulletin,* 52(8), 900–915, doi:10.1248/cpb.52.900.
76. van der Veen BA, van Alebeek G-JWM, Uitdehaag JCM, Dijkstra BW and Dijkhuizen L (2000) The three transglycosylation reactions catalyzed by Cyclodextrin glycosyltransferase from Bacillus circulans (strain 251) proceed via different kinetic mechanisms, *European Journal of Biochemistry*, 267(3):658–665, doi:10.1046/j.1432-1327.2000.01031.x.
77. Fenyvesi E, Jicsinszky L, Szejtli J, Schwarzenbach R (1999) Inclusion Complexes of UV Filters in Solution and in Solid State. In: Labandeira JJT, Vila-Jato JL (eds) *Proceedings of the Ninth International Symposium on Cyclodextrins.* Springer, Dordrecht, https://doi.org/10.1007/978-94-011-4681-4_151.
78. Costa R and Santos L (2017) Delivery systems for cosmetics – From manufacturing to the skin of natural antioxidants, *Powder Technology*, 322:402–416, doi:10.1016/j.powtec.2017.07.086.
79. Anselmi C, Centini M, Maggiore M, Gaggelli N, Andreassi M, Buonocore A, Beretta G and Facino RM (2008) Non-covalent inclusion of ferulic acid with alpha-cyclodextrin improves photo-stability and delivery: NMR and modeling studies, *Journal of Pharmaceutical and Biomedical Analysis*, 46(4):645–652, doi:10.1016/j.jpba.2007.11.037.

80. Cassez A, Ponchel A, Bricout H, Fourmentin S, Landy D and Monflier E (2006) Eco-efficient catalytic hydrodechlorination of carbon tetrachloride in aqueous cyclodextrin solutions, *Catalysis Letters*, 108(3–4):209–214, doi:10.1007/s10562-006-0045-7.
81. Crini G (2014) Review: A history of cyclodextrins, *Chemical Reviews,* 114(21):10940–10975, doi:10.1021/cr500081p.
82. Lin L, Piao S, Choi Y, Lyu L, Hong H, Kim D, Lee J, Zhang W and Piao Y (2022) Nanostructured transition metal nitrides as emerging electrocatalysts for water electrolysis: Status and challenges, *EnergyChem*, 4(2):100072, doi:10.1016/j.enchem.2022.100072.
83. Chodankar D, Vora A and Kanhed A (2022) β-Cyclodextrin and its derivatives: Application in wastewater treatment, *Environmental Science and Pollution Research International*, 29(2):1585–1604, doi:10.1007/s11356-021-17014-3.
84. Irie T and Uekama K (1999) Cyclodextrins in peptide and protein delivery, *Advanced Drug Delivery Reviews,* 36(1), 101–123, doi:10.1016/s0169-409x(98)00057-x.
85. Matsuda H and Arima H (1999) Cyclodextrins in transdermal and rectal delivery. *Advanced Drug Delivery Reviews*, 36(1), 81–99, doi:10.1016/s0169-409x(98)00056-8.
86. Perassinoto NL, Raponi MRB, Shitara JLY and Kumayama TM (2014) Composition comprising cyclodextrin as Uv- and Ir-Radiation screen agent, https://patentscope.wipo.int/search/en/detail.jsf?docId=WO2014165471, accessed 10 May 2024.
87. Chisvert A, Pascual-Martí MC and Salvador A (2001) Determination of the UV filters worldwide authorised in sunscreens by high-performance liquid chromatography: Use of cyclodextrins as mobile phase modifier, *Journal of Chromatography A*, 921(2):207–215, doi:10.1016/S0021-9673(01)00866-4.
88. Godwin DA, Wiley CJ and Felton LA (2006) Using cyclodextrin complexation to enhance secondary photoprotection of topically applied ibuprofen, *European Journal of Pharmaceutics and Biopharmaceutics*, 62(1):85–93, doi:10.1016/j.ejpb.2005.06.002.
89. Preeti RV, Ignacy C.and Josef H., Pereva S., Nikolova V., Angelova S., Spassov T and Dudev T (2019) Stability constants of the inclusion complexes of β-Cyclodextrin with various adamantane derivatives. A UV-Vis study, *Beilstein Journal of Organic Chemistry*, 15:1592–1600, doi:10.3762/bjoc.15.163.
90. Ferreira L, Filipa M, Rabaça S, Mathur A, Sharma A, Giram PS, Pawar KD, Rahdar A, Raza F, Veiga F, Mazzola PG, Paiva-Santos, AC (2003) Cyclodextrin-based dermatological formulations: Dermopharmaceutical and cosmetic applications, *Colloids and Surfaces B: Biointerfaces*, 221, 113012, ISSN 0927-7765, https://doi.org/10.1016/j.colsurfb.2022.113012.
91. Garland CF, Garland FC and Gorham ED (2003) Epidemiologic evidence for different roles of ultraviolet A and B radiation in melanoma mortality rates, *Annals of Epidemiology*, 13(6):395–404, doi:10.1016/s1047-2797(02)00461-1.
92. Pfeifer GP and Besaratinia A (2012) UV wavelength-dependent DNA damage and human non-melanoma and melanoma skin cancer, *Photochemical & Photobiological Sciences: Official Journal of the European Photochemistry Association and the European Society for Photobiology*, 11(1):90–97, doi:10.1039/c1pp05144j.
93. Lhiaubet-Vallet V, Marin M, Jiménez O, Gorchs O, Trullas C and Miranda M (2010) Filter-filter interactions. Photostabilization, triplet quenching and reactivity with singlet oxygen, *Photochemical & Photobiological Sciences: Official Journal of the European Photochemistry Association and the European Society for Photobiology*, 9:552–558, doi:10.1039/b9pp00158a.
94. Sharma N and Baldi A (2016) Exploring versatile applications of Cyclodextrins: An overview, *Drug Delivery*, 23(3):739–757, doi:10.3109/10717544.2014.938839.

95. Kryczyk-Poprawa A, Kwiecień A and Opoka W (2019) Photostability of topical agents applied to the skin: A review, *Pharmaceutics*, 12(1):10, doi:10.3390/pharmaceutics12010010.
96. Cawthray JF (2010) *UVA chemical filters: A systematic study.* [Thesis], https://digital.library.adelaide.edu.au/dspace/handle/2440/61508, accessed 11 July 2024.
97. Murphy RB, Staton J, Rawal A and Darwish TA (2020) The effect of deuteration on the keto–enol equilibrium and photostability of the sunscreen agent avobenzone, *Photochemical & Photobiological Sciences*, 19(10):1410–1422, doi:10.1039/d0pp00265h.
98. Nasr M, Mansour S, Mortada ND and El Shamy AA (2008) Lipospheres as carriers for topical delivery of Aceclofenac: Preparation, characterization and in vivo evaluation, *AAPS PharmSciTech*, 9(1):154–162, doi:10.1208/s12249-007-9028-2.
99. Rasheed A (2008) Cyclodextrins as drug carrier molecule: A review, *Scientia Pharmaceutica*, 76(4):567–598, doi:10.3797/scipharm.0808-05.
100. dos Santos C, Buera P and Mazzobre F (2017) Novel trends in Cyclodextrins encapsulation. Applications in food science, *Current Opinion in Food Science*, 16:106–113, doi:10.1016/j.cofs.2017.09.002.
101. Li NB, Luo HQ and Liu SP (2005) Resonance Rayleigh scattering study of the inclusion complexation of chloramphenicol with beta-Cyclodextrin, *Talanta*, 66(2):495–500, doi:10.1016/j.talanta.2004.11.022.
102. Rácz CP, Borodi G, Pop MM, Kacso I, Sánta S and Tomoaia-Cotisel M (2012) Structure of the inclusion complex of β-Cyclodextrin with lipoic acid from laboratory powder diffraction data, *Acta Crystallographica. Section B, Structural Science*, 68(Pt 2):164–170, doi:10.1107/S0108768112004284.
103. Jani P, D'Souza S, and Rajora MA (2017) Cyclodextrins: Pharmaceutical applications, *Drug Development and Industrial Pharmacy*, 43(6), 872–886.
104. Pawar S, Shende P and Trotta F (2019) Diversity of β-Cyclodextrin-based nanosponges for transformation of actives, *International Journal of Pharmaceutics*, 565:333–350, doi:10.1016/j.ijpharm.2019.05.015.
105. Adeoye O, and Cabral-Marques H (2017) Cyclodextrin nanosystems in oral drug delivery: A mini review, *International Journal of Pharmaceutics*, 531(2): 521–531, https://doi.org/10.1016/j.ijpharm.2017.04.050.
106. Gonzalez Pereira A, Carpena M, García Oliveira P, Mejuto JC, Prieto MA and Simal Gandara J (2021) Main applications of cyclodextrins in the food industry as the compounds of choice to form Host–Guest complexes, *International Journal of Molecular Sciences*, 22(3):1339, doi:10.3390/ijms22031339.
107. Loftsson T, Jarho P, Másson M and Järvinen T (2005) Cyclodextrins in drug delivery. *Expert Opinion on Drug Delivery,* 2(2), 335–351, doi:10.1517/17425247.2.2.335.
108. Hirayama F, and Uekama K (1999) Cyclodextrin-based controlled drug release system, *Chemical Reviews*, 36(1):125–141, doi:10.1016/s0169-409x(98)00058-1.
109. Shimpi S, Chauhan B and Shimpi P (2005) Cyclodextrins: Application in different routes of drug administration, *Acta Pharmaceutica (Zagreb, Croatia)*, 55(2):139–156.
110. Buchwald P and Bodor N (2016) Brain-targeting chemical delivery systems and their cyclodextrin-based formulations in light of the contributions of Marcus E. Brewster. *Journal of Pharmaceutical Sciences,* 105(9):2589-260, https://doi.org/10.1016/j.xphs.2016.04.007.
111. Conceição J, Adeoye O, Cabral-Marques HM and Lobo JMS (2018) Cyclodextrins as excipients in tablet formulations, *Drug Discovery Today,* 23(6);1274-1284, https://doi.org/10.1016/j.drudis.2018.04.009.

112. Conceicao J, Adeoye O, Cabral-Marques HM and Lobo JMS (2018) Cyclodextrins as drug carriers in pharmaceutical technology: The state of the art. *Current Pharmaceutical Design*, 24(13):1405–1433, doi: 10.2174/1381612824666171218125431.
113. Kali G, Haddadzadegan S, and Bernkop-Schnürch A (2024) Cyclodextrins and derivatives in drug delivery: New developments, relevant clinical trials, and advanced products. *Carbohydrate Polymers,* 324:121500, https://doi.org/10.1016/j.carbpol.2023.121500.
114. Singh M, Sharma R and Banerjee UC (2002) Biotechnological applications of Cyclodextrins, *Biotechnology Advances*, 20(5–6):341–359, doi:10.1016/s0734-9750(02)00020-4.
115. Stella VJ and He Q (2008) Cyclodextrins, *Toxicologic Pathology*, 36(1):30–42, doi:10.1177/0192623307310945.
116. Yoshii H, Furuta T, Yasunishi A and Hirano H (1994) Minimum number of water molecules required for inclusion of d-limonene in the Cyclodextrin cavity, *Journal of Biochemistry*, 115(6):1035–1037, doi:10.1093/oxfordjournals.jbchem.a124452.
117. Zhang W, Kong Y, Jin X, Yan B, Diao G and Piao Y (2020) Supramolecule-assisted synthesis of Cyclodextrin polymer functionalized polyaniline/carbon nanotube with core-shell nanostructure as high-performance supercapacitor material, *Electrochimica Acta*, 331:135345, doi:10.1016/j.electacta.2019.135345.

8 Applications of UVR Absorbers

Cyclodextrin Complexes

Sagarika Deepthy Tallapragada, Shouvik Kumar Nandy, Sanjukta Das, and Saikat Bhattacharya

8.1 INTRODUCTION

Cyclodextrins (CDs) stand as synthetic marvels derived from the enzymatic degradation of starch, one of nature's most fundamental polysaccharides. One characteristic that sets CD molecules apart is their core hydrophobic cavity, which serves as a molecular cage capable of ensnaring and encapsulating a wide range of compounds. Even though CDs date back to the late 19th century, their discovery in the 1980s was a game-changer because of the exponential growth in their use across numerous industries [1]. The discovery of cyclodextrins began with Antoine Villiers in 1891, who identified the enzymatic breakdown of potato starch, revealing two acid-resistant, non-reducing molecules likely to be α-CD and β-CD, initially named "cellulosines." Franz Schardinger later renamed them "dextrins," forming the basis for the term "cyclodextrins." Larger-ring cyclodextrins, such as γ-cyclodextrins, were later identified. In sunscreens, cyclodextrins enable controlled release of active ingredients by encapsulating hydrophobic UV filters within stable inclusion complexes, ensuring prolonged and consistent UV protection [2, 3].

Cyclodextrins (CDs) have been recognized since the late 1970s for forming inclusion complexes with active molecules, enhancing their stability, solubility, skin permeation, and masking odours. They provide a biocompatible delivery system for targeting specific skin sites, showing potential in treating skin diseases and improving cosmetic formulations. József Szejtli (2004) studied the complexation of insoluble UV filters with CDs, identifying randomly methylated β-cyclodextrin (RAMEB) as the best solubilizer, improving both solubility and photostability of UV filters and model drugs, thereby advancing dermopharmaceutical and cosmetic delivery systems [2–4].

Cyclodextrins (CDs) enhance the solubility of poorly soluble compounds by encapsulating hydrophobic molecules, such as UV filters or antioxidants, within their hydrophobic cavities in aqueous solutions. This process improves water solubility

 DOI: 10.1201/9781003492757-8

and facilitates product formulation. When applied to the skin, the cyclodextrin-UV filter complex gradually releases the UV filter, with the release rate influenced by temperature, humidity, and interaction with the skin's natural oils [4].

8.1.1 Brief Introduction to Cyclodextrins and Their Properties

In 1939, Karl Freudenberg's work set the stage for Friedrich Cramer's 1954 discoveries on cyclodextrins and their interactions with guest molecules. The introduction of α-, β-, and γ-CDs made them affordable and widely used in industries like food, pharmaceuticals, and textiles [5].

Cyclodextrins' versatility lies in their selective cavities, accommodating molecules based on size—α-CD for smaller molecules and γ-CD for larger ones [6]. Their solubility varies with ring strain and hydrogen bonding, with γ-CD being the most soluble [2, 7].

Cyclodextrins resist acid hydrolysis but degrade under strong acids or oxidative reactions. They are also enzymatically biodegradable and more resistant to α-amylase than starch [3, 8].

8.1.2 Overview of Ultraviolet Radiation and Its Harmful Effects on Skin and Materials

Solar energy impacts the environment and species, with UV radiation, especially UV-B and UV-C, posing mutagenic risks. The atmosphere filters most UVC and some UVB, while UVA penetrates deeper, causing cellular damage [6].

UVB induces sunburn and inflammation via DNA damage, but treatments like NSAIDs and corticosteroids show limited efficacy, favouring conservative options like ice compresses [6].

Natural UV absorbers like ferulic acid face formulation challenges due to poor solubility. Cyclodextrins, such as HP-γ-CD, enhance solubility and UVA protection, offering potential for safer sunscreens [9–11].

8.2 FORMATION AND CHARACTERIZATION OF CYCLODEXTRIN COMPLEXES

8.2.1 Preparation of Complex

Advances in cyclodextrin (CD) research have clarified their enzymatic synthesis, separation, and ability to form inclusion complexes due to their toroidal structure. The term "inclusion complex" was first used in 1956, highlighting CDs' reversible complexation and non-covalent interactions within their hydrophobic cavity [10, 11].

Studies on inclusion complex formation emphasize hydrophobic and van der Waals interactions, with guest molecules aligning to maximize contact with CDs' apolar cavity. Stability and formation constants have been extensively analyzed, as water replacement by hydrophobic substrates drives complexation [12, 13].

Solid-state CD complexes require tailored preparation methods, including microwave irradiation, co-grinding, and techniques like spray-drying or colyophilization. Characterization techniques such as XRPD, DSC, FTIR, and SEM are critical for optimizing host–guest interactions and ensuring desired outcomes [8].

8.3 ENHANCED STABILITY AND SOLUBILITY OF CYCLODEXTRIN COMPLEXES

8.3.1 Cyclodextrin Complexes Improve the Stability and Solubility of UVR Absorbers

Cyclodextrin complexes enhance the stability and solubility of UV radiation (UVR) absorbers by encapsulating them within their cavities, improving their efficiency in absorbing UV rays. Encapsulation also enables controlled and prolonged release of UV absorbers, making them more effective in applications like sunscreens and cosmetics.

This synergistic interaction between cyclodextrins and UV absorbers paves the way for advanced formulations with improved photoprotective properties. Research focuses on optimizing cyclodextrin-based complexes, refining formulation techniques, and understanding complexation mechanisms to address challenges in UV defense and photostability [6].

8.3.2 Comparison of the Properties of Free UVR Absorbers and Cyclodextrins Complexes

Cyclodextrins enhance the solubility, stability, and photostability of UVR absorbers, which are otherwise limited in aqueous formulations. Free UVR absorbers often have low solubility and are susceptible to UV-induced degradation, reducing their effectiveness. When complexed with cyclodextrins, UVR absorbers become more soluble, stable, and resistant to photodegradation, providing more consistent and prolonged UV protection. This improvement makes cyclodextrin-complexed UVR absorbers more effective in sunscreens and cosmetics, offering enhanced performance compared to free UVR absorbers.

8.3.3 Applications in Sunscreen Formulations

Solar UVR can damage skin DNA, leading to skin cancer. Sunscreens with UV filters protect the skin by absorbing, scattering, or reflecting UV radiation. These formulations often include antioxidants that help prevent DNA damage and free radical production. However, both UV filters and antioxidants are prone to degradation when exposed to oxygen and sunlight [14, 15]. Natural antioxidants, in particular, are difficult to produce due to low solubility and limited skin penetration. Cyclodextrins (CDs) can enhance the solubility, stability, and bioavailability of these compounds, protect them from oxidative stress, and modify their release. CD inclusion complexes, such as those with UV filter benzophenone-3, can improve sunscreen performance and increase SPF value [16, 17].

8.4 UTILIZATION OF CYCLODEXTRIN COMPLEXES IN SUNSCREEN FORMULATIONS AS UVR ABSORBERS

UV filters are categorized into physical filters (inorganic, reflecting/scattering light) and chemical filters (absorbing UVB radiation). Their efficacy is measured by sun protection factor (SPF). Chemical filters are typically used at 5–15% concentrations, while physical filters can reach up to 25%. Benzophenone-3 is a cost-effective UV filter, absorbing both UVA and UVB light. It can penetrate the skin up to 2%, with a 6% concentration limit. A study showed that Benzophenone-3, complexed with HP-β-CD, improved solubility and photostability while reducing skin permeability [16]. CD-UV filter complexes have been shown to slightly improve SPF values [18]. A study by Srinivasan and Bauer (2013) found that incorporating β-CD complexes into sunscreens led to a 19.6% increase in SPF. Additionally, an HP-β-CD-oxybenzone complex provided UV protection equivalent to an SPF 30 sunscreen [19, 20].

8.4.1 Enhancement of Sunscreen Efficacy and Stability Through Cyclodextrin Complexation

Cyclodextrins (CDs) enhance the solubility, stability, and antioxidant activity of compounds through complex formation. Studies show that UV irradiation reduces the scavenging capacity of pure chrysin, whereas the chrysin-β-CD complex maintains stability. Similarly, the rutin-β-CD complex shows improved stability and scavenging capacity. In anticancer research, saikosaponin-d (SSD) showed higher cytotoxicity and bioavailability when complexed with HP-β-CD, achieving maximum efficacy at a 1:5 molar ratio [16]. Sunscreen chemicals like oxybenzone, octocrylene, and OMC exhibit enhanced photostability when complexed with β-CD, reducing degradation upon UV exposure [21]. Studies have demonstrated that β-CD improves stability and mitigates undesirable interactions within sunscreen formulations, extending UV filter absorption times and reducing systemic absorption, which may decrease toxicity and allergic reactions. This strategy enhances sunscreen efficacy and reduces skin cancer risk [18].

8.4.2 Photoprotective Properties

The use of excipients like ascorbic acid, α-tocopherol, and supramolecular matrices has gained attention for enhancing drug photochemical stability and solubility. These matrices, formed via host–guest inclusion chemistry, improve drug photoprotection. A study on 1,4-dihydropyridine derivatives revealed that lacidipine, manidipine, nifedipine, and nimodipine were highly sensitive to light, degrading within 5 minutes in an aqueous medium. In contrast, amlodipine, felodipine, nisoldipine, and nitrendipine showed improved photostabilization when complexed with cyclodextrins

(CDs), achieving a drug recovery rate of 90% after 30 minutes of light exposure [22]. Additionally, tretinoin demonstrated light sensitivity, but incorporation of α-CD and hydroxypropyl-β-CD improved its stability, although β-CD had minimal photoprotective effects.

8.5 BIOCOMPATIBILITY AND SAFETY PROFILE OF CYCLODEXTRIN COMPLEXES

Cyclodextrins (CDs) are macrocyclic molecules with a hydrophobic cavity, enabling the formation of inclusion complexes with various molecules. They offer advantages in biocompatibility, safety, and stability, and are used in pharmaceuticals to improve drug bioavailability. However, factors like the type of CD, complex formation method, molar ratio, and the drug involved affect their safety and efficacy. Despite their benefits, CDs have limitations such as low solubility, slow degradation, and weak drug encapsulation. Modifications like covalent incorporation into polymer structures can enhance properties such as solubility and drug release. However, their biological applications remain limited due to hydrophobicity and lack of active sites [22–24].

8.6 ASSESSMENT OF THE BIOCOMPATIBILITY AND SAFETY PROFILE OF CYCLODEXTRIN COMPLEXES FOR TOPICAL APPLICATION

Cyclodextrins (CDs) are cyclic oligosaccharides with a hydrophilic outer surface and a lipophilic central cavity, allowing them to form water-soluble inclusion complexes with lipophilic, water-insoluble drugs. These complexes improve the solubility and stability of hydrophobic drugs in aqueous solutions. When applied topically, CDs facilitate drug release from the cavity, allowing the drug to penetrate the skin's lipophilic barrier, thus enhancing drug delivery. The process is both diffusion-controlled and membrane-controlled. CDs rely on aqueous solutions for effective drug delivery, and while α-CD is highly soluble (20 g/100 mL), β-CD has lower solubility (1.8 g/100 mL) due to strong intermolecular hydrogen bonds, limiting its performance in drug delivery [22, 23].

Random substitution of the hydrogen bond forming hydroxy groups can dramatically increase aqueous solubility. Approximately 100 different cyclodextrin derivatives are commercially available as fine chemicals, but only a few have gone through toxicological evaluations and are available as bulk chemicals for pharmaceutical applications. Current derivatives used in drug formulations include HP-CD, randomly methylated CD (RM-CD), sulfobutylether CD (SBE-CD), maltosyl CD (ML-CD), and (2-hydroxypropyl)-CD (HP-CD). The aqueous solubility of these derivatives is typically 50–70% (w/v). In general, cyclodextrins can only enhance topical drug delivery in the presence of water. They solubilize lipophilic water-insoluble drugs in aqueous vehicle systems and deliver the drug molecules to the barrier surface. A novel compound, retinaldehyde γ-cyclodextrin complex (RCC), has been developed for topical skin therapy, showing unique skin care attributes and less topical irritation.

Cyclodextrin-conjugated nanoparticles combine the advantages of cyclodextrins and nanoparticles, offering enhanced water solubility, targeted drug delivery, greater surface area, improved drug loading, and higher stability [24].

8.6.1 Advantages for the Use of Cyclodextrins in Cosmetics

Cyclodextrins offer a range of benefits in cosmetic formulations, making them valuable additives for enhancing product stability and performance. First, cyclodextrins provide effective protection for guest molecules, preventing decomposition from light or heat, oxidation, and hydrolysis, as well as reducing unwanted reactions with other organic compounds and minimizing loss through evaporation. They also facilitate solubilization of active ingredients in water, increasing both solubility and the rate of solubilization while avoiding the need for organic solvents and enabling viscosity adjustments. Additionally, cyclodextrins can eliminate undesired odours, tastes, and hygroscopicity, thus improving the sensory qualities of products. By converting liquid or oily substances into powder form, cyclodextrins also improve the ease of handling and enhance the stability of emulsions, ultimately contributing to a more refined and stable end product.

Cyclodextrins are considered safe for human use, as they generally pose minimal risk at typical concentrations. Toxicity studies indicate that only extremely high concentrations may be harmful, with no signs of acute toxicity observed in animal experiments. Supporting its safety profile, β-cyclodextrin received approval in Germany as a food additive in November 2000, further establishing its acceptance in products intended for human consumption. This licensing underscores the compound's low toxicity and suitability in food and cosmetic applications, where it provides functional benefits without compromising safety.

The safety and effectiveness of cyclodextrin complexes depend on several factors, such as the specific cyclodextrin used, the method used to produce the complex, the resulting molar ratio, and the specific medicine being complexed. Considering this, it is crucial to evaluate the safety and effectiveness of each cyclodextrin-drug complex individually [24].

8.7 OTHER APPLICATIONS OF CYCLODEXTRIN COMPLEXES IN DIFFERENT INDUSTRIES

8.7.1 Exploration of Additional Applications of Cyclodextrin Complexes in Industries Beyond Cosmetics, Such as Textiles, Plastics, and Pharmaceuticals

Complexes of cyclodextrin have several distinguishing characteristics that make them extremely versatile and suitable for a wide variety of applications across a number of different sectors. Cyclodextrin (CD) complexes, with their molecular inclusion and encapsulation properties, have been widely used in the cosmetics industry. Cyclodextrins are non-toxic cyclic polysaccharides that form inclusion complexes with various organic molecules, enhancing their stability against light and oxygen, reducing vapour pressure, and improving solubility. These properties make cyclodextrins valuable in cosmetic formulations, enabling effects not achievable with conventional methods.

They can be used in textiles for odour control, colour fastness, moisture management, and polymer blends. In plastics, CD complexes can improve the compatibility between polymers, control the release of additives, and enhance flame retardancy [1, 3]. They can also be used in pharmaceuticals for drug delivery, improving solubility, stability, and bioavailability. They can also mask the bitter taste of drugs, making them more palatable and improving patient compliance. CD complexes can also be used in food and beverage to improve the stability, solubility, and bioavailability of food additives and ingredients [2, 25, 26]. They can also be used in agriculture to enhance the delivery and efficacy of pesticides and fertilizers. Additionally, CD complexes can be used in environmental remediation to remove pollutants from the environment [27].

The increasing awareness of the environmental and health risks associated with synthetic UV filters has prompted research into natural alternatives. The study focused on extracting and microencapsulating bioactive compounds from onion peel (OP) using β-cyclodextrin. High-performance liquid chromatography (HPLC-DAD) identified quercetin and resveratrol as key components. The resulting microparticles demonstrated strong photoprotection, comparable SPF to synthetic filters, high antioxidant activity, and 91.8% encapsulation efficiency. Stability tests of oil-in-water emulsions confirmed their sustained SPF efficacy, suggesting these OP microparticles could serve as eco-friendly, natural UV filters in sunscreen formulations [28].

8.7.2 Cyclodextrin in Enhanced UV Absorbers

In the context of UV filters, cyclodextrins can be used to improve the performance of UV-filtering agents in several ways. Many UV filters can degrade when exposed to sunlight, losing their effectiveness. Cyclodextrins can encapsulate these UV filters, protecting them from photodegradation and extending their effective lifespan.

Researchers explore how cyclodextrins (CDs) can enhance sunscreen formulations. Unprotected skin exposure to UV radiation can damage DNA and lead to skin cancer. Sunscreens protect by using UV filters and natural antioxidants, but these components are prone to degradation and have poor water solubility. CDs can form inclusion complexes with these ingredients, improving their solubility, stability, and controlled release. Incorporating CDs in sunscreens can potentially increase their effectiveness and improve skin cancer prevention [29].

UV filters are crucial in sunscreens as they protect the skin from harmful UV radiation. However, many effective UV filters, such as avobenzone and oxybenzone, have poor water solubility. This characteristic limits their effectiveness and complicates their formulation into aqueous-based sunscreens, which are popular for their lighter, non-greasy feel and ease of application.

Natural antioxidants, such as vitamin E, ferulic acid, and various botanical flavonoids, are often added to sunscreens for their additional skin benefits. They help to neutralize free radicals generated by UV exposure, thereby providing a secondary layer of protection. However, like many UV filters, these antioxidants also tend to be poorly soluble in water. This solubility issue limits their bioavailability and effectiveness when included in aqueous-based formulations.

Traditional sunscreens can lose effectiveness as the active ingredients degrade or are removed through perspiration, water exposure, or rubbing. The controlled release mechanism provided by cyclodextrins ensures that a steady amount of UV filter is available on the skin surface over an extended period, maintaining effective UV protection for longer durations. This reduces the need for frequent reapplication, which is beneficial for users who are active outdoors or spend extended periods in the sun.

By maintaining a consistent level of UV filters on the skin, cyclodextrin-based sunscreens ensure continuous protection against harmful UV radiation. This is particularly important for preventing cumulative skin damage caused by prolonged exposure to the sun.

The gradual release of UV filters can also minimize the potential for skin irritation. High concentrations of UV filters applied all at once can sometimes cause irritation or sensitivity, especially in people with sensitive skin. Controlled release helps to avoid such issues by delivering the active ingredients in a more measured manner.

UV filters encapsulated within cyclodextrins are protected from premature degradation caused by environmental factors like light, heat, and oxygen. This encapsulation not only prolongs the effectiveness of the sunscreen during use but also extends the shelf life of the product. The controlled release mechanism ensures that the UV filters remain effective even in varying environmental conditions. Whether in water, under sweat, or during physical activity, the gradual release maintains a steady protective barrier on the skin. Cyclodextrins can improve the photostability of both synthetic and natural UV filters. This is particularly beneficial for natural compounds like botanical flavonoids and ferulic acid, which are otherwise challenging to use due to their instability in light.

Ferulic acid, a natural antioxidant, is known for its poor water solubility. When complexed with hydroxypropyl-γ-cyclodextrin (HP-γ-CD), ferulic acid's solubility and stability in water are significantly improved. This complex can then be easily incorporated into aqueous-based sunscreen formulations, enhancing the product's overall effectiveness in protecting the skin from UV damage. Avobenzone is a commonly used UVA filter with poor water solubility. By forming an inclusion complex with β-cyclodextrin, avobenzone's solubility is increased, making it easier to formulate into water-based sunscreens. The cyclodextrin complex also helps to stabilize avobenzone, preventing its rapid degradation when exposed to sunlight.

Mori et al. (2019) investigated natural alternatives to synthetic UV absorbers like octylmethoxy cinnamate (OMC) and avobenzone due to their high skin irritation. Natural compounds such as botanical flavonoids and ferulic acid have poor solubility, making formulation challenging. The study found that cyclodextrins can form inclusion complexes with these natural UV absorbers. Specifically, hydroxypropylated-γ-cyclodextrin (HP-γ-CD) formed a 1:2 complex with ferulic acid, enhancing its solubility and UV protection. Cosmetic formulations containing the ferulic acid/HP-γ-CD complex showed superior sun protection, especially against UVA, compared to those with OMC, indicating their potential for safer and more effective sunscreens [16].

The study delves into the intricate physics of cyclodextrins, examining their dynamic interactions and structural details, aiming to uncover the underlying principles that govern their remarkable adaptability [3, 27].

8.7.3 Challenges and Potential Solutions in Optimizing the Formulation and Application of These Complexes

Cyclodextrin complexes have significant potential in various applications, but they face challenges in formulation and application. These include solubility, stability, toxicity, and regulatory issues. Solutions include inclusion complexes with essential oils, nanoparticles, derivatives, and regulatory compliance. Inclusion complexes with essential oils can overcome solubility and stability issues [30, 31]. Nanoparticles can enhance water solubility and drug loading, provide targeted drug delivery with minimal toxicity, and offer greater surface area and stability. Chemically modified cyclodextrins can improve solubility and stability. Regulatory compliance is crucial when using cyclodextrins in pharmaceutical formulations. The safety and effectiveness of cyclodextrin complexes depend on factors like the type of cyclodextrin used, the method of complex formation, the resulting molar ratio, and the specific compound being complexed. Each cyclodextrin complex should be evaluated individually for its safety and effectiveness stability [27, 32–34].

8.8 CONCLUSION

The cyclodextrin complexes play an important role in UVR protection because they are able to dissolve ultraviolet absorbers, which is advantageous for their usage in cosmetics, and they interact with organic UV absorbers through the formation of inclusion complexes. Such complexes may strengthen the solubility and stability of UV absorbers and antioxidants, thus enhancing sunscreen efficiency and preventing skin cancer. UVR was proven to cause skin cell DNA damage when exposed with no protection and leads to the development of skin cancer. Suntan lotions are topical preparations that possess natural antioxidants as an added layer of protection against UVR. These antioxidants repair DNA damage and prevent the formation of free radicals. These filters may be chemical or physical and must be part of sunscreen formulation to protect the skin from damage caused by UV radiation. Generally, antioxidants are added to the sunscreen formulation to neutralize adverse effects from UVR that can pass through UV filters inadequately. However, antioxidants like UV filters are susceptible to UV radiation or other environmental factors causing degradation. Organic and inorganic UV filters are a number of components of sunscreens. Organic filters absorb the UV light, thus converting it into a safer form of radiation. Oxides of titanium dioxide and zinc oxide are examples of inorganic filters, which work by reflecting and scattering the ultraviolet radiation away from the skin. Some chemical UV filters include oxybenzone, avobenzone, octisalate, octocrylene, homosalate, and octinoxate. This means it is either an organic or inorganic compound that absorbs or reflects UV radiation to prevent the harmful effects such as sunburn, early ageing, and even skin cancers it can bring. However, for them to be effective, they have to cross the barrier of the skin to reach the deeper layers of the dermis. Among others affecting the percutaneous absorption of UV filters are the chemical properties of the filter, condition of the skin, and substances present on the skin. Cyclodextrin complexes, like UVA filters, can enhance sunscreen effectiveness and longevity by making UV absorbers more soluble and ensuring slow release. Such complexes of cyclodextrin might, therefore improve performance and stability of the sunscreen through the formation of inclusion complexes that increase the solubility of UV absorbers and ensure controlled release. Some of the potential applications of cyclodextrin complexes in sunscreens, textiles, and personal care products are thus promising and expectant [32, 34].

REFERENCES

1. Crini, G., (2014). 'A history of cyclodextrins' *Chem. Rev.*, Vol 114(21), pp.10940–10975.
2. Poulson, B.G., Alsulami, Q.A., Sharfalddin, A., El Agammy, E.F., Mouffouk, F., Emwas, A.H., Jaremko, L. and Jaremko, M., (2021). 'Cyclodextrins: Structural, chemical, and physical properties, and applications' *Polysaccharides*, Vol 3(1), pp.1–31.
3. Loftsson, T., Sigurdsson, H.H. and Jansook, P., (2023) 'Anomalous properties of cyclodextrins and their complexes in aqueous solutions' *Materials*, Vol 16(6), p.2223.
4. Szejtli, J., (2004) 'Past, present and future of cyclodextrin research' *Pure Appl. Chem.*, Vol 76(10), pp. 1825–1845.
5. Loftsson, T. and Duchene, D., (2007) 'Cyclodextrins and their pharmaceutical applications' *Int. J. Pharm.*, Vol 329(1–2), pp.1–11.
6. Rincón-López, J., Almanza-Arjona, Y.C., Riascos, A.P. and Rojas-Aguirre, Y., (2021) 'Technological evolution of cyclodextrins in the pharmaceutical field' *J. Drug Deliv. Sci. Technol.*, Vol 61, p.102156.
7. Matencio, A., Navarro-Orcajada, S., García-Carmona, F. and López-Nicolás, J.M., (2020) 'Applications of cyclodextrins in food science. A review' *Trends Food. Sci. Technol.*, Vol 104, pp.132–143.
8. Pélingre, M., Koffi Teki, D.S.E., El-Abid, J., Chagnault, V., Kovensky, J. and Bonnet, V., (2021) 'Ring-opening of Cyclodextrins: An efficient route to pure Maltohexa-, Hepta-, and Octaoses' *Organics*, Vol 2(3), pp.287–305.
9. Dhiman, P. and Bhatia, M., (2020) 'Pharmaceutical applications of cyclodextrins and their derivatives' *J. Incl. Phenom. Macrocycl. Chem.*, Vol 98(3), pp. 171–86.
10. Lu, Q., (2023) 'Bioresponsive and multifunctional cyclodextrin-based non-viral nanocomplexes in cancer therapy: Building foundations for gene and drug delivery, immunotherapy and bioimaging' *Environ. Res.*, Vol 234, p.116507.
11. Jug, M. and Mura, P.A., (2018) 'Grinding as solvent-free green chemistry approach for cyclodextrin inclusion complex preparation in the solid state' *Pharmaceutics*, Vol 10(4), p.189.
12. Rekharsky, M. and Inoue, Y., (2000). '1: 1 and 1: 2 Complexation thermodynamics of γ-cyclodextrin with N-carbobenzyloxy aromatic amino acids and ω-phenylalkanoic acids' *J. Am. Chem. Soc.*, Vol 122(44), pp.10949–55.
13. Liu, M., Guo, Q., Shi, Y., Cai, C., Pei, W., et al., (2019) 'Studies on pH and temperature dependence of inclusion complexes of bisdemethoxycurcumin with β-cyclodextrin derivatives' *J. Mol. Struct.*, Vol 5(1179), pp.336–46.

14. Scalia, S., Coppi, G. and Iannuccelli V., (2011) 'Microencapsulation of a cyclodextrin complex of the UV filter, butyl methoxydibenzoylmethane: In vivo skin penetration studies' *J. Pharm. Biomed. Anal.*, Vol 54(2) pp.345–50.
15. Simeoni, S., Scalia, S., Tursilli, R. and Benson, H., (2006) 'Influence of cyclodextrin complexation on the in vitro human skin penetration and retention of the sunscreen agent, oxybenzone' *J. Incl. Phenom. Macrocycl. Chem.*, Vol 54, pp. 275–82.
16. Mori, T., Tsuchiya, R., Doi, M., Nagatani, N. and Tanaka, T., (2019) 'Solubilization of ultraviolet absorbers by cyclodextrin and their potential application in cosmetics' *J. Incl. Phenom.*, Vol 93(1), pp.91–96.
17. Mura, P., (2015) 'Analytical techniques for characterization of cyclodextrin complexes in the solid state: A review' *J. Pharm. Biomed. Anal.*, Vol 113, pp.226–238.
18. Mangolim, C.S., Moriwaki, C., Nogueira, A.C., Sato, F., Baesso, M.L., Neto, A.M. and Matioli, G., (2014) 'Curcumin–β-cyclodextrin inclusion complex: Stability, solubility, characterisation by FT-IR, FT-Raman, X-ray diffraction and photoacoustic spectroscopy, and food application' *Food Chem.*, Vol 153, pp.361–370.
19. Srinivasan, R. and Bauer, K.M., (2013) 'Safety performance function development guide: Developing jurisdiction-specific SPFs'. United States. Federal Highway Administration. Office of Safety.
20. Felton, L.A., Wiley, C.J. and Godwin, D.A., (2004) 'Influence of cyclodextrin complexation on the in vivo photoprotective effects of oxybenzone' *Drug Dev. Ind. Pharm.*, Vol 30(1), pp.95–102.
21. Tian, L., Wu, Y., Hou, Y., Dong, Y., Ni, K. and Guo, M., (2024) 'Environmentally friendly UV absorbers: Synthetic characterization and biosecurity studies of the Host–Guest supramolecular complex' *Int. J. Mol. Sci.*, Vol 25(15), pp. 1–14.
22. Rajamohan, R., Kamaraj, E., Govindasamy, C., Subramanian, S.A., Kim, S.J., Krishnan, M.M., Murugavel, K., Murugan, M. and Lee, Y.R., (2024) 'Formulation of ensulizole with beta-cyclodextrins for improved sunscreen activity and biocompatibility' *J. Pharm. Sci.*, Vol 113(6), pp.1536–45.
23. Hădărugă, N.G., Bandur, G.N., David, I. and Hădărugă, D.I., (2019) 'A review on thermal analyses of cyclodextrins and cyclodextrin complexes' *Environ. Chem. Lett.*, Vol 17, pp.349–373.
24. Rajamohan, R., Muthuraja, P., Govindasamy, C., Subramanian, S.A., Kim, S.J., Krishnan, M.M., Murugavel, K. and Lee, Y.R., (2023) 'Enhanced photostability and biocompatibility of sunscreen formulation of 2-phenylbenzimidazole-5-sulfonic acid with methyl-beta-cyclodextrin' *J. Mol. Liq.*, Vol 390, p.123013.
25. Al-Rawashdeh, N.A., Al-Sadeh, K.S. and Al-Bitar, M.B., (2013) 'Inclusion complexes of sunscreen agents with β-cyclodextrin: Spectroscopic and molecular modeling studies' *J Spectrosc.* Vol 2013, pp.1-11
26. Jicsinszky, L. and Cravotto, G., (2021) 'Toward a greener world—cyclodextrin derivatization by mechanochemistry' *Molecules*, Vol 26(17), p.5193.
27. Ioele, G., De Luca, M., Garofalo, A. and Ragno, G., (2017) 'Photosensitive drugs: A review on their photoprotection by liposomes and cyclodextrins' *Drug Deliv.*, Vol 24(2), pp.33–44.
28. Wadhwa, G., Kumar, S., Chhabra, L., Mahant, S. and Rao, R., (2017) 'Essential oil–cyclodextrin complexes: An updated review' *Incl. Phenom. Macrocycl. Chem.*, Vol 89, pp.39–58.
29. Dahabra, L., Broadberry, G., Le Gresley, A., Najlah, M. and Khoder, M., (2021) 'Sunscreens containing cyclodextrin inclusion complexes for enhanced efficiency: A strategy for skin cancer prevention' *Molecules*, Vol 26(6), pp.1698.

30. Tian, B., Hua, S., Tian, Y. and Liu, J., (2021) 'Cyclodextrin-based adsorbents for the removal of pollutants from wastewater: A review' *Environ. Sci. Pollut. Res.*, Vol 28(2), pp.1317–1340.
31. Liu, Z., Dai, X., Sun, Y. and Liu, Y., (2020) 'Organic supramolecular aggregates based on water-soluble cyclodextrins and calixarenes' *Aggregate*, Vol 1(1), pp.31–44.
32. Yadav, M., Thakore, S. and Jadeja, R., (2022) 'A review on remediation technologies using functionalized Cyclodextrin' *Environ. Sci. Pollut. R*es., Vol 29(1), pp.236–250.
33. Healy, B., Yu, T., C. da Silva Alves, D., Okeke, C. and Breslin, C.B., (2021) 'Cyclodextrins as supramolecular recognition systems: Applications in the fabrication of electrochemical sensors' *Materials*, Vol 14(7), p.1668.
34. Karimi-Maleh, H., Darabi, R., Baghayeri, M., Karimi, F., Fu, L., Rouhi, J., Niculina, D.E., Gündüz, E.S. and Dragoi, E.N., (2023) 'Recent developments in carbon nanomaterials-based electrochemical sensors for methyl parathion detection' *J. Food Meas. Charact.*, Vol 17(5), pp.5371–5389.

9 Cyclodextrins in UV Protection

Advanced Absorbers for Versatile Materials

Saptadwipa Bhattacharjee, Arif Hossain Patowari, Barnali De Sarkar, Taniya Das, Tamalika Maitra, and Prof. Pranab Ghosh

9.1 INTRODUCTION

Cyclodextrin is a naturally occurring, ring-like molecule that has the unique capacity to envelope other molecules [1]. Its hydrophobic core and hydrophilic outer make it a perfect host for a diverse set of guest molecules [2]. Cyclodextrin's high water solubility and biocompatibility have made it a handy tool in a variety of sectors [3]. The sun's ultraviolet (UV) radiation poses a serious hazard to human health, including skin damage, wrinkles, and an increased risk of skin cancer [4]. Typical UV filters utilized by sunscreens have disadvantages, such as toxicity, instability, and adverse effects on the environment [5]. As a result, there is an increasing demand for alternative UV filtration systems that are reliable, efficient, and sustainable. Cyclodextrins (CDs), a cyclic polysaccharide family, have emerged as a viable UV filtering option [6]. CDs have the unusual ability to generate inclusion complexes with UV-absorbing compounds, which improves their stability, solubility, and UV absorption capacity [7]. This feature makes CDs an appealing element for sunscreen compositions, since they provide greater UV protection, lower toxicity, and ecological benefits [8].

This chapter examines CDs' potential as UV filters, examining their mechanism of action, benefits, limits, and recent research on CD-based UV filters [6]. We will investigate the safety and efficacy of CD-based UV filters and propose future research and development directions, emphasizing CDs' potential to revolutionize UV screening in sunscreen applications.

9.2 PRINCIPLES OF CYCLODEXTRIN-INCLUSION COMPLEXES

Cyclodextrin-inclusion complexes form when guest molecules fit into the hydrophobic cavity of cyclodextrins, influenced by factors like size, shape, and polarity [2, 3]. These complexes alter the properties of guest molecules, improving solubility, stability, and reactivity, while controlling release [9]. This makes them valuable in pharmaceuticals, food, cosmetics, and environmental clean-up [10]. Advanced

 DOI: 10.1201/9781003492757-9

techniques like NMR and X-ray crystallography help study the complex structures and dynamics [11].

9.2.1 Structure and Properties of Cyclodextrins

Cyclodextrins (CDs) are non-toxic, water-soluble sugar polymers with unique physicochemical properties, making them versatile in various applications. They are stable in alkaline conditions and certain organic solvents, but susceptible to acid hydrolysis [1, 12]. Natural CDs have been demonstrated to be highly stable in alkaline conditions; their reported pKa values range from 12.1 to 13.5 [13]. As a result, CDs are susceptible to acid hydrolysis at low pH, which opens the ring and yields a variety of linear oligosaccharides and glucose units [14]. CDs are distinguished by their hydrophilic circular, truncated cone shape having 7.9 Å deep hydrophobic, hollow conical chambers and allow the hydrophobic guest molecules to encapsulate them. This opening enables the insertion of appropriately sized hydrophobic guest molecules [1, 15–18]. The arrangement of three natural cyclodextrins: alpha CD (α-CD), beta CD (β-CD), and gamma CD (γ-CD) is shown in Figure 9.1. Each has six, seven, or eight D-glucose subunits [12].

Cyclodextrins (CDs) have a hydroxyl group lining on their exterior, which forms hydrogen bonds with water molecules, allowing for some aqueous solubility. The three native CDs, α-CD, β-CD, and γ-CD, have varying hydrophobic cavity sizes and numbers of glucose units, with α-CD being the smallest and γ-CD the largest [19]. Table 9.1 outlines the size characteristics of α-CD, β-CD, and γ-CD, with inner cavity diameters typically reported as a single value [12, 20, 21], although some studies provide a range [22]. Each CD type has distinct physical properties, including molecular weight (972–1297 g/mol), water solubility (1.85–23.2 g/100 mL), and melting temperature (255–280°C) [27].

9.2.2 Mechanisms of Inclusion Complex Formation with UVR Absorbers

The formation of inclusion complexes between cyclodextrins (CDs) and UVR absorbers involves a multi-step process [2]. Initially, the UVR absorber molecule enters the CD cavity through hydrophobic interactions, releasing water molecules [1]. As it penetrates deeper, van der Waals forces and hydrogen bonds reinforce the complex, with the CD ring adapting to the shape of the UVR absorber [3]. Factors like concentration, pH, and temperature influence the complex's stability and release rates [28]. The resulting inclusion complex protects the UVR absorber from degradation, enhances its solubility, and controls its release, making it a stable and effective UVR shielding agent for various applications.

9.2.3 Factors Influencing Complex Stability and Release Kinetics

The stability and release kinetics of cyclodextrin-insertion ensembles with UVR absorbers are influenced by several factors. Temperature, pH, concentration of CD and UVR absorbers, size and structure of the UVR absorber molecule, and type of

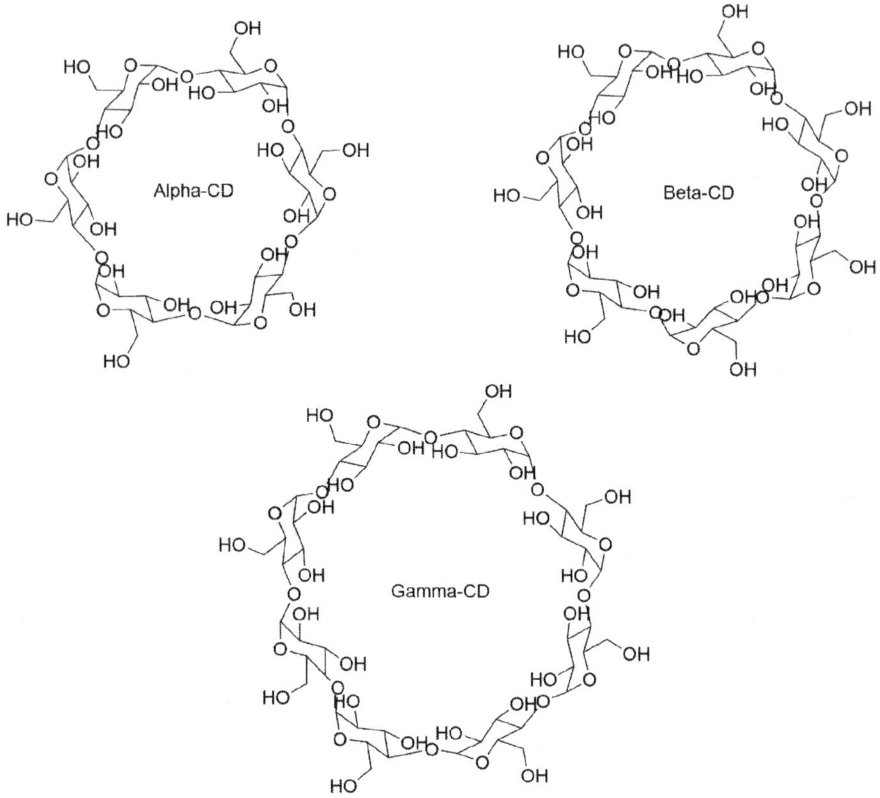

FIGURE 9.1 Structure of cyclodextrin.

TABLE 9.1
Characteristics of α-CD, β-CD, and γ-CD

Cyclodextrin	No. of Glucose Units	Molecular Weight	Height (Å) [23]	Cavity Volume (Å^3) [24]	Outer Diameter (Å) [3, 25]	Cavity Diameter (Å) [26]
α-CD	6	972	7.8	174	14.6	4.7–5.2
β-CD	7	1135	7.8	262	15.4	6.0–6.4
γ-CD	8	1297	7.8	427	17.5	7.5–8.3

CD used (α-, β-, or γ-CD) all play a crucial role [28–32]. For instance, highlights that the concentration of CD and UVR absorbers determines the complex's synthesis and stability, with larger concentrations favouring complexation [31]. Understanding these parameters is essential to optimize complex formation and release kinetics for specific applications.

a. *Temperature:* The stability of cyclodextrin-insertion complexes containing UVR absorbers is highly dependent on temperature [29].
   - Thermal energy: High temperatures can disrupt hydrophobic interactions, causing complex dissociation.
   - Conformational changes: Temperature fluctuations can alter CD or UVR absorber structures, impacting complex stability and release.
   - Optimal temperature range: The ideal temperature for complex formation and stability varies depending on the CD and UVR absorber used.
b. *pH:* pH is another important element that influences the stability of cyclodextrin-insertion complexes containing UVR absorbers [30].
   - Ionization state: pH changes alter CD or UVR absorber ionization, impacting complex stability.
   - Hydrophobicity and electrostatic interactions: pH variations influence hydrophobic and electrostatic interactions, affecting complex formation and stability.
   - Buffering capacity: A solution's buffering capacity maintains pH stability, minimizing the impact of pH changes on complex stability.
c. *Concentration:* The stability of cyclodextrin-insertion complexes containing UVR absorbers is highly dependent on concentration. Here is a more detailed explanation [31].
   - CD concentration: Higher concentrations promote complex formation, but excessive CD molecules can destabilize the complex.
   - UVR absorber concentration: Higher concentrations can enhance stability, but excessive absorbers can compete for CD cavities, reducing stability.
   - Optimal concentration ratio: The ideal CD-to-UVR absorber ratio varies depending on the specific CD and absorber used, and finding this ratio is crucial for maximizing complex stability.
d. *Size and shape of the UVR absorber molecule:* The shape and size of the UVR absorber molecule have a substantial impact on the stability of the cyclodextrin-inclusion complex. Here is a more detailed explanation [28].
   - Molecular size: Smaller UVR absorbers form more stable complexes due to better hydrophobic contacts and fit within the CD cavity.
   - Molecular structure: Linear or cylindrical UVR absorber shapes facilitate stable complex formation.
   - Optimal dimensions and shape: The ideal UVR absorber size and shape vary depending on the CD used, and finding this optimal combination is crucial for maximizing complex stability.
e. *Type of CD used (α-, β-, or γ-CD)*: The shape and dimensions of the CD cavity influence the stability of the complex. α-CD has a smaller cavity, β-CD has a medium-sized cavity, and γ-CD has a bigger one [32].
   - Substitution pattern: The type of CD substitution (e.g., hydroxypropyl, methyl) influences complex stability by altering hydrophobicity and electrostatic interactions.
   - Optimal CD type: The best CD type (α, β, or γ) depends on the UVR absorber used, with smaller absorbers potentially benefiting from α-CD and larger ones from γ-CD.

f. Presence of other molecules or ions [28]:

- Competitive binding: Other molecules or ions can compete with the UVR absorber for CD binding, reducing complex stability.
- Electrostatic interactions: Additional molecules or ions can alter electrostatic interactions between the CD and UVR absorber, affecting complex stability.
- Optimal conditions: The ideal conditions to minimize competition and maximize complex stability vary depending on the CD, UVR absorber, and other present molecules or ions.

These factors interact with each other and with the specific CD and UVR absorber molecules to influence complex stability and release kinetics.

9.3 ORGANIC UVR ABSORBERS WITH CYCLODEXTRINS

9.3.1 Overview of Common Organic UVR Absorbers (Oxybenzone, Avobenzone, etc.)

Common organic UVR absorbers in sunscreens and cosmetics include oxybenzone (OBZ), avobenzone (ABZ), octinoxate (OMC), octocrylene (OCR), and ethylhexyl salicylate (EHS) [33]. These chemicals provide strong UV protection, but have raised concerns about hormone disruption, skin irritation, photoinstability, and environmental impact, such as coral reef destruction [34–37]. Despite these issues, they remain widely used due to their effectiveness and aesthetic compatibility. However, ongoing research aims to find alternative absorbers with improved safety and environmental profiles [38].

9.3.2 Cyclodextrin-Based Inclusion Complexes for Fabric and Textile Applications

Cyclodextrin-based inclusion complexes are being researched for textile applications, enabling controlled release of functional compounds like perfumes, antibacterial agents, and UV protectants [39, 40]. These complexes enhance fabric functionality, offering benefits such as durability, washability, and thermal stability [41]. They can also improve comfort and hygiene by releasing perfumes and antibacterial compounds, while providing UV protection and antioxidant properties [42, 43]. This technology presents a versatile and effective approach to enhancing textile performance, opening up new possibilities for the textile industry [44].

9.3.3 UV-Protective Clothing, Upholstery, and Industrial Textiles [45, 46]

Case studies have demonstrated the effectiveness of cyclodextrin-based inclusion complexes in various textile applications. For instance, cyclodextrin complexes with UV absorbers like oxybenzone and avobenzone have maintained high SPF ratings after repeated washing and wear. Furthermore, cyclodextrin complexes with antimicrobial agents like triclosan have inhibited bacterial growth and odour in upholstery, improving indoor air quality. These complexes have also been used in industrial textiles to enhance fire resistance and safety, and in temperature-regulating textiles to provide comfort. Overall, these studies showcase the versatility and benefits of cyclodextrin-based inclusion complexes in textiles, leading to improved performance, efficiency, and sustainability.

9.4 INORGANIC UVR ABSORBERS WITH CYCLODEXTRINS

9.4.1 Overview of Common Inorganic UVR Absorbers (Zinc Oxide, Titanium Dioxide, etc.)

Conventional inorganic UVR absorbers, especially zinc oxide (ZnO) and titanium dioxide (TiO_2), are frequently utilized in sunscreens, coatings, and textiles due to their good UV absorption and photostability. Zinc oxide, a physical sunscreen, absorbs both UVA and UVB light, ensuring broad-spectrum protection [47]. Titanium dioxide, a physical sunscreen, absorbs UVB light and is commonly used in conjunction with other absorbers to provide UVA protection [48]. Additional inorganic UVR absorbers include cerium oxide (CeO_2), iron oxide (Fe_2O_3), and silicon dioxide (SiO_2), all of which provide variable degrees of UV protection and are employed in certain applications [49]. Inorganic UVR absorbers are widely regarded as secure and gentle on the skin, making them appropriate for use in delicate applications like as baby care items and medical devices [50].

9.4.2 Cyclodextrin-Based Coatings for Glass and Film Applications

Cyclodextrin-based coatings have been developed for glass and film, offering benefits like UV protection, self-cleaning, and controlled release of functional molecules [39]. These coatings, composed of cyclodextrin molecules attached to surfaces, provide a thin layer for absorbing and encapsulating chemicals [51]. They can be applied using various methods, including dip-coating and chemical vapour deposition, and have a range of applications, including UV protection for glassware, self-cleaning films that release antibacterial agents or perfumes, and improved surface hardness and scratch resistance [41–43]. Overall, cyclodextrin-based coatings provide a versatile and effective way to modify glass and film surfaces, expanding their applications in industries like automotive, construction, and consumer products.

9.4.3 UV-Protective Glass, Automotive Windshields, and Optical Films

The integration of cyclodextrin-based coatings with UV-absorbing molecules has led to the creation of innovative UV-protective materials, including glass, car windshields, and optical films [39]. These advanced polymers provide superior UV protection, minimizing the risk of skin damage, fading, and discolouration [43]. They can filter up to 99% of UV rays, ensuring a safer driving experience, and can also be used in architectural and eyewear applications [51, 41]. Additionally, these materials can be designed to be self-cleaning, reducing maintenance and increasing durability [42]. This unique combination has opened up new opportunities for the development of high-performance, UV-protective materials across various industries, including automotive, construction, and healthcare.

9.5 NANOPARTICLE UVR ABSORBERS WITH CYCLODEXTRINS

9.5.1 Overview of Nanoparticle UVR Absorbers (Nano-Zinc Oxide, Nano-Titanium Dioxide, etc.)

UVR-absorbing nanoparticles, such as nano-zinc oxide and nano-titanium dioxide, have gained popularity due to their excellent UV absorption and cosmetic appeal [52]. Their small size (10–100 nanometers) enables superior UV protection, high surface areas, and improved dispersion [53]. These nanoparticles are widely used in sunscreens, cosmetics, and coatings, offering broad-spectrum UV protection and photostability [54]. Researchers are also exploring other nanoparticle UVR absorbers, like nano-cerium oxide and nano-iron oxide [55]. The tiny size of these nanoparticles allows for transparent and non-whitening formulations, making them suitable for clear sunscreens, cosmetics, and optical devices [56]. However, concerns about potential environmental and physiological impacts require careful consideration and ongoing research.

9.5.2 Cyclodextrin-Based Dispersion and Stabilization of Nanoparticles

Cyclodextrin-based dispersion and stabilization of nanomaterials has emerged as a promising method for enhancing the properties and applications of nanoparticle systems [57]. By forming inclusion complexes with cyclodextrins, nanoparticles such as zinc oxide, titanium dioxide, and cerium oxide can be distributed and stabilized in various media, including water, oils, and polymers [58]. This technique improves the solubility, stability, and compatibility of nanoparticles, enabling their use in diverse applications like cosmetics, medicines, and coatings [59]. Additionally, cyclodextrin-based dispersion and stabilization enhance the UV absorption and drug delivery properties of nanoparticles, reduce toxicity and adverse ecological effects, and increase biocompatibility [42, 43, 60]. Overall, this method provides a flexible and effective approach to unlocking the full potential of nanoparticle-based technology.

9.5.3 UV-Protective Nanotextiles, Plastics, and Composites

UV-protective nanotextiles, polymers, and composites have been developed by incorporating nanoparticles like zinc oxide and titanium dioxide, which absorb and scatter UV light, preventing damage [61]. These materials are used in outdoor apparel, automotive assemblies, and construction materials, offering improved UV resistance, durability, and mechanical properties [62, 63]. Additionally, they can be made self-cleaning, antibacterial, and flame-resistant, expanding their applications across various industries such as textiles, automotive, construction, and medicine.

9.6 MECHANISM OF ACTION

UV-protective nanomaterials, containing nanoparticles like zinc oxide and titanium dioxide [52], which are incorporated into fabrics, plastics, and composites, absorb and scatter UV radiation, creating a physical barrier that prevents UV rays from penetrating the material [33]. The nanoparticles capture UVA and UVB rays, converting them into harmless heat, and also scatter UV radiation, reducing its intensity [46, 49]. This dual process provides broad-spectrum UV protection, effectively preventing UV-induced harm, degradation, and discoloration [64].

9.7 ADVANTAGES AND LIMITATIONS [65]

Advantages

- High UV protection efficiency
- Broad-spectrum protection (UVA and UVB)
- Improved durability and resistance to degradation
- Enhanced mechanical properties
- Self-cleaning and antimicrobial properties
- Versatile applications (textiles, plastics, composites)
- Non-toxic and environmentally friendly

Limitations

- Potential toxicity of nanoparticles (depending on size and material)
- Limited UV protection at high concentrations of nanoparticles
- Aggregation and sedimentation of nanoparticles can reduce effectiveness
- Interference with other material properties (e.g., color, texture)
- High cost of production and processing
- Potential environmental impact (e.g., nanoparticle release, accumulation)
- Need for standardized testing and regulation

9.8 RECENT STUDIES AND APPLICATIONS [66]

UV-protective nanomaterials have led to innovative solutions across various industries. They are used in textiles to create nano-coated fabrics for outdoor apparel [67], in healthcare for self-cleaning surfaces and antimicrobial coatings [68], and in aerospace and automotive [69, 70] for UV-resistant components. Additionally, they are employed in biomedicine for surgical dressings and instruments [71], agriculture for nano-coated seeds [72, 73], cosmetics for enhanced UV protection in personal care products [74, 75], and architecture for energy-efficient windows and coatings in buildings [76]. Future research is expected to unlock even more novel uses.

Some of the applications are given below.

9.8.1 Investigating CD-UV Absorber Interactions and Complexation Mechanisms [6, 64]

In order to better understand cyclodextrin-based UVR absorbers, further research is needed on CD-UV absorber interactions. Factors like cyclodextrin type, UV absorber, and solvents affect complexation. Interactions are driven by hydrogen bonding, hydrophobic forces, and van der Waals forces. Advanced techniques like NMR, X-ray crystallography, and molecular modelling help optimize CD-UV absorber design.

9.8.2 Optimizing CD Cavity Size and Shape for Enhanced UV-Absorbing Capacity [77, 78]

Optimizing cyclodextrin (CD) cavity size and shape enhances UV absorption. Larger cavities (γ-CD) accommodate bulkier UV absorbers, increasing complexation efficiency. Smaller cavities (α-CD) offer better control. Chemical modifications, like methylations, and using CD derivatives with flexible structures improve binding affinity, stoichiometry, and UV absorption. Tailoring CD cavity size and shape boosts UV protection and expands applications across industries.

9.8.3 Developing CD-Based UV Filters with Controlled Release and Targeted Delivery Properties [79]

Designing cyclodextrin (CD)-based UV filters with controlled release and targeted delivery can enhance efficacy and safety. Researchers can create systems that release UV absorbers in a regulated manner, providing sustained protection and reduced application frequency. Modified CDs can target specific skin layers or cells, and respond to environmental changes, releasing UV absorbers as needed. This approach enables adaptive UV protection systems tailored to individual skin types and situations.

9.8.4 Exploring CD-Containing Formulations for Other Applications (e.g., Cosmetics, Textiles) [80, 40]

Research on cyclodextrin (CD)-containing formulations is expanding beyond UV protection. CDs are being explored as carriers for fragrances, antioxidants, and anti-ageing agents in cosmetics, and as treatments for textiles to provide moisture management, antibacterial properties, and colour protection. CDs are also being used to encapsulate dyes, pesticides, and medications, enhancing stability, solubility, and targeted delivery. Additionally, CDs are being applied in agriculture, food packaging, and other industries, showcasing their versatility and potential for innovative applications.

9.8.5 Potential Applications in Biomedical, Agricultural, and Space Exploration Contexts [25, 81]

Cyclodextrin-based formulations have diverse applications beyond UV protection, including biomedicine, agriculture, and space research. In biomedicine, CDs enhance drug delivery, solubility, and stability, and can create antimicrobial coatings. In agriculture, CDs improve pesticide and fertilizer delivery, reducing environmental impact and boosting crop yields. CDs also enable new biosensors and diagnostic tools. In space exploration, CDs aid in developing advanced medical equipment, air and water purification systems, and radiation-absorbing materials, supporting long-term space missions. CDs' versatility drives innovation across these fields.

9.9 SAFETY AND EFFICACY [50, 82]

The development and use of UV-protective nanomaterials require careful consideration of safety and efficacy. While they offer benefits, their small size and unique properties raise concerns about toxicity, environmental impact, and human health risks. Regulatory agencies have deemed some nanomaterials, like nano-zinc oxide and nano-titanium dioxide, safe for cosmetic use. To ensure safety, manufacturers must design and test these materials carefully, minimizing risks through techniques like encapsulation. Efficacy is also crucial and standardized testing and labelling can help ensure products meet safety and efficacy standards.

9.10 OUTLOOK FOR FUTURE RESEARCH AND DEVELOPMENT IN CYCLODEXTRIN-BASED UVR ABSORBERS

Future research on cyclodextrin-based UVR absorbers will focus on enhancing stability, solubility, and bioavailability, as well as exploring new compounds and multifunctional formulations. Advances in nanotechnology and materials science will enable targeted release and smart UVR protection systems. Applications in healthcare, agriculture, and space exploration will drive innovation, ultimately protecting human health, the environment, and materials from UV radiation.

9.11 CONCLUSION

To summarize, cyclodextrin-based UVR absorbers are a cutting-edge technology for guarding against the detrimental effects of ultraviolet radiation. Cyclodextrins increase the stability, solubility, and bioavailability of active substances by forming inclusion complexes with UV absorbers, resulting in greater efficacy and safety. Ongoing research and development are aimed at improving CD-UV absorber complexes, investigating novel derivatives and applications, and capitalizing on breakthroughs in nanotechnology and materials science. Cyclodextrin-based UVR absorbers are expected to play an important role in a variety of industries, including cosmetics and textiles, biomedicine, agriculture, and rocketry, eventually leading to improved human health, sustainable development, and technological innovation.

REFERENCES

1. Szejtli, J. (1998) 'Introduction and General Overview of Cyclodextrin Chemistry' *Chemical Reviews*, Vol. 98, pp. 1743–1754.

2. Connors, K. A. (1997) 'The Stability of Cyclodextrin Complexes in Solution' *Chemical Reviews*, Vol. 97, pp. 1325–1358.
3. Del Valle, E. M. M. (2004) 'Cyclodextrins and Their Uses: A Review' *Process Biochemistry*, Vol. 39, pp. 1033–1046.
4. Gallagher, R. P.; Lee, T. K. (2006) 'Adverse Effects of Ultraviolet Radiation: A Brief Review' *Progress in Biophysics and Molecular Biology*, Vol. 92, pp. 119–131.
5. Chatzigianni, M.; Pavlou, P.; Siamidi, A.; Vlachou, M.; Varvaresou, A.; Papageorgiou, S. (2022) 'Environmental Impacts Due to the Use of Sunscreen Products: A Mini-Review' *Ecotoxicology*, Vol. 31, pp. 1331–1345.
6. Fenyvesi, É.; Otta, K.; Kolbe, I.; Novák, Cs.; Szejtli, J. (2004) 'Cyclodextrin Complexes of UV Filters' *Journal of Inclusion Phenomena*, Vol. 48, pp. 117–123.
7. Fenyvesi, E.; Jicsinszky, L.; Szejtli, J.; Schwarzenbach, R. (1999) 'Inclusion Complexes of UV Filters in Solution and in Solid State' In *Proceedings of the Ninth International Symposium on Cyclodextrins*; Springer Netherlands: Dordrecht, pp. 639–643.
8. Dahabra, L.; Broadberry, G.; Le Gresley, A.; Najlah, M.; Khoder, M. (2021) 'Sunscreens Containing Cyclodextrin Inclusion Complexes for Enhanced Efficiency: A Strategy for Skin Cancer Prevention' *Molecules*, Vol. 26, pp. 1698.
9. Assaf, K. I.; Gabel, D.; Zimmermann, W.; Nau, W. M. (2016) 'High-Affinity Host–Guest Chemistry of Large-Ring Cyclodextrins' *Organic & Biomolecular Chemistry*, Vol. 14, pp. 7702–7706.
10. Carneiro, S.; Costa Duarte, F.; Heimfarth, L.; Siqueira Quintans, J.; Quintans-Júnior, L.; Veiga Júnior, V.; Neves de Lima, Á. (2019) 'Cyclodextrin–Drug Inclusion Complexes: In Vivo and In Vitro Approaches' *International Journal of Molecular Sciences*, Vol. 20, pp. 642.
11. Kfoury, M.; Landy, D.; Fourmentin, S. (2018) 'Characterization of Cyclodextrin/Volatile Inclusion Complexes: A Review' *Molecules*, Vol. 23, pp. 1204.
12. Martin, J.; Díaz-Montaña, E. J.; Asuero, A. G. (2018) 'Cyclodextrins: Past and Present' In Poonam Arora and Neelima Dhingra (eds.), *Cyclodextrin – A Versatile Ingredient*; InTech.
13. Saokham, P.; Muankaew, C.; Jansook, P.; Loftsson, T. (2018) 'Solubility of Cyclodextrins and Drug/Cyclodextrin Complexes' *Molecules*, Vol. 23, pp. 1161.
14. Gaidamauskas, E.; Norkus, E.; Butkus, E.; Crans, D. C.; Grincienė, G. (2009) 'Deprotonation of β-Cyclodextrin in Alkaline Solutions' *Carbohydrate Research*, Vol. 344, pp. 250–254.
15. Szejtli, J. (2004) 'Past, Present and Futute of Cyclodextrin Research' *Pure and Applied Chemistry*, Vol. 76, pp. 1825–1845.
16. Bender, M. L.; Komiyama, M. (2012) '*Cyclodextrin Chemistry*' Springer Science & Business Media.
17. Tegge, G.; Szejtli, J. (1982) 'Cyclodextrins and Their Inclusion Complexes' *Starch – Stärke*, Vol. 34, pp. 395–395.
18. Dodziuk, H. (2006) 'Molecules with Holes – Cyclodextrins' In *Cyclodextrins and Their Complexes*, pp. 1–30.
19. Irie, T.; Uekama, K. (1997) 'Pharmaceutical Applications of Cyclodextrins. III. Toxicological Issues and Safety Evaluation' *Journal of Pharmaceutical Sciences*, Vol. 86, pp. 147–162.
20. Noreña-Caro, D.; Alvarez-Láinez, M.(2014) 'Experimental Design as a Tool for the Manufacturing of Filtering Media Based on Electrospun Polyacrylonitrile/ β - Cyclodextrin Fibers' *International Journal on Interactive Design and Manufacturing (IJIDeM)*, Vol. 10, pp. 153–164.

21. Cova, T. F.; Murtinho, D.; Pais, A. A. C. C.; Valente, A. J. M. (2018) 'Combining Cellulose and Cyclodextrins: Fascinating Designs for Materials and Pharmaceutics' *Frontiers in Chemistry*, Vol. 6, pp. 271.
22. Challa, R.; Ahuja, A.; Ali, J.; Khar, R. K. (2005) 'Cyclodextrins in Drug Delivery: An Updated Review' *AAPS PharmSciTech*, Vol. 6, pp. E329–E357.
23. Loftsson, T.; Brewster, M. E. (2010) 'Pharmaceutical Applications of Cyclodextrins: Basic Science and Product Development' *Journal of Pharmacy and Pharmacology*, Vol. 62, pp. 1607–1621.
24. Cal, K. (2008) 'Use of Cyclodextrins in Topical Formulations: Practical Aspects' *European Journal of Pharmaceutics and Biopharmaceutics*, Vol. 68, pp. 467–478.
25. Sharma, N.; Baldi, A. (2014) 'Exploring Versatile Applications of Cyclodextrins: An Overview' *Drug Delivery*, Vol. 23, pp. 729–747.
26. Atwood, J. L. (2003) 'Inclusion (Clathrate) Compounds' In Robert A. Meyers (ed.) *Encyclopedia of Physical Science and Technology*; Elsevier, pp. 717–729.
27. Esteso, M. A.; Romero, C. M. (2024) 'Cyclodextrins: Properties and Applications' *International Journal of Molecular Sciences*, Vol. 25, pp. 4547.
28. Linde, G. A.; Laverde, A., Jr; Colauto, N. B. (2011) 'Changes to Taste Perception in the Food Industry: Use of Cyclodextrins' In Victor R. Preedy, Ronald Ross Watson, and Colin R. Martin (eds.), *Handbook of Behavior, Food and Nutrition*; Springer New York: New York, NY, pp. 99–118.
29. Sakai, S.; Hirano, Y.; Kobayashi, Y.; Arai, N. (2023) 'Effect of Temperature on the Structure and Drug-Release Behaviour of Inclusion Complex of β-Cyclodextrin with Cyclophosphamide: A Molecular Dynamics Study' *Soft Matter*, Vol. 19, pp. 2902–2907.
30. Cedillo-Flores, O. E.; Rodríguez-Laguna, N.; Hipólito-Nájera, A. R.; Nivón-Ramírez, D.; Gómez-Balderas, R.; Moya-Hernández, R. (2022) 'Effect of the pH on the Thermodynamic Stability of Inclusion Complexes of Thymol and Carvacrol in β-Cyclodextrin in Water' *Food Hydrocolloids*, Vol. 124, pp. 107307.
31. da Rocha Neto, A. C.; de Oliveira da Rocha, A. B.; Maraschin, M.; Di Piero, R. M.; Almenar, E. (2018) 'Factors Affecting the Entrapment Efficiency of β-Cyclodextrins and Their Effects on the Formation of Inclusion Complexes Containing Essential Oils' *Food Hydrocolloids*, Vol. 77, pp. 509–523.
32. Yin, H.; Wang, C.; Yue, J.; Deng, Y.; Jiao, S.; Zhao, Y.; Zhou, J.; Cao, T. (2021) 'Optimization and Characterization of 1,8-Cineole/Hydroxypropyl-β-Cyclodextrin Inclusion Complex and Study of Its Release Kinetics' *Food Hydrocolloids*, Vol. 110, pp. 106159.
33. Diffey, B. (2009) 'Sunscreens: Expectation and Realization' *Photodermatology, Photoimmunology & Photomedicine*, Vol. 25, pp. 233–236.
34. Wang, S. Q.; Balagula, Y.; Osterwalder, U. (2010) 'Photoprotection: A Review of the Current and Future Technologies' *Dermatologic Therapy*, Vol. 23, pp. 31–47.
35. Berardesca, E.; Zuberbier, T.; Sanchez Viera, M.; Marinovich, M. (2019) 'Review of the Safety of Octocrylene Used as an Ultraviolet Filter in Cosmetics' *Journal of the European Academy of Dermatology and Venereology*, Vol. 33, pp. 25–33.
36. Krause, M.; Klit, A.; Blomberg Jensen, M.; Søeborg, T.; Frederiksen, H.; Schlumpf, M.; Lichtensteiger, W.; Skakkebaek, N. E.; Drzewiecki, K. T. (2012) 'Sunscreens: Are They Beneficial for Health? An Overview of Endocrine Disrupting Properties of UV-filters' *International Journal of Andrology*, Vol. 35, pp. 424–436.
37. Downs, C. A.; Kramarsky-Winter, E.; Segal, R.; Fauth, J.; Knutson, S.; Bronstein, O.; Ciner, F. R.; Jeger, R.; Lichtenfeld, Y.; Woodley, C. M.; et al. (2015) 'Toxicopathological Effects of the Sunscreen UV Filter, Oxybenzone (Benzophenone-3), on Coral Planulae

and Cultured Primary Cells and Its Environmental Contamination in Hawaii and the U.S. Virgin Islands' *Archives of Environmental Contamination and Toxicology*, Vol. 70, pp. 265–288.

38. Aguilera, J.; Gracia-Cazaña, T.; Gilaberte, Y. (2023) 'New Developments in Sunscreens' *Photochemical & Photobiological Sciences*, Vol. 22, pp. 2473–2482.
39. Voncina, B.; Vivo, V. (2013) 'Cyclodextrins in Textile Finishing' In Melih Günay (ed.), *Eco-Friendly Textile Dyeing and Finishing*; InTech.
40. Voncina, B. (2011) 'Application of Cyclodextrins in Textile Dyeing' In Peter Hauser (ed.), *Textile Dyeing*; InTech.
41. Haji, A. (2020) 'Functional Finishing of Textiles with β-Cyclodextrin'. In Mohd Shabbir, Shakeel Ahmed & Javed N. Sheikh (eds.) *Frontiers of Textile Materials*, LLC, pp. 87–116.
42. Bezerra, F. M.; Lis, M. J.; Firmino, H. B.; Dias da Silva, J. G.; Curto Valle, R. de C. S.; Borges Valle, J. A.; Scacchetti, F. A. P.; Tessaro, A. L. (2020) 'The Role of β-Cyclodextrin in the Textile Industry—Review' *Molecules*, Vol. 25, pp. 3624.
43. Singh, M.; Sharma, R.; Banerjee, U. C. (2002) 'Biotechnological Applications of Cyclodextrin' *Biotechnology Advances*, Vol. 20, pp. 341–359.
44. Singh, N.; Sahu, O. (2019) 'Sustainable Cyclodextrin in Textile Applications' In Shahid-ul-Islam and B.S. Butola (eds.), *The Impact and Prospects of Green Chemistry for Textile Technology*; Elsevier, pp. 83–105.
45. Shahidi, S.; Wiener, J.; Ghoranneviss, M. (2013) 'Surface Modification Methods for Improving the Dyeability of Textile Fabrics' In Melih Gunay (ed.), *Eco-Friendly Textile Dyeing and Finishing*; InTech.
46. Scacchetti, F. A. P.; Pinto, E.; Soares, G. M. B. (2017) 'Functionalization and Characterization of Cotton with Phase Change Materials and Thyme Oil Encapsulated in Beta-Cyclodextrins' *Progress in Organic Coatings*, Vol. 107, pp. 64–74.
47. Mitchnick, M. A.; Fairhurst, D.; Pinnell, S. R. (1999) 'Microfine Zinc Oxide (Z-Cote) as a Photostable UVA/UVB Sunblock Agent' *Journal of the American Academy of Dermatology*, Vol. 40, pp. 85–90.
48. Trivedi, M.; Murase, J. (2017) 'Titanium Dioxide in Sunscreen' In Magdalena Janus (ed.), *Application of Titanium Dioxide*; InTech.
49. Schneider, S. L.; Lim, H. W. (2018) 'A Review of Inorganic UV Filters Zinc Oxide and Titanium Dioxide' *Photodermatology, Photoimmunology & Photomedicine*, Vol. 35, pp. 442–446.
50. Smijs, T.; Pavel. (2011) 'Titanium Dioxide and Zinc Oxide Nanoparticles in Sunscreens: Focus on Their Safety and Effectiveness' *Nanotechnology, Science and Applications*, Vol. 2011, pp. 95–112.
51. Cannavale, A.; Fiorito, F.; Manca, M.; Tortorici, G.; Cingolani, R.; Gigli, G. (2010) 'Multifunctional Bioinspired Sol-Gel Coatings for Architectural Glasses' *Building and Environment*, Vol. 45, pp. 1233–1243.
52. Rapaport, M. J. (1983) 'Sunscreening Agents and Sun Protective Factors' *International Journal of Dermatology*, Vol. 22, pp. 293–294.
53. Mondal, S. (2021) 'Nanomaterials for UV Protective Textile' *Journal of Industrial Textiles*, Vol. 51, pp. 5592S–5621S.
54. Wang, X. (2023) 'The Comparison of Titanium Dioxide and Zinc Oxide Used in Sunscreen Based on Their Enhanced Absorption' *Applied and Computational Engineering*, Vol. 24, pp. 237–245.
55. Silva, M. R. F.; Alves, M. F. R. P.; Cunha, J. P. G. Q.; Costa, J. L.; Silva, C. A.; Fernandes, M. H. V.; Vilarinho, P. M.; Ferreira, P. (2023) 'Nanostructured Transparent

Solutions for UV-Shielding: Recent Developments and Future Challenges' *Materials Today Physics*, Vol. 35, pp. 101131.

56. Fytianos, G.; Rahdar, A.; Kyzas, G. Z. (2020) 'Nanomaterials in Cosmetics: Recent Updates' *Nanomaterials*, Vol. 10, pp. 979.
57. Noël, S.; Léger, B.; Ponchel, A.; Philippot, K.; Denicourt-Nowicki, A.; Roucoux, A.; Monflier, E. (2014) 'Cyclodextrin-Based Systems for the Stabilization of Metallic(0) Nanoparticles and Their Versatile Applications in Catalysis' *Catalysis Today*, Vol. 235, pp. 20–32.
58. Gómez-Graña, S.; Pérez-Juste, J.; Hervés, P. (2021) 'Cyclodextrins and Inorganic Nanoparticles: Another Tale of Synergy' *Advances in Colloid and Interface Science*, Vol. 288, pp.102338.
59. Pandey, A. (2021) 'Cyclodextrin-Based Nanoparticles for Pharmaceutical Applications: A Review' *Environmental Chemistry Letters*, Vol. 19, pp. 4297–4310.
60. Gadade, D. D.; Pekamwar, S. S. (2020) 'Cyclodextrin Based Nanoparticles for Drug Delivery and Theranostics' *Advanced Pharmaceutical Bulletin*, Vol. 10, pp. 166–183.
61. Hossain, M. A.; Rahman, M. (2015) 'A Review of Nano Particle Usage on Textile Material against Ultra Violet Radiation' *Journal of Textile Science and Technology*, Vol. 01, pp. 93–100.
62. Tsuzuki, T.; Wang, X. (2010) 'Nanoparticle Coatings for UV Protective Textiles' *Research Journal of Textile and Apparel*, Vol. 14, pp. 9–20
63. Zhu, Q.; Chua, M. H.; Ong, P. J.; Cheng Lee, J. J.; Le Osmund Chin, K.; Wang, S.; Kai, D.; Ji, R.; Kong, J.; Dong, Z.; et al. (2022) 'Recent Advances in Nanotechnology-Based Functional Coatings for the Built Environment' *Materials Today Advances*, Vol. 15, pp.100270.
64. Yang, H.; Zhu, S.; Pan, N. (2004) 'Studying the Mechanisms of Titanium Dioxide as Ultraviolet blocking Additive for Films and Fabrics by an Improved Scheme' *Journal of Applied Polymer Science*, Vol. 92, pp. 3201–3210.
65. Noreen, S.; Maqbool, I.; Ijaz, M.; Tanveer, S. (2016) 'Cyclodextrin Inclusion Complexes: Novel Techniques to Improve Solubility of Poorly Soluble Drugs: A Review' *Global Pharmaceutical Sciences Review*, Vol. 1, pp.29–34.
66. Rajasekar, M.; Mary, J.; Sivakumar, M.; Selvam, M. (2024) 'Recent Developments in Sunscreens Based on Chromophore Compounds and Nanoparticles' *RSC Advances*, Vol. 14 (4), pp. 2529–2563.
67. Xing, Y.; Tang, J.; Li, X.; Huang, R.; Wu, L.; Xu, Q.; Liu, X.; Bi, X. (2022) 'Photo-Induced Antifungal Activity of Chitosan Composite Film Solution with Nano–Titanium Dioxide and Nano-Silver' *Journal of Food Protection*, Vol. 85, pp. 597–606.
68. Fan, X.; Yahia, L.; Sacher, E. (2021) 'Antimicrobial Properties of the Ag, Cu Nanoparticle System' *Biology*, Vol. 10, pp. 137.
69. Mathew, J.; Joy, J.; George, S. C. (2019 'Potential Applications of Nanotechnology in Transportation: A Review' *Journal of King Saud University –Science*, Vol. 31, pp. 586–594.
70. Salla, J.; Pandey, K. K.; Srinivas, K. (2012) 'Improvement of UV Resistance of Wood Surfaces by Using ZnO Nanoparticles' *Polymer Degradation and Stability*, Vol. 97, pp. 592–596.
71. Placido, F.; McLean, A.; Ogwu, A. A.; Ademosu, W. (2016) 'Titanium Dioxide Coatings for Medical Devices' In Waqar Ahmed and Mark J. Jackson (eds.), *Surgical Tools and Medical Devices*; Springer International Publishing: Cham, pp. 81–91.
72. Donia, D. T.; Carbone, M. (2023) 'Seed Priming with Zinc Oxide Nanoparticles to Enhance Crop Tolerance to Environmental Stresses' *International Journal of Molecular Sciences*, Vol. 24, pp. 17612.

73. Shang, Y.; Hasan, Md. K.; Ahammed, G. J.; Li, M.; Yin, H.; Zhou, J. (2019) 'Applications of Nanotechnology in Plant Growth and Crop Protection: A Review' *Molecules*, Vol. 24, pp. 2558.
74. Singha, K.; Maity, S.; Pandit, P. (2020) 'UV Protection via Nanomaterials' *Frontiers of Textile Materials*, pp. 153–166.
75. Pandey, A. S.; Bawiskar, D.; Wagh, V. (2024) 'Nanocosmetics and Skin Health: A Comprehensive Review of Nanomaterials in Cosmetic Formulations' *Cureus*, Vol. 16, pp. e52754.
76. Bao, Y.; Guo, R.; Kang, Q.; Liu, C.; Zhang, W.; Zhu, Q. (2021) 'Transparent, Thermal Insulation and UV-Shielding Coating for Energy Efficient Glass Window' *Ceramics International*, Vol. 47, pp. 24597–24606.
77. Cid-Samamed, A.; Rakmai, J.; Mejuto, J. C.; Simal-Gandara, J.; Astray, G. (2022) 'Cyclodextrins Inclusion Complex: Preparation Methods, Analytical Techniques and Food Industry Applications' *Food Chemistry*, Vol. 384, pp. 132467.
78. Sandilya, A. A.; Natarajan, U.; Priya, M. H. (2020) 'Molecular View into the Cyclodextrin Cavity: Structure and Hydration' *ACS Omega*, Vol. 5, pp. 25655–25667.
79. Feng, Y.; Chen, S.; Li, Z.; Gu, Z.; Xu, S.; Ban, X.; Hong, Y.; Cheng, L.; Li, C. (2021) 'A Review of Controlled Release from Cyclodextrins: Release Methods, Release Systems and Application' *Critical Reviews in Food Science and Nutrition*, Vol. 63, pp. 4744–4756.
80. Ferreira, L.; Mascarenhas-Melo, F.; Rabaça, S.; Mathur, A.; Sharma, A.; Giram, P. S.; Pawar, K. D.; Rahdar, A.; Raza, F.; Veiga, F.; et al. (2023) 'Cyclodextrin-Based Dermatological Formulations: Dermopharmaceutical and Cosmetic Applications' *Colloids and Surfaces B: Biointerfaces*, Vol. 221, pp. 113012.
81. Yousaf, R.; Abdul Razzaq, F.; Asghar, S.; Irfan, M.; Ullah Khan, I.; Haroon Khalid, S. (2022) 'Cyclodextrins: An Overview of Fundamentals, Types, and Applications' In Rashid Ali (ed.), *Cyclodextrins – Core Concepts and New Frontiers*; IntechOpen, pp. 1–13.
82. Oliveira, C. A. de; Dario, M. F.; Sarruf, F. D.; Mariz, I. F. A.; Velasco, M. V. R.; Rosado, C.; Baby, A. R. (2016) 'Safety and Efficacy Evaluation of Gelatin-Based Nanoparticles Associated with UV Filters' *Colloids and Surfaces B: Biointerfaces*, Vol. 140, pp. 531–537.

10 Cyclodextrin-UVR-Absorbing Molecular Complexes for Photodynamic Therapy

K Ganesh Kadiyala and Kanakadurga Bandaru

10.1 INTRODUCTION

The non-invasive treatment method known as photodynamic therapy (PDT) has gained considerable interest in recent years because of its distinctive mode of action and its ability to specifically target damaged tissues while preserving healthy cells [1]. PDT is the application of a photosensitizer, which is a light-absorbing polymer that, when activated by a particular wavelength of light, produces reactive oxygen species (ROS) capable of causing cell death. The proposed therapeutic approach has demonstrated efficacy in the treatment of several pathological ailments, encompassing malignancies, infectious diseases, age-related macular degeneration, psoriasis, and autoimmune disorders. Notwithstanding its encouraging therapeutic advantages, PDT has certain obstacles that restrict its wider use, including the inadequate solubility, limited bioavailability, and possible systemic toxicity of photosensitizers [2]. In order to address these constraints, current studies have concentrated on the advancement of innovative delivery systems and molecular complexes to enhance the safety, effectiveness, and specificity of PDT. Cyclodextrins (CDs) have been identified as a highly promising category of carriers for photosensitizers and UVR-absorbing compounds. They provide remarkable improvements in solubility, stability, and targeted administration.

CDs are cyclic oligosaccharides composed of α-(1,4)-linked glucopyranose units, widely used in pharmaceuticals for their ability to form inclusion complexes with hydrophobic guest molecules [3]. Their unique structure, with a hydrophobic core and hydrophilic outer surface, allows CDs to encapsulate hydrophobic drugs and photosensitizers, enhancing solubility and stability in water. In PDT, collidal dispersions [4] improve the pharmacokinetics, reduce toxicity, and enable targeted delivery of photosensitizers to affected tissues. They also protect photosensitizers from degradation due to light and oxygen, enhancing their therapeutic efficacy in PDT applications.

DOI: 10.1201/9781003492757-10

Integrating UVR-absorbing molecules into cyclodextrin complexes improves PDT by protecting photosensitizers from UVR-induced degradation, preserving their stability and effectiveness. Cyclodextrin encapsulation ensures targeted delivery to tissues while UVR absorbers can be customized to absorb wavelengths distinct from the photosensitizer's activation wavelength, enabling precise control. This approach enhances PDT's specificity, minimizing side effects and damage to healthy tissues, ultimately improving the safety and efficacy of PDT for various therapeutic applications.

The integration of CDs with UVR-absorbing compounds offers numerous advantages for PDT. CDs enhance the solubility and bioavailability of hydrophobic photosensitizers, facilitating their incorporation into aqueous formulations suitable for intravenous delivery. This is particularly important for poorly water-soluble photosensitizers, which may aggregate, reducing bioavailability and therapeutic efficacy. CDs also improve the photophysical properties of photosensitizers by modifying their microenvironment, enhancing light absorption and ROS production. Additionally, CDs provide thermal and enzymatic stability, protecting photosensitizers from metabolic degradation.

Incorporating UVR-absorbing molecules into CD complexes allows multi-wavelength PDT, activating photosensitizers with different light wavelengths for sequential or simultaneous treatment of multiple targets. Early studies reveal promising results, showing that CD complexes enhance photosensitizer solubility, stability, and photodynamic activity while protecting against degradation. This approach also improves activation precision, minimizing unintended side effects and addressing limitations of conventional photosensitizer formulations.

Cyclodextrin-UVR-absorbing molecular complexes hold significant promise for PDT beyond cancer treatment, with potential applications in dermatology for conditions like psoriasis, acne, and infections. These complexes enhance photosensitizer transport, stability, and efficacy, while UVR-absorbing molecules protect against premature degradation and enable precise activation. However, several challenges remain, such as optimizing complex designs to improve therapeutic effectiveness while minimizing toxicity. A thorough understanding of cyclodextrin, UVR-absorbing compounds, and photosensitizer interactions is crucial to enhance their stability, solubility, and bioavailability.

10.2 UVR ABSORBERS

UVR absorbers are compounds that absorb ultraviolet light in UV-A (320–400 nm), UV-B (280–320 nm), and UV-C (200–280 nm) regions, offering protection against UV-induced damage (Figure 10.1).

Common UVR absorbers include benzophenones, which absorb both UV-A and UV-B light, commonly found in sunscreens; cinnamates and salicylates, which absorb UV-B light and are used in cosmetics and skincare; oxybenzone and octocrylene, which are broad-spectrum UV filters used in sunscreens; and avobenzone, which absorbs UV-A light and is often combined with other UV filters for comprehensive protection against harmful UV radiation.

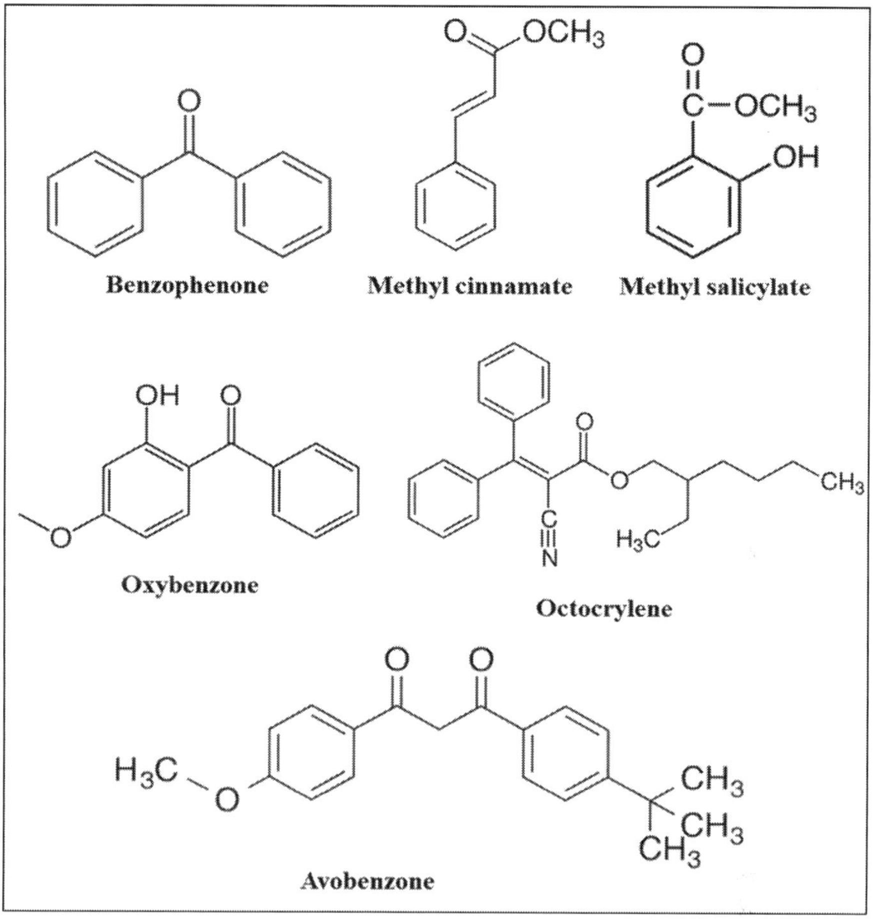

FIGURE 10.1 Structure of UV ray absorbers.

These UVR absorbers can be sensitive to environmental conditions such as light, heat, and oxygen, leading to their degradation and reduced effectiveness. Encapsulation within cyclodextrins can provide a protective environment for these molecules, enhancing their stability and effectiveness (Table 10.1).

10.3 CYCLODEXTRIN COMPLEXES WITH UVR ABSORBERS: ENHANCING STABILITY AND EFFICACY

Incorporating UVR absorbers into cyclodextrin complexes enhances their stability, solubility, and efficacy in photoprotection and photodynamic therapy. Cyclodextrins protect UVR absorbers from environmental degradation, including light-induced photodegradation, by encapsulating them in their hydrophobic cavities. This shields the UVR absorbers from direct contact with light and oxygen, preserving

TABLE 10.1
Applications of Various Photosensitizers in Different Fields

Photosensitizer	Trademark	$\lambda_{exc.}$ (nm)	Application
Porfimer sodium	Photofrin	632	Bladder, esophogeal, lung, cervical, endobronchial, and gastric cancers
5-ALA	Levulan/Ameluz	635	Actinic keratosis
MAL	Metvix/Metvixia	570–670	Actinic keratosis and basal cell carcinoma
HAL	Cysview/Hexvix	635	Colon and bladder diagnosis cancer
mTHPC	Foscan	652	Head and neck cancer
NPe6	Laserphyrin	664	Lung cancer
Synthetic hypericin	SGX301 [8]	570–650	Cutaneous T-cell lymphoma
LUZ11	Redaporfin [8]	749	Biliary tract cancer
AlPcS4	Photosens	675	Stomach, skin, lips, oral cavity, tongue, and breast cancers
Pd-Bpheid	Tookad	538/762	Prostate cancer

their effectiveness. Cyclodextrins also improve the solubility of hydrophobic compounds, making them suitable for water-based formulations like creams and lotions. Furthermore, these complexes offer controllable release, providing prolonged UV protection without frequent reapplication, ensuring consistent and effective photoprotection.

Cyclodextrin-UVR absorber complexes improve the solubility and effectiveness of UV filters. In order to enhance their solubility in aqueous solutions and increase their photostability, cyclodextrins, namely hydroxypropylated-γ-cyclodextrin (HP-γ-CD) and randomly methylated β-cyclodextrin, form inclusion complexes with UV absorbers [5, 6]. The synthesis of these complexes is distinguished by particular stoichiometries, such as 1:2 for ferulic acid and HP-γ-CD, which enhances the efficiency of UV protection delivery.

Furthermore, the interaction between cyclodextrins and guest molecules results in notable modifications in UV-vis absorption characteristics, including shifts towards blue wavelengths in absorption peaks. This suggests modified electronic conditions and decreased blockage in the guest molecules [7]. The findings indicate that cyclodextrins have the ability to increase the solubility and alter the photophysical characteristics of UV absorbers, which may result in enhanced UV protection in cosmetic formulations. In general, cyclodextrin complexes function as efficient carriers that enhance the efficiency of UV filters in diverse applications.

Srinivasan et al. [8] studied the UV absorption bands of 2,4-dinitrophenol (2,4-DNP), 2,4-dinitroaniline (2,4-DNA), 2,6-dinitroaniline (2,6-DNA), and 2,6-dinitrobenzoic acid (2,6-DNB), which showed broad absorption in the UV region below 400 nm. The inclusion complexes formed between these compounds and β-cyclodextrin (β-CD) enhanced absorption intensities, improving UV protection. FT-IR and ^{1}H NMR spectroscopy characterized the complexes, and SPF calculations confirmed their UV protective properties, ensuring the stability of ballpoint pen ink against photoinduced degradation.

Cellulose-polysaccharide (CD-PS) systems have been developed through three distinct binding modes: non-covalent interactions (CD-PS inclusion complexes), covalent bonding (CD-PS conjugates), and non-specific external interactions (CD-PS nanoassemblies). Porphyrinoid photosystems (PSs) have been shown to have several beneficial effects when included in CD-PS complexes. These effects include (1) enhancing the efficiency of photodynamic therapy (PDT) compared to free PSs [9–11], (2) mitigating certain limitations of PSs such as their limited water solubility [12, 13], and tendency to self-aggregate [14]; (3) reducing cytotoxicity in some cases; (4) promoting improved or different cellular uptake [15], even in spheroids; modifying the intracellular distribution; and (6) promoting faster diffusion of PSs into biological media. Comparable results have also been observed with non-porphyrinoid PSs.

In addition, various CD–PS conjugates and nanoassemblies have proven effective in encapsulating (1) chemotherapeutic drugs, facilitating a synergistic interaction between photodynamic therapy (PDT) and chemotherapy [16–21], or (2) NO photodonor, which enhances the PDT effect by achieving simultaneous photo release of two cytotoxic species ($^1O_2$ and $NO^\bullet$) [22, 23, 24].

Besides, the integration of CD–PS systems with NPs results in passive-targeting PDT due to the enhanced photoinduced extinction (EPR) action of NPs. The enhanced penetration of NIR excitation-type nanoparticles (UCNPs) into biological tissues [24–27], allows for deeper tumor penetration. Additionally, AuNPs enable the simultaneous performance of photothermal therapy (PTT) and PDT [28]. With the objective of enhancing the selectivity of all these CD–PS systems, the researchers anticipate the emergence of novel CD–PS systems integrated with active targeting agents in the near future. Limited literature currently documents the process of attaching active targeting agents onto CD-PS systems. Notably, there are reports on the grafting of various targeting molecules, including the cell-penetrating peptide R6H4 [29], the membrane-permeable and mitochondria-targeting peptide RLA, CD44-targeting hyaluronic acid [30], folate-targeting folic acid [31], and asialoglycoprotein-targeting sugars such as galactose [32] and lactobionic acid [33].

10.4 NANOPARTICLE-BASED CYCLODEXTRIN COMPLEXES FOR PDT

Recent advances in nanotechnology have further expanded the potential applications of cyclodextrin-UVR-absorbing molecular complexes in PDT [34]. Nanoparticle-based cyclodextrin complexes offer several advantages over traditional formulations, including enhanced stability, controlled release, and targeted delivery of photosensitizers [35]. By incorporating cyclodextrins into nanoparticles, researchers can create multifunctional platforms that combine the benefits of CDs with the unique properties of nanoparticles, such as EPR effect, which facilitates the accumulation of the therapeutic agents in tumour tissues [36].

Several studies have explored the use of cyclodextrin-based nanoparticles for PDT. For example, researchers have developed beta-cyclodextrin-conjugated gold nanoparticles loaded with photosensitizers for enhanced photodynamic therapy of cancer [37].

To date, a substantial amount of research has focused on exploring single-component nanoparticles for cancer detection and treatment [38, 39]. The use of single-component nanoparticles, whether metallic or non-metallic, can be traced back to ancient Roman times in the fourth century AD and has been extensively studied due to their versatile applications. Gold nanoparticles (Au NPs) have attracted significant interest due to their distinctive optical characteristics and strong biocompatibility, making them well-suited for biomedical applications [40]. These nanoparticles have demonstrated their utility not only in diagnostics but also in therapeutic applications such as photodynamic and photothermal therapy [41], drug and gene delivery [42], and bioimaging [43]. Additionally, magnetic nanoparticles (MNPs) have attracted considerable attention in biomedical applications, including cell therapy [44], magnetic resonance imaging (MRI) [45], and targeted drug delivery [46], due to their unique properties, excellent biocompatibility, and low toxicity. Magnetite (Fe_3O_4) and maghemite (γ-Fe_2O_3) are among the most commonly studied magnetic nanoparticles. Gold and magnetic nanoparticles are often used in combination with coatings or surface modifications to enhance their efficiency and performance [47–49]. Janus nanoparticles (JNPs), named after the Roman god Janus, who is depicted with two faces, offer a novel alternative to single-component nanoparticles. Unlike single-component nanoparticles, JNPs overcome the limitation of having only one set of properties by combining two distinct types of characteristics, making them particularly intriguing in current research.

This study describes the effective synthesis of lipid-based JNPs with both a hydrophilic and a hydrophobic side by the Benrabah group [50]. Using a novel nanocarrier, these nanoparticles show potential as vehicles for delivering active compounds with different solubilities. Furthermore, Yang et al. [51] produced Janus nanoparticles composed of polyvinylpyrrolidone-ciprofloxacin/ethyl cellulose-silver nanoparticles (PVP-CIP/EC-AgNPs). These nanoparticles demonstrated exceptional bactericidal properties against both Gram-positive *S. aureus* and Gram-negative E. *coli*, making them a highly promising material for the development of efficient wound dressings. PLGA/AcDex JNPs with pH-responsive bicompartmental structure were synthesised by the Gregory group [52], enabling autonomous drug loading and release. The aforementioned results illustrate the considerable potential for application of JNPs in several biomedical domains, by utilizing their capacity to capture several characteristics inside a single nanoparticle.

Chen et al. [53] demonstrated a significant photothermal effect in their synthesis of multifunctional branching PEG-Au–Fe_3O_4 JNPs. Additionally, Reguera et al. [54] developed Au–Fe_3O_4 JNPs with potential applications in X-ray computed tomography (CT), T_2-weighted nuclear magnetic resonance imaging (NMRI), photoacoustic imaging (PAI), and surface-enhanced Raman spectroscopy (SERS). Moreover, Abedin et al. [55] successfully synthesized polymer-coated Au–Fe_3O_4 superparamagnetic Janus-like nanoparticles, confirming the viability of this material for cancer diagnostics and photothermal ablation.

β-cyclodextrin (β-CD)-conjugated JNPs combining gold and iron oxide have been developed for magneto-plasmonic properties and loaded with 5-FU (anticancer) and IBU (anti-inflammatory). These nanoparticles offer optimal size, biocompatibility, and safety, enabling effective photothermal and photodynamic therapy in cancer treatment. The successful β-CD conjugation facilitates drug incorporation, enhancing pharmacological properties. Laboratory and animal studies confirm that β-CD-conjugated JNPs eliminate cancer cells through synergistic effects. With a high photothermal conversion efficiency of 32.88% [37], these nanostructures integrate chemotherapy, photothermal, and photodynamic therapies, improving efficacy and safety while ensuring high drug loading.

10.5 CYCLODEXTRIN-BASED HYBRID MATERIALS FOR CONTROLLED RELEASE AND ACTIVATION IN PDT

Researchers have also explored hybrid materials that combine cyclodextrins with other polymers or materials to create controlled-release systems for photosensitizers. These systems can provide a more controlled activation of photosensitizers, enhancing the specificity and reducing the side effects of PDT.

Dong et al. [56] developed dual fluorescent CpG nanoparticle-loaded IR820-hydrogel (IR820-α-CD) for synergistic photothermal radiation immunotherapy. This system combines photothermal therapy and immunotherapy to enhance tumour antigen release, improving immune activation and therapeutic efficacy. The CpG nanoparticles increase loading capacity and immunological activation, eliminating the need for additional carriers. The IR820-hydrogel also provides imaging guidance. In a separate study, Semeraro et al. incorporated Chlorophyll a (Chl a), a natural photosensitizer, into modified cyclodextrins (CDs) like 2-HP-β-CD, enhancing solubility and stability for photodynamic therapy. These complexes dissolved Chl a effectively in cell culture medium.

Subsequently, the cytotoxicity and intracellular localization of the Chl a/2-HP-β-CD and Chl a/2-HP-γ-CD complexes were quantitatively assessed in vitro using the HT-29 human colorectal cancer cell line. Further investigations focused specifically on the Chl a/2-HP-β-CD system, evaluating phototoxicity, ROS production, intracellular distribution, and mechanisms of cell death. The results revealed no dark toxicity and significant phototoxicity towards HT-29 cells, leading to cell death through a necrotic mechanism. Thus, the Chl a/2-HP-β-CD supramolecular complex shows great promise as a formulation for photodynamic therapy (Figure 10.2) [57].

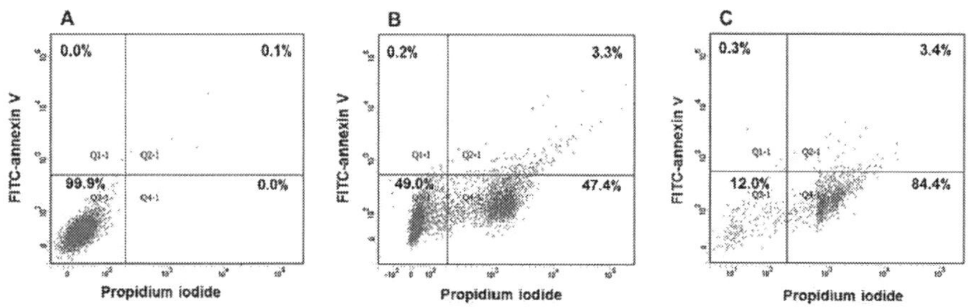

FIGURE 10.2 (A) Cells incubated with Chl a/2-HP-β-CD for 4 h and analyzed before irradiation. (B) Cells incubated with Chl a/2-HP-β-CD for 4 h and analyzed after irradiation. (C) Cells incubated with Chl a/2-HP-β-CD for 24 h and analyzed after irradiation [57].

Significant research has focused on improving the water solubility of fullerenes for biological applications. Ikeda et al. demonstrated the formation of water-soluble fullerene-liposome systems, while Altaf et al. [58] synthesized C_{60}-HSA nanoparticles by transferring C_{60} moieties from C_{60}/HP-β-CD nanoparticles to human serum albumin (HSA). The resulting C_{60}-HSA nanoparticles showed excellent dispersion stability and enhanced ROS scavenging under visible light. Both C_{60}-HSA and C_{60}-HP-β-CD nanoparticles exhibited similar photodynamic therapy (PDT) effects on A549 cells, with HSA alone showing minimal activity. These findings highlight the crucial role of β-CD in developing effective PDT drug delivery systems.

Experimental studies were performed on C_{60}/HP-β-CD NPs both in vitro and in vivo to evaluate their photosensitizing activity for photodynamic therapy [59]. Within those nanoparticles, C_{60} partially filled the hydrophobic cavity of the carbon dots. Upon exposure to visible light, this system produced a significant amount of $^1O_2$ ($\phi = 0.96$ in benzene) that surpassed the amount generated by P_pIX ($\phi = 0.56$ in PBS). In addition, the C_{60} component of the nanoparticles also generated O_2. Both forms of ROS were generated in a dose-dependent manner, continuously and proportionally increasing as the irradiation time and light intensity rose. Under visible light illumination, C_{60}-HP-β-CD NPs exhibited significant photoinduced toxicity towards HeLa and A549 cells in vitro, with IC_{50} values of 10 M and 60 M, respectively.

The enhanced efficacy of C_{60}-HP-β-CD nanoparticles (NPs) as a PDT photosensitizer compared to 5-ALA can be attributed to their greater generation of ROS. Additionally, the authors investigated the optimal parameters for effective in vivo PDT. Administering C_{60}-HP-β-CD NPs (2 mg/kg of C_{60}) along with 12 cycles of visible light exposure (λ = 400–700 nm, 350 mW/cm²) for a brief duration of 15 seconds, resulting in a total light dose of 63 J/cm², effectively inhibited the proliferation of S-180 sarcoma cells in ddY mice. Notably, this suppression occurred without causing any skin damage or significant hyperthermia.

To address the issue of water solubility in fullerene C60 and enhance its phototoxicity, Zhang et al. utilized a CD-functionalization strategy to modify C60. Instead of unmodified γ-CD, they employed a β-CD polymer (β-CD-P) to form the inclusion complex (C60-γ-CD-P). This approach involved the hydrophobic cavities of γ-CD-P being occupied by monomeric C_{60}. Similar to β-CD and γ-CD, γ-CD-P effectively solubilized C_{60} in water, and the absorption spectra of the inclusion complex displayed the characteristic peaks of C_{60}, albeit with a red shift compared to those of C_{60}-γ-CD-P. It appears that γ-CD-P has a superior capacity for forming inclusion complexes compared to β-CD. Furthermore, the C_{60} atom within the C_{60}-γ-CD-P complex remained well-dispersed in water without aggregating, even at high concentrations. This stability can be attributed to the supramolecular interactions with γ-CD-P, giving it an advantage over previous C60 inclusion complexes with CDs.

The C_{60}-γ-CD-P complex demonstrated high efficiency in producing singlet oxygen ($^1O_2$) exclusively under UVA light, while C_{60}-γ-CD could generate $^1O_2$ under both UVA and visible light. In vitro studies using the MTT assay with B16-F10 murine melanoma cell lines showed that γ-CD-P enhanced the biocompatibility of C_{60}, as it did not exert harmful effects on cells exposed to various concentrations of C_{60}-γ-CD-P ($[C_{60}]$ = 0.5–20 μM) after 48 hours of incubation in the dark. The non-aggregated structure of C60 encapsulated in γ-CD-P, combined with the substantial generation of $^{1}O_2$, resulted in C_{60}-γ-CD-P exhibiting a stronger phototoxic effect on tumor cells when exposed to UVA radiation compared to C60 alone.

The application of cyclodextrin-UVR absorber complexes in the food industry is a promising area that enhances food safety, stability, and sensory properties. CDs form inclusion complexes that can encapsulate UV absorbers, protecting food from degradation caused by ultraviolet radiation.

10.6 APPLICATION OF CYCLODEXTRIN-UVR ABSORBER COMPLEXES IN THE FOOD INDUSTRY

CDs offer multiple benefits in food applications by enhancing stability, sensory properties, and safety. They encapsulate UV absorbers, protecting sensitive food components from UV degradation and extending shelf life. CDs also improve sensory properties by masking undesirable flavours and preserving the visual appeal of food by protecting natural pigments. Furthermore, they enhance food safety by selectively capturing target contaminants, ensuring safer products. Despite their advantages, challenges remain in optimizing their use in diverse food matrices and ensuring regulatory compliance. Further research is needed to fully explore the potential of cyclodextrin-UVR absorber complexes in the food industry.

10.7 CONCLUSIONS

Cyclodextrins have emerged as versatile carriers in PDT due to their ability to form inclusion complexes with UV-absorbing molecules, such as photosensitizers. These macrocyclic compounds enhance the solubility, stability, and bioavailability of photosensitizers, which are often hydrophobic and prone to degradation. By encapsulating the photosensitizer within the cyclodextrin cavity, these complexes provide a protective environment that prevents premature degradation under UV radiation, allowing more efficient activation by light at the targeted site.

The formation of cyclodextrin-UVR absorbing molecular complexes enhances the selective delivery of photosensitizers to cancer cells or other diseased tissues, reducing damage to surrounding healthy cells. Additionally, cyclodextrins can modulate the release rate of the photosensitizer, improving its therapeutic window. This controlled release minimizes systemic toxicity and enhances the precision of PDT.

In conclusion, cyclodextrin-based UVR-absorbing molecular complexes represent a significant advancement in the field of photodynamic therapy. Their ability to improve photosensitizer properties, reduce side effects, and enhance treatment efficacy positions them as promising tools in PDT. Continued research into these complexes could lead to more refined therapies for cancer and other diseases requiring localized treatment with minimal collateral damage.

REFERENCES

1. Alvarez, N., and Sevilla, A. (2024b) 'Current advances in photodynamic therapy (PDT) and the future potential of PDT-Combinatorial cancer therapies' *Int. J. Mol. Sci.*, Vol 25, pp 1023. https://doi.org/10.3390/ijms25021023
2. Sun, W., Luo, L., Feng, Y., Qiu, Y., Shi, C., Meng, S., Chen, X., and Chen, H. (2020) 'Gadolinium–Rose Bengal coordination polymer nanodots for MR-/Fluorescence-Image-Guided radiation and photodynamic therapy' *Adv. Mater.*, Vol 32, pp e2000377. doi: 10.1002/adma.202000377
3. Cheirsilp, B., and Rakmai, J. (2017) 'Inclusion complex formation of cyclodextrin with its guest and their applications' *Biol. Eng. Med.*, Vol 2, pp 1–6. https://doi.org/10.15761/bem.1000108
4. Bruchez, M., Moronne, M., Gin, P., Weiss, S., and Alivisatos, A. P. (1998) 'Semiconductor nanocrystals as fluorescent biological labels' *Science*, Vol 281, pp 2013–2016. https://doi.org/10.1126/science.281.5385.2013
5. Mori, T., Tsuchiya, R., Doi, M., Nagatani, N., and Tanaka, T. (2018) 'Solubilization of ultraviolet absorbers by cyclodextrin and their potential application in cosmetics' *J. Incl. Phenom. Macrocycl. Chem.*, Vol 93, pp 91–96. https://doi.org/10.1007/s10847-018-0846-5
6. Fenyvesi, É., Otta, K., Kolbe, I., Novák, C., and Szejtli, J. (2004) 'Cyclodextrin complexes of UV filters' *J. Incl. Phenom. Mol. Recognit. Chem.*, pp 117–123. https://doi.org/10.1023/b:jiph.0000022518.38097.55
7. Jiang, H., Xu, Y., Na, L., Jin, R., and Zhang, S. (2008) 'UV-vis Spectral Analysis of Inclusion Complexes between β-Cyclodextrin and Aromatic/Aliphatic Guest Molecules' *Curr. Drug Discov. Technol.*, Vol 5, pp 173–176. https://doi.org/10.2174/157016308784746283
8. Srinivasan, K., Radhakrishnan, S., and Stalin, T. (2014) 'Inclusion complexes of β-cyclodextrin-dinitrocompounds as UV absorber for ballpoint pen ink' *Spectrochim. Acta A Mol. Biomol. Spectrosc.*, Vol 129, pp 551–564. https://doi.org/10.1016/j.saa.2014.03.021
9. Kolárová, H., Mosinger, J., Lenobel, R., Kejlová, K., JíRová, D., and Strnad, M. (2003) 'In vitro toxicity testing of supramolecular sensitizers for photodynamic therapy' *Toxicol. Vitro*, Vol 17, pp 775–778. https://doi.org/10.1016/s0887-2333(03)00094-8
10. Kolarova, H., Macecek, J., Nevrelova, P., Huf, M., Tomecka, M., Bajgar, R., Mosinger, J., and Strnad, M. (2005) 'Photodynamic therapy with zinc-tetra(p-sulfophenyl) porphyrin bound to cyclodextrin induces single strand breaks of cellular DNA in G361 melanoma cells' *Toxicol. Vitro*, Vol 19, pp 971–974. https://doi.org/10.1016/j.tiv.2005.06.015
11. Paul, S., Heng, P. W. S., and Chan, L. W. (2016) 'pH-dependent complexation of hydroxypropyl-beta-cyclodextrin with chlorin e6: Effect on solubility and aggregation in relation to photodynamic efficacy' *JPP*, Vol 68, pp 439–449. https://doi.org/10.1111/jphp.12535
12. Lu, S., Ma, J. Y., Xuan, H. Y., Wang, A., Zhao, B., Li, X. D., Zhou, J. H., Lin, Y., Zhou, L., and Wei, S. H. (2014) 'A novel assembling complex of hydrobobic phthalocyanine-cyclodextrin: Preparation, characterization, molecular modeling, and in vitro activity' *RSC Adv.*, Vol 4, pp 59759–59763. https://doi.org/10.1039/c4ra12654h
13. Lu, S., Wang, A., Ma, Y., Xuan, H., Zhao, B., Li, X., Zhou, J., Zhou, L., and Wei, S. (2016) 'Cyclodextrin type dependent host-guest interaction mode with phthalocyanine

and their influence on photodynamic activity to cancer' *Carbohydr. Polym.*, Vol 148, pp 236–242. https://doi.org/10.1016/j.carbpol.2016.04.062

14. Yankovsky, I., Bastien, E., Yakavets, I., Khludeyev, I., Lassalle, H., Gräfe, S., Bezdetnaya, L., and Zorin, V. (2016) 'Inclusion complexation with β-cyclodextrin derivatives alters photodynamic activity and biodistribution of meta-tetra(hydroxyphenyl)chlorin' *Eur. J. Pharm. Sci.*, Vol 91, pp 172–182. https://doi.org/10.1016/j.ejps.2016.06.012
15. Ikeda, A., Satake, S., Mae, T., Ueda, M., Sugikawa, K., Shigeto, H., Funabashi, H., and Kuroda, A. (2017) 'Photodynamic activities of porphyrin derivative–Cyclodextrin complexes by photoirradiation' *ACS Med. Chem. Lett.*, Vol 8, pp 555–559. https://doi.org/10.1021/acsmedchemlett.7b00098
16. Yakavets, I., Yankovsky, I., Millard, M., Lamy, L., Lassalle, H., Wiehe, A., Zorin, V., and Bezdetnaya, L. (2017) 'The alteration of temoporfin distribution in multicellular tumor spheroids by β-cyclodextrins' *Int. J. Pharm.*, Vol 529, pp 568–575. https://doi.org/10.1016/j.ijpharm.2017.07.037
17. Kejík, Z., Bříza, T., Králová, J., Poučková, P., Král, A., Martásek, P., and Král, V. (2011) 'Coordination conjugates of therapeutic proteins with drug carriers: A new approach for versatile advanced drug delivery' *Bioorg. Med. Chem. Lett.*, Vol 21, pp 5514–5520. https://doi.org/10.1016/j.bmcl.2011.06.101
18. Aggelidou, C., Theodossiou, T. A., and Yannakopoulou, K. (2013) 'Protoporphyrin IX–Β-Cyclodextrin bimodal conjugate: Nanosized drug transporter and potent phototoxin' *J. Photochem. Photobiol.*, Vol 89, pp 1011–1019. https://doi.org/10.1111/php.12127
19. Conte, C., Scala, A., Siracusano, G., Sortino, G., Pennisi, R., Piperno, A., Miro, A., Ungaro, F., Sciortino, M. T., Quaglia, F., and Mazzaglia, A. (2016) 'Nanoassemblies based on non-ionic amphiphilic cyclodextrin hosting Zn(II)-phthalocyanine and docetaxel: Design, physicochemical properties and intracellular effects' *Colloids Surf. B: Biointerfaces*, Vol 146, pp 590–597. https://doi.org/10.1016/j.colsurfb.2016.06.047
20. Zhang, W., Li, Y., Sun, J., Tan, C., Ji, L., and Mao, Z. (2015) 'Supramolecular self-assembled nanoparticles for chemo-photodynamic dual therapy against cisplatin resistant cancer cells' *Chem Comm.*, Vol 51, pp 1807–1810. https://doi.org/10.1039/c4cc08583c
21. Xu, T., Li, J., Cheng, F., Zhang, Y., Cao, J., Gao, W., and He, B. (2017) 'Fabrication of a polypseudorotaxane nanoparticle with synergistic photodynamic and chemotherapy' *Chin. Chem. Lett.*, Vol 28, pp 1885–1888. https://doi.org/10.1016/j.cclet.2017.07.029
22. Lourenço, L. M. O., Pereira, P. M. R., Maciel, E., Válega, M., Domingues, F. M. J., Domingues, M. R. M., Neves, M. G. P. M. S., Cavaleiro, J. A. S., Fernandes, R., and Tomé, J. P. C. (2014) 'Amphiphilic phthalocyanine–cyclodextrin conjugates for cancer photodynamic therapy' *Chem Comm.*, Vol 50, pp 8363–8366. https://doi.org/10.1039/c4cc02226b
23. Fraix, A., Gref, R., and Sortino, S. (2014) 'A multi-photoresponsive supramolecular hydrogel with dual-color fluorescence and dual-modal photodynamic action' *J. Mater. Chem. B.*, Vol 2, pp 3443–3449. https://doi.org/10.1039/c4tb00257a
24. Tian, G., Ren, W., Yan, L., Jian, S., Gu, Z., Zhou, L., Jin, S., Yin, W., Li, S., and Zhao, Y. (2012) 'Red-Emitting upconverting nanoparticles for photodynamic therapy in cancer cells under Near-Infrared excitation' *Small*, Vol 9, pp 1929–1938. https://doi.org/10.1002/smll.201201437
25. Wang, A., Jin, W., Chen, E., Zhou, J., Zhou, L., and Wei, S. (2016) 'Drug delivery function of carboxymethyl-β-cyclodextrin modified upconversion nanoparticles for

adamantine phthalocyanine and their NIR-triggered cancer treatment' *Dalton Trans.*, Vol 45, pp 3853–3862. https://doi.org/10.1039/c5dt04900h

26. Wang, H., Han, R., Yang, L., Shi, J., Liu, Z., Hu, Y., Wang, Y., Liu, S., and Gan, Y. (2016) 'Design and synthesis of Core–Shell–Shell upconversion nanoparticles for NIR-Induced drug release, photodynamic therapy, and cell imaging' *ACS Appl. Mater.*, Vol 8(7), pp 4416–4423. https://doi.org/10.1021/acsami.5b11197
27. Han, R., Shi, J., Liu, Z., Wang, H., and Wang, Y. (2017) 'Fabrication of Mesoporous-Silica-Coated upconverting nanoparticles with ultrafast photosensitizer loading and 808 nm NIR-Light-Triggering capability for photodynamic therapy' *Chem. Asian J.*, Vol 12, pp 2197–2201. https://doi.org/10.1002/asia.201700836
28. Trapani, M., Romeo, A., Parisi, T., Sciortino, M. T., Patanè, S., Villari, V., and Mazzaglia, A. (2013) 'Supramolecular hybrid assemblies based on gold nanoparticles, amphiphilic cyclodextrin and porphyrins with combined phototherapeutic action' *RSC Adv.*, Vol 3, pp 5607–5614. https://doi.org/10.1039/c3ra40204e
29. Tong, H., Wang, Y., Li, H., Jin, Q., and Ji, J. (2016) 'Dual pH-responsive 5-aminolevulinic acid pseudopolyrotaxane prodrug micelles for enhanced photodynamic therapy' *Chem Comm.*, Vol 52, pp 3966–3969. https://doi.org/10.1039/c6cc00450d
30. Chen, W., Luo, G., Qiu, W., Lei, Q., Liu, L., Wang, S., and Zhang, X. (2017) 'Mesoporous silica-based versatile theranostic nanoplatform constructed by layer-by-layer assembly for excellent photodynamic/chemo therapy' *Biomaterials*, Vol 117, pp 54–65. https://doi.org/10.1016/j.biomaterials.2016.11.057
31. Chai, S., Guo, Y., Zhang, Z., Chai, Z., Ma, Y., and Qi, L. (2017) 'Cyclodextrin-gated mesoporous silica nanoparticles as drug carriers for red light-induced drug release' *Nanotechnology*, Vol 28, pp 145101. https://doi.org/10.1088/1361-6528/aa5e74
32. Hu, X., Zang, Y., Li, J., Chen, G., James, T. D., He, X., and Tian, H. (2016) 'Targeted multimodal theranostics via biorecognition controlled aggregation of metallic nanoparticle composites' *Chemical Science*, Vol 7, pp 4004–4008. https://doi.org/10.1039/c6sc01463a
33. Luo, G., Chen, W., Lei, Q., Qiu, W., Liu, Y., Cheng, Y., and Zhang, X. (2016) 'A Triple-Collaborative strategy for High-Performance tumor therapy by multifunctional mesoporous Silica-Coated Gold nanorods' *Adv. Func. Mater.*, Vol 26, pp 4339–4350. https://doi.org/10.1002/adfm.201505175
34. Mihoub, A. B., Larue, L., Moussaron, A., Youssef, Z., Colombeau, L., Baros, F., Frochot, C., Vanderesse, R., and Acherar, S. (2018) 'Use of cyclodextrins in anticancer photodynamic therapy treatment' *Molecules*, Vol 23, pp 1936. https://doi.org/10.3390/molecules23081936
35. Gadade, D. D., and Pekamwar, S. S. (2020) 'Cyclodextrin based nanoparticles for drug delivery and theranostics' *Adv. Pharm. Bull.*, Vol 10, pp 166–183. https://doi.org/10.34172/apb.2020.022
36. Karthic, A., Roy, A., Lakkakula, J., Alghamdi, S., Shakoori, A., Babalghith, A. O., Emran, T. B., Sharma, R., Lima, C. M. G., Kim, B., Park, M. N., Safi, S. Z., De Almeida, R. S., and Coutinho, H. D. M. (2022) 'Cyclodextrin nanoparticles for diagnosis and potential cancer therapy: A systematic review' *Front. Cell Dev. Biol.*, Vol 10, pp 1–26. https://doi.org/10.3389/fcell.2022.984311
37. Park, S., Choi, J., Ko, N., Mondal, S., Pal, U., Lee, B., and Oh, J. (2024) 'Beta Cyclodextrin conjugated Au-Fe3O4 Janus nanoparticles with enhanced chemo-photothermal therapy performance' *Acta Biomater.*, Vol 182, pp 213–227. https://doi.org/10.1016/j.actbio.2024.05.008
38. Hu, A., Pu, Y., Xu, N., Yang, H., Hu, X., Sun, R., Jin, R., and Nie, Y. (2024) 'Hierarchically decorated magnetic nanoparticles amplify the oxidative stress and

promote the chemodynamic/magnetic hyperthermia/immune therapy' *Acta Biomater.*, Vol 173, pp 457–469. https://doi.org/10.1016/j.actbio.2023.11.023

39. Niu, S., Zhang, X., Williams, G. R., Wu, J., Gao, F., Fu, Z., Chen, X., Lu, S., and Zhu, L. (2021) 'Hollow mesoporous silica nanoparticles gated by Chitosan-Copper sulfide composites as theranostic agents for the treatment of breast cancer' *Acta Biomater.*, Vol 126, pp 408–420. https://doi.org/10.1016/j.actbio.2021.03.024
40. Elahi, N., Kamali, M., and Baghersad, M. H. (2018) 'Recent biomedical applications of gold nanoparticles: A review' *Talanta*, Vol 184, pp 537–556. https://doi.org/10.1016/j.talanta.2018.02.088
41. Wei, X., Chen, H., Tham, H. P., Zhang, N., Xing, P., Zhang, G., and Zhao, Y. (2018) 'Combined photodynamic and photothermal therapy using cross-linked polyphosphazene nanospheres decorated with gold nanoparticles' *ACS Appl. Nano Mater.*, Vol 1, pp 3663–3672. https://doi.org/10.1021/acsanm.8b00776
42. Xiong, Z., Alves, C. S., Wang, J., Li, A., Liu, J., Shen, M., Rodrigues, J., Tomás, H., and Shi, X. (2019) 'Zwitterion-functionalized dendrimer-entrapped gold nanoparticles for serum-enhanced gene delivery to inhibit cancer cell metastasis' *Acta Biomater.*, Vol 99, pp 320–329. https://doi.org/10.1016/j.actbio.2019.09.005
43. García-Álvarez, R., Chen, L., Nedilko, A., Sánchez-Iglesias, A., Rix, A., Lederle, W., Pathak, V., Lammers, T., Von Plessen, G., Kostarelos, K., Liz-Marzán, L. M., Kuehne, A. J. C., and Chigrin, D. N. (2020) 'Optimizing the geometry of photoacoustically active gold nanoparticles for biomedical imaging' *ACS Photonics*, Vol 7(3), pp 646–652. https://doi.org/10.1021/acsphotonics.9b01418
44. Kolosnjaj-Tabi, J., Wilhelm, C., Clément, O., and Gazeau, F. (2013) 'Cell labeling with magnetic nanoparticles: Opportunity for magnetic cell imaging and cell manipulation' *J. Nanobiotechnology*, Vol 11(S1). https://doi.org/10.1186/1477-3155-11-s1-s7
45. Wei, Y., Liao, R., Mahmood, A. A., Xu, H., and Zhou, Q. (2017) 'pH-responsive pHLIP (pH low insertion peptide) nanoclusters of superparamagnetic iron oxide nanoparticles as a tumor-selective MRI contrast agent' *Acta Biomater.*, Vol 55, pp 194–203. https://doi.org/10.1016/j.actbio.2017.03.046
46. Arruebo, M., Fernández-Pacheco, R., Ibarra, M. R., and Santamaría, J. (2007) 'Magnetic nanoparticles for drug delivery' *Nano Today*, Vol 2, pp 22–32. https://doi.org/10.1016/s1748-0132(07)70084-1
47. Lv, S., Yuan, X., Xiao, J., and Jiang, X. (2023) 'Hemostasis-osteogenesis integrated Janus carboxymethyl chitin/hydroxyapatite porous membrane for bone defect repair' *Carbohydr. Polym.*, Vol 313, pp 120888. https://doi.org/10.1016/j.carbpol.2023.120888
48. Tong, H., Wang, J., Qi, L., and Gao, Q. (2023) 'Starch-based Janus particle: Fabrication, characterization and interfacial properties in stabilizing Pickering emulsion' *Carbohydr. Poly.*, Vol 313, pp 120867. https://doi.org/10.1016/j.carbpol.2023.120867
49. Qian, S., Zhao, B., Mao, J., Liu, Z., Zhao, Q., Lu, B., Mao, X., Zhang, L., Cheng, L., Zhang, Y., Cui, W., and Sun, X. (2022) 'Biomedical applications of Janus membrane' *Biomed. Technol.*, Vol 2, pp 58–69. https://doi.org/10.1016/j.bmt.2022.11.003
50. Benrabah, L., Kemel, K., Twarog, C., Huang, N., Solgadi, A., Laugel, C., and Faivre, V. (2020) 'Lipid-based Janus nanoparticles for pharmaceutical and cosmetic applications: Kinetics and mechanisms of destabilization with time and temperature' *Colloids Surf. B: Biointerfaces*, Vol 195, pp 111242. https://doi.org/10.1016/j.colsurfb.2020.111242
51. Yang, J., Wang, K., Yu, D., Yang, Y., Bligh, S. W. A., and Williams, G. R. (2020) 'Electrospun Janus nanofibers loaded with a drug and inorganic nanoparticles as an effective antibacterial wound dressing' *MSEC*, Vol 111, pp 110805. https://doi.org/10.1016/j.msec.2020.110805

52. Gregory, J. V., Vogus, D. R., Barajas, A., Cadena, M. A., Mitragotri, S., and Lahann, J. (2020) 'Programmable delivery of synergistic cancer drug combinations using bicompartmental nanoparticles' *Adv. Healthcare Mater.*, Vol 9. https://doi.org/10.1002/adhm.202000564
53. Chen, X., Li, G., Han, Q., Li, X., Li, L., Wang, T., and Wang, C. (2017) 'Rational design of branched AU–Fe3O4 Janus nanoparticles for simultaneous trimodal imaging and photothermal therapy of cancer cells' *Chem. Eur. J.*, Vol 23, pp 17204–17208. https://doi.org/10.1002/chem.201704514
54. Reguera, J., De Aberasturi, D. J., Henriksen-Lacey, M., Langer, J., Espinosa, A., Szczupak, B., Wilhelm, C., and Liz-Marzán, L. M. (2017) 'Janus plasmonic–magnetic gold–iron oxide nanoparticles as contrast agents for multimodal imaging' *Nanoscale*, Vol 9, pp 9467–9480. https://doi.org/10.1039/c7nr01406f
55. Abedin, M. R., Umapathi, S., Mahendrakar, H., Laemthong, T., Coleman, H., Muchangi, D., Santra, S., Nath, M., and Barua, S. (2018) 'Polymer coated gold-ferric oxide superparamagnetic nanoparticles for theranostic applications' *J. Nanobiotechnology*, Vol 16. https://doi.org/10.1186/s12951-018-0405-7
56. Dong, X., Liang, J., Yang, A., Qian, Z., Kong, D., and Lv, F. (2019) 'Fluorescence imaging guided CpG nanoparticles-loaded IR820-hydrogel for synergistic photothermal immunotherapy' *Biomaterials*, Vol 209, pp 111–125. https://doi.org/10.1016/j.biomaterials.2019.04.024
57. Semeraro, P., Chimienti, G., Altamura, E., Fini, P., Rizzi, V., and Cosma, P. (2018) 'Chlorophyll a in cyclodextrin supramolecular complexes as a natural photosensitizer for photodynamic therapy (PDT) applications' *MSEC*, Vol 85, pp 47–56. https://doi.org/10.1016/j.msec.2017.12.012
58. Abdulmalik, A., Hibah, A., Zainy, B. M., Makoto, A., Daisuke, I., Masaki, O., Kaneto, U., and Fumitoshi, H. (2013) 'Preparation of soluble stable C60/human serum albumin nanoparticles via cyclodextrin complexation and their reactive oxygen production characteristics' *Life Sci.*, Vol 93, pp 277–282. https://doi.org/10.1016/j.lfs.2013.06.021
59. Altaf, A., Aldawsari, H., Banjar, Z. M., Iohara, D., Anraku, M., Uekama, K., and Hirayama, F. (2014) 'Potential use ofC60/2-Hydroxypropyl-B-cyclodextrin nanoparticles as a new photosensitizer in the treatment of cancer' *Int. J. Photoenergy*, Vol 10, pp 1–8. https://doi.org/10.1155/2014/570506

11 Cyclodextrins

Revolutionizing Sunscreen Performance and Protection

Khan Aejaz Ahmed, Wan Nurhidayah Wan Hanaffi, Usama Ahmad, and Mohd Muazzam Khan

11.1 INTRODUCTION

The sun emits UVR, with UV-A (315–400 nm) reaching the Earth's surface the most, while the ozone layer partially absorbs UV-B (280–315 nm), and UVC (100–280 nm) is completely absorbed. UVR has biological impacts like vitamin D synthesis in humans, but overexposure can lead to harmful effects like skin damage, DNA mutations, and increased cancer risk [1]. Sunscreens are essential in mitigating these effects by blocking or absorbing harmful UVR. These are divided into organic sunscreens that absorb UV light and inorganic that deflect or disperse it. Sunscreen efficacy is measured by SPF, primarily reflecting UVB protection, with higher SPF values indicating better protection [2].

UVR absorbers are essential in sun protection, preventing harm from prolonged UV exposure. In sunscreens, they reflect, absorb, or disperse UV rays, minimizing skin radiation. Chemical filters like oxybenzone, ensulizole, and avobenzone block UVA and UVB rays. However, concerns exist about their potential systemic absorption and negative consequences. To address these issues, CDs are used as carriers for UVR absorbers, increasing their concentration and stability in the skin's top layers while reducing direct skin contact. This encapsulation technique preserves sun protection effectiveness and enhances UVR absorbers' photostability without endangering skin health. Combining CDs with UVR absorbers advances sun protection technology [3]. UVR absorbers enclosed in CDs have improved photostability, ensuring effective long-term UV protection. CD complexes with UVR absorbers provide a more consistent protective factor at the skin's surface and lower the risk of adverse side effects by preventing deeper skin penetration. Dispersed in natural waters, UVR absorbers must be environmentally friendly and meet regulatory standards for safety and efficacy to prevent harm to wildlife and the environment [4]. Figure 11.1 shows UV penetration without protection on the left and sunscreen blocking UV rays on the right.

DOI: 10.1201/9781003492757-11

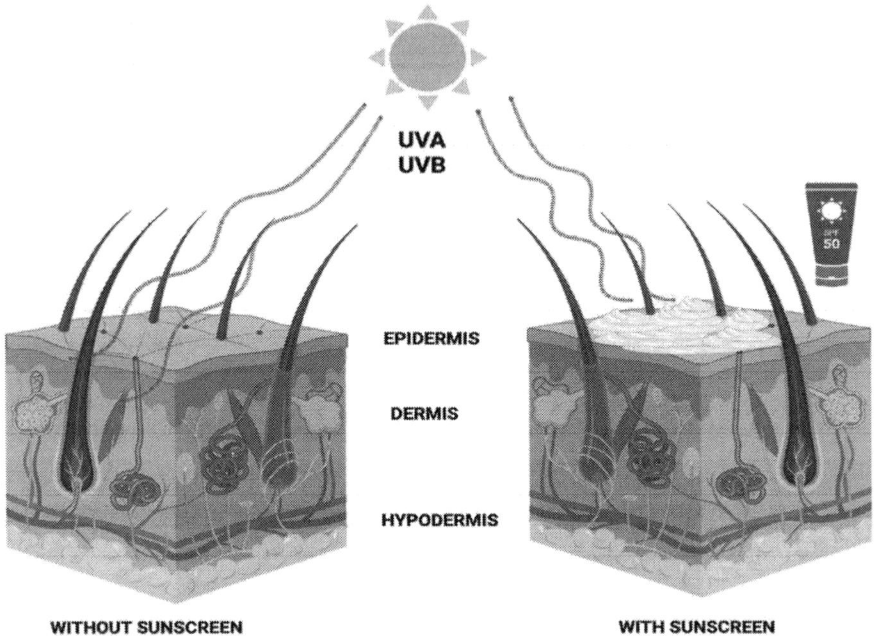

FIGURE 11.1 Impact of CD-based sunscreen on UVR penetration through skin layers.

11.2 UVR ABSORBERS: MECHANISM OF ACTION

Mechanism of action of various commonly used UVR absorbers is described in Table 11.1.

11.3 INTERACTION OF CDS AND UVR ABSORBERS

CDs form inclusion complexes with guest molecules through molecular recognition. Understanding CD-UVR absorber interaction is crucial for developing effective UV protection systems. CD inclusion complexes are formed when macrocyclic CDs encapsulate guest molecules within their hydrophobic inner cavity through non-covalent interactions, enhancing the solubility and stability of these molecules [22]. CDs are derived from starch, consisting of glucose monomers linked by α-(1-4) glycosidic bonds, resulting in a hydrophilic exterior and a hydrophobic interior that facilitates the encapsulation of lipophilic substances, improving their solubility and bioavailability. This property is crucial for enhancing drug permeability and absorption in biological systems [23]. The cavity size of CDs determines their ability to accommodate different molecules, with γ-cyclodextrin capable of binding larger compounds compared to α- and β-CD, influencing their selectivity in applications such as electrochemical sensors [24]. CDs exhibit exceptional molecular recognition abilities, distinguishing guest molecules based on size, polarity, and chirality, which is critical for applications in drug delivery and molecular separation [25]. Their chiral

TABLE 11.1
Mechanism of Action of Various Commonly Used UVR Absorbers

UV Filter	Mechanism of Action	Remarks
Avobenzone	Converts absorbed UVA energy into harmless heat. Neutralizes ROS.	84.6–95.5% of the absorbed UV filter was found in the stratum corneum, with only a small percentage reaching the epidermis [5]. Its antioxidant properties neutralize ROS, reducing their harmful effects on skin cells and DNA.
Ensulizole	Absorbs UVB and UVA rays. Reduces ROS.	Its absorption spectrum shows strong absorption at around 303 nm, blocking considerable harmful UVB radiation [6]. The PL spectrum shows a pronounced blue shift in polar solvents, moving from 445 nm in non-polar solvents to 412 nm in polar solvents like water and ethanol.
Padimate O	Absorbs UVB rays. Neutralizes ROS.	Upon exposure to UVB radiation, Padimate O absorbs the energy and undergoes a photochemical reaction, preventing UVR from penetrating the skin [7]. By reducing ROS levels, Padimate O helps prevent chain reactions leading to further cellular damage and inflammation.
Homosalate	Absorbs UVB and UVA rays. Converts UV energy into heat. Reduces ROS	HMS undergoes ultrafast excited state intramolecular proton transfer. This minimizes the formation of harmful photoproducts from UV exposure [8]. HMS has low dermal permeability and does not cause endocrine disruption, which are significant concerns for many existing UV filters.
Meradimate	Absorbs UVA rays. Reduces ROS.	Meradimate absorbs UVR, converting it into less harmful energy. The ultrafast dynamics of meradimate reveal complex energy behaviors influencing its protective capabilities [9].
Cinoxate	Absorbs UVB rays. Minimizes ROS formation.	Cinoxate effectively absorbs UVB radiation, reducing the risk of sunburn and skin damage [10]. Like nitroxides, cinoxate scavenges ROS generated by UV exposure, protecting skin cells from oxidative damage [11].
Aminobenzoic acid	Absorbs UVB rays. Reduces ROS	PABA effectively absorbs UVR, reducing its skin penetration. It can also photosensitize singlet oxygen formation, a reactive species implicated in oxidative stress and skin ageing [7].

(continued)

TABLE 11.1 (Continued)
Mechanism of Action of Various Commonly Used UVR Absorbers

UV Filter	Mechanism of Action	Remarks
Dioxybenzone	Absorbs UVB and UVA rays. Reduces ROS.	Dioxybenzone acid absorbs UVR, preventing DNA damage like thymine dimer formation, linked to skin cancer. Its antioxidant properties reduce oxidative stress, inhibiting apoptosis and matrix metalloproteinase expression associated with skin ageing [12].
Octocrylene	Absorbs UVB and UVA rays. Neutralizes free radicals.	Octocrylene's structure includes a chromophore that absorbs UVR, providing broad-spectrum protection. It undergoes ultrafast non-radiative relaxation after UV-B excitation, quickly returning to its ground state, enhancing its protective capabilities [13].
Oxybenzone	Absorbs UVB and UVA rays. Minimal antioxidant effect. Neutralizes some free radicals	Oxybenzone absorbs UVR across a broad spectrum, particularly in UVA and UVB ranges, crucial for its protective function [14]. The molecule undergoes ESIPT, facilitating energy dissipation through pathways, including internal conversion to its ground state [15].
Octinoxate	Absorbs UVB and UVA rays. Minimal antioxidant effect. Neutralizes some free radicals.	Octinoxate absorbs UVB radiation, converting harmful UV energy into less harmful heat [16]. It modulates AhR signalling, inhibiting CYP1A1 and CYP1B1, potentially affecting skin health and metabolism [17].
Sulisobenzone	Absorbs UVB and UVA rays. Minimal antioxidant effect. Neutralizes some free radicals.	Sulisobenzone effectively absorbs UVB and UVA radiation, preventing skin damage by converting harmful UV energy into heat [18].
Titanium dioxide	Reflects, scatters, and absorbs UVB and UVA rays. Stays on the skin surface.	TiO_2 is widely used in sunscreens as it absorbs and scatters UV radiation, preventing skin penetration. The absorption of UV radiation by TiO_2 excites electrons from the valence band to the conduction band, creating electron-hole pairs [19].
Zinc Oxide	Reflects and scatters UVA and UVB rays, providing broad-spectrum protection. Remains on the skin surface, minimizing absorption and irritation.	ZnO absorbs UVR and converts it into less harmful infrared radiation, dissipating away from the skin [20]. ZnO nanoparticles enhance the antioxidant defense system by increasing antioxidant enzyme activity such as superoxide dismutase and catalase [21].

recognition capability allows them to differentiate between enantiomers, enhancing their specificity in pharmaceutical applications [26]. Additionally, functionalized CDs with reactive groups can further improve molecular recognition and catalytic efficiency by facilitating tailored interactions [27].

11.4 IMPACT OF INCLUSION COMPLEXES OF UVR ABSORBERS WITH CDS

CDs have received much interest in sunscreen formulation due to their capacity to form inclusion complexes with UVR absorbers. These complexes have shown promising results in enhancing sunscreen efficiency and photostability (Figure 11.2). CDs have emerged as a key tool in optimizing sun protection solutions, improving overall SPF.

CDs form inclusion complexes with lipophilic UV filters, like oxybenzone, to increase solubility by up to 1049-fold, ensuring uniform distribution and better skin protection [28, 29]. This solubility improvement aids in skin penetration, enhances absorption, and reduces the risk of phototoxicity by controlling the release and minimizing percutaneous absorption. CDs also stabilize UV filters, protecting them from environmental degradation, extending sunscreen's effectiveness, and ensuring longer protection [30, 31].

CDs significantly boost the bioavailability and skin penetration of UVR absorbers by enhancing solubility and enabling gradual release. This controlled release helps maintain prolonged sun protection, reducing the need for frequent reapplications and protecting UV filters from photodegradation [32]. CDs also minimize potential side effects, such as phototoxicity, by reducing skin penetration of UV filters and

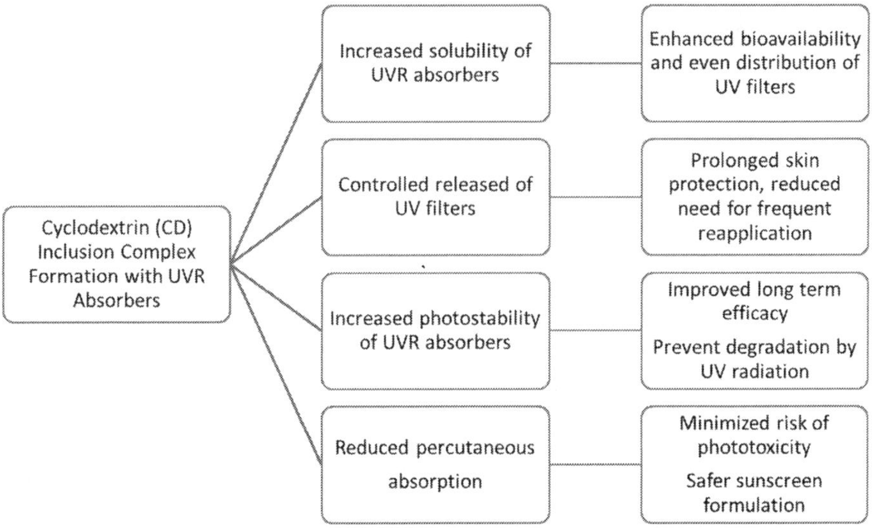

FIGURE 11.2 Benefits of cyclodextrin inclusion complex formation.

ensuring an even distribution of active ingredients, which guarantees consistent UV protection [33]. CDs improve the stability and efficacy of UVR absorbers, raising SPF values in sunscreen formulations. By forming inclusion complexes with UV filters, like avobenzone and ferulic acid, CDs reduce photodegradation, ensuring longer-lasting protection [32, 34]. Research has shown that CDs enhance SPF and UVA protection factors by improving solubility, stability, and retention of active ingredients, making sunscreens more effective. Additionally, CDs prevent UV filters from penetrating deeper layers of skin, reducing the risk of adverse reactions like photoallergenicity [28].

11.5 CASE STUDIES OF INCLUSION COMPLEX OF CYCLODEXTRINS WITH VARIOUS UVR ABSORBERS

The ensuing case studies examine different UV absorbers and their complexation with CDs, explaining techniques used, analyzing effects on sunscreen formulations, and evaluating improvement results. Table 11.2 summarizes the inclusion complexes formed between UVR absorbers and CDs.

Inclusion complexes between UVR absorbers, such as octyl p-methoxycinnamate (OMC) and CDs, increase the solubility and epidermal penetration. One study found that OMC can form stable inclusion complexes with *β*-CD. Despite having a larger molecular volume (367.53 $Å^3$) than the *β*-CD cavity (262 $Å^3$), inclusion complexes formed between UVR absorbers and CDs effectively accommodate UVR absorbers with an inclusion efficiency of 89.0% [30, 35, 36]. The *β*-CD/OMC complex's characterization revealed that OMC's aromatic protons had shifted in the NMR spectrum, indicating that the aromatic part of OMC is inside the *β*-CD cavity. This encapsulation enhances OMC's stability and solubility in formulations. Encapsulating OMC in *β*-CD significantly enhances its solubility in aqueous formulations, addressing its hydrophobic nature and improving its dispersion in sunscreen products to achieve the desired SPF [37]. *β*-CD also enhances OMC's penetration into the skin, facilitating its absorption into deeper layers. Monteiro et al. (2012) reported higher OMC concentrations in the dermis with *β*-CD inclusion, improving UV absorption and skin protection. The interaction of CDs with skin lipids further promotes OMC release and surface availability, enhancing its effectiveness. Compared to non-complexed formulations, *β*-CD/OMC complexes exhibited better skin distribution and UV absorption, resulting in increased SPF [30].

Avobenzone (AVZ), also known as butyl methoxydibenzoylmethane, is a chemical derivative structurally different from the benzophenone family. AVZ has a methoxydibenzoylmethane group, while benzophenones, like oxybenzone, have a benzene ring attached to a carbonyl group (C=O) ortho to another benzene ring [38]. AVZ absorbs UVA rays more efficiently than benzophenone derivatives, aiding UVB protection. Encapsulation of AVZ in *γ*-CD enhances its solubility, stability, and photostability, resulting in higher SPF values compared to non-complexed formulations [39]. The *γ*-CD complex protects AVZ from UV-induced degradation, reducing free radical generation and improving skin protection [40, 41]. It prevents

TABLE 11.2
Summary of Inclusion Complexes of UVR Absorbers with Cyclodextrins

UVR Absorber	CD Type	Key Findings	Stability Mechanism	Reference
Methoxy Cinnamic Acid Derivatives	α-CD, β-CD, γ-CD	Formation of five specific complexes (4-MCA/αCD, 4-MCA/βCD, etc.). Enhanced stability via intermolecular hydrogen bonds. Absorbs UVB/UVA.	Hydrogen bonds (stronger in orientation B)	[50]
Benzimidazolium Bromide Derivatives	β-CD, Sulfobutylether β-CD	Two inclusion modes observed. Strongest interactions with sulfobutylether β-CD.	Versatile molecular orientations	[51]
Vanillin	β-CD	Enhances solubility and stability of vanillin. Association constant: 179–187 M^{-1}.	Host–guest interactions	[52]
Thiourea Derivatives	α-CD, β-CD	Stability enhanced by cyclodextrin dimers.	Supramolecular structure	[53]
Betulonic Acid	γ-CD, 2-hydroxypropyl-γ-CD	Improves solubility and bioavailability of betulonic acid. Stability constant: 1855.55 M^{-1}.	Hydrophobic interactions	[54]
Ferulic Acid	Hydroxypropyl-γ-CD	Higher sun-cut effect than octylmethoxy cinnamate. Exceptionally high UVA protection.	Enhanced solubility in cosmetics.	[46]
Curcumin + UV Filters	β-CD Polymers	Improved water solubility and photostability of curcumin and UV filters (EHMC, DBM).	Enhanced by cross-linking with epichlorohydrin and citric acid.	[55]
Benzophenone	Modified β-CD (ODA-bis-β-CD, ABS-β-CD)	Increased solubility and thermal stability. ODA-bis-β-CD forms more stable complexes.	Hydrophobic interactions and hydrogen bonds.	[56]
2-Hydroxybenzophenone	β-CD	Enhances solubility and thermal stability of oxybenzone.	Hydrophobic interactions and hydrogen bonds.	[57]

(continued)

TABLE 11.2 (Continued)
Summary of Inclusion Complexes of UVR Absorbers with Cyclodextrins

UVR Absorber	CD Type	Key Findings	Stability Mechanism	Reference
Avobenzone	β-CD, γ-CD	Dual emission observed. Improved photostability with β-CD inclusion.	Excimer modulation.	[45]
Titanium Dioxide	β-CD	Grafting CDs onto TiO_2 nanoparticles improves photocatalytic degradation efficiency and sunscreen performance. Enhances solubility, stability, and structural properties.	Host–guest interactions, self-assembly.	[3, 31]
Zinc Oxide	β-CD	Prevents agglomeration, ensuring better dispersion and transparency. Enhances stability, solubility, and skin compatibility.	Host–guest interactions, molecular dispersion.	[3, 31]

harmful excimer states, enhancing photostability and extending UV protection. Additionally, γ-CD broadens AVZ's UV absorption spectrum, providing better UVA defense, while reducing crystallization for improved formulation performance [42–44]. Encapsulation in HP-β-CD further enhances AVZ's stability while minimizing skin penetration, with over 70% remaining on the skin's surface, reducing systemic absorption and improving safety [45]. Studies show β-CD complexes significantly improve AVZ's photostability by modulating excimer formation, enhancing sunscreen efficacy [45]. Balancing synthetic and natural components is essential to address potential skin sensitivity and environmental concerns [46].

Mori et al. demonstrated the ability of various CDs to form inclusion complexes with organic UV absorbers, improving the solubility of weakly soluble natural chemicals like ferulic acid (FA) and botanical flavonoids, which have potential as natural UV absorbers but are challenging to formulate. They investigated the solubilization of UV absorbers in aqueous solutions, focusing on FA, using HP-γ-CD. They showed that the FA/HP-γ-CD combination had a better sun-cut effect than OMC, providing noticeably better UVA protection and highlighting its potential as a safer and more effective substitute for sunscreen formulations [46]. HP-γ-CD and ferulic acid together significantly increase water solubility, overcoming a barrier to applying these natural UV absorbers in cosmetic compositions. According to Ferreira et al. (2023) and Mori et al. (2019), the inclusion complex of tFA/HP-γ-CD shows higher SPF and UVA protection factor (UVA-PF) values than formulations containing ferulic acid alone. The SPF values were noticeably higher, suggesting complexation increases the sunscreen formulations' solubility and protective efficiency. The results support consumer demands for safer and more environmentally friendly cosmetics. Bathochromic shifts seen in UV-Vis spectroscopy enhance the UV absorption characteristics of the inclusion complexes. The findings indicated that CDs can change the photophysical properties of UV absorbers, crucial knowledge for future research and development [4, 30].

In a study, β-CD was used to encapsulate three UV absorbers—oxybenzone, octocrylene, and ethylhexyl methoxycinnamate. The resulting inclusion complexes were analyzed using DSC, 13C NMR, and UV-Vis spectroscopy, with molecular mechanics computations providing structural insights [44, 47]. The findings revealed a significant enhancement in the photostability of sunscreen agents encapsulated within β-CD, reducing their photodecomposition and potentially improving long-term SPF effectiveness [47]. β-CD's hydrophobic cavity shields UV filters from degradation while potentially limiting skin penetration, enhancing both efficacy and safety. Higher β-CD concentrations can further minimize systemic absorption, making it valuable for safer formulations. Encapsulation in HP-β-CD increased oxybenzone's water solubility by up to 1049 times while limiting its percutaneous absorption, enhancing its UV protection capabilities with reduced skin exposure risks [45]. Similarly, SBE-β-CD complexation significantly improved oxybenzone's solubility and reduced transdermal absorption, emphasizing its potential for safer sunscreen formulations [29]. Octocrylene and ethylhexyl methoxycinnamate also benefit from β-CD complexation, which enhances their stability and effectiveness during prolonged sun exposure, ensuring consistent UV protection [48].

When combined with 4-MBC, RM-β-CD shows the best solubilizing and photostabilizing properties; β-CD significantly reduces 4-MBC degradation under UV light, maintaining its effectiveness in sunscreen formulations as a UVB filter [49]. The RM-β-CD and 4-MBC relationship improves the UV filter's stability, highlighting β-CD's significance in developing effective photoprotective products. Ensulizole's stability and efficacy in sunscreen formulations increase when complexed with β-CD, ensuring it provides adequate UV protection and prevents quick degradation in sunlight. This complexation is essential for enhancing the active ingredient's photostability and the overall effectiveness of sunscreen products [48]. Methoxy cinnamic acid (MCA) derivatives, powerful UVR absorbers, can combine with CDs to create stable inclusion complexes. Promkatkaew et al. investigated complex host–guest interactions between MCA derivatives and different cyclodextrins using density functional theory (DFT) with time-dependent DFT (TD-DFT) calculations. Five complexes were clarified: 4-MCA/αCD, 4-MCA/βCD, 245-MCA/βCD, 246-MCA/βCD, and 246-MCA/γCD. The study highlighted that intermolecular hydrogen bonds are primarily responsible for these complexes' stability, with orientation B being the most stable configuration. These complexes can absorb UVB and UVA light, replicating the parent MCA's absorptive characteristics and enhancing UV filters' durability and effectiveness [50].

Porcarasu et al. studied inclusion complex formation between CDs and benzoimidazolium bromide derivatives. They analyzed interactions using UV-Vis and NMR spectroscopy. Findings showed 1:1 stoichiometry, with sulfobutylether-substituted β-cyclodextrin showing the most significant interactions. The investigation revealed two unique inclusion modes, demonstrating cyclodextrins' adaptability to accommodate different molecular orientations [51]. Vanillin, known for UVR absorption, combines with β-CD to produce a stable inclusion complex. Eteer's investigation used UV-Vis spectroscopy to clarify the complex's stoichiometry and stability constant. Results show 1:1 stoichiometry and association constants of 179 and 187 M^{-1}. This complexation improves vanillin's solubility and stability, making it viable for UVR protection applications [52].

Thiourea derivatives form effective inclusion complexes with CDs. Stoicescu et al. (2020) studied synthesizing these complexes using α- and β-CD. Their research showed CD dimer interaction increases complex stability. This supramolecular structure development highlights the complexes' importance in chemical and pharmaceutical domains, indicating promising future uses in improving thiourea derivatives' solubility and stability [53]. Low water solubility triterpenoid betulonic acid forms inclusion complexes with γ-CD and 2-HP-γ-CD. Sbârcea et al.'s study reveals these complexes increase betulonic acid's solubility and bioavailability. They display 1:1 stoichiometry and a stability constant of 1855.55 M^{-1}. This discovery highlights CDs' ability to enhance bioactive substances' physicochemical characteristics, enabling medicinal uses [54]. Curcumin and commercial UV filters, including 4-tert-butyl-4′-methoxydibenzoylmethane (DBM) and 2-ethylhexyl-p-methoxycinnamate (EHMC), have limited water solubility, restricting their use in watery situations. Photostability is critical for UV filters and curcumin effectiveness. Karpkird et al. researched if inclusion complexes with β-CD improve the photostability in ethylene glycol and water solutions. Inclusion complexes of curcumin, EHMC, and DBM with β-CD polymers cross-linked with citric acid and epichlorohydrin, showed improved water solubility. Inclusion complexes with pbCDE in aqueous and ethylene glycol solutions markedly enhanced curcumin, EHMC, and DBM photostability [55].

Deng et al. investigated inclusion complexes of benzophenone (BP) with two modified β-CD derivatives: p-aminobenzenesulfonic acid-β-cyclodextrin (ABS-β-CD) and 4,4′-diaminodiphenyl ether-bridged-bis-β-cyclodextrin (ODA-bis-β-CD). These complexes, synthesized via co-precipitation, formed in a 1:1 stoichiometric ratio. Stability constants, calculated using the Benesi–Hildebrand equation, indicated a spontaneous and exothermic inclusion process, primarily driven by entropy. ODA-bis-β-CD formed more stable complexes than ABS-β-CD, highlighting the impact of structural modifications [56]. Oxybenzone, a benzophenone derivative known as 2-hydroxybenzophenone (2OHBP), is widely used in sunscreens for UVB and UVA II absorption. However, its low water solubility and thermal stability limit its application. Sancho et al. (2015) studied the inclusion of 2OHBP in β-CD to enhance these properties. The freeze-dried complex showed improved solubility and thermal stability. Thermodynamic analysis revealed an exothermic process in methanol-water and endothermic in water, with intermolecular hydrogen bonding and hydrophobic interactions key in complexation. These findings suggest β-CD inclusion can significantly enhance 2OHBP's solubility and stability, promising for advanced formulations [57].

CDs enhance TiO_2-based sunscreen formulations by improving solubility, stability, and photostability of active ingredients. TiO_2 is widely used in sunscreens to scatter and reflect UVR, offering protection against solar exposure [3]. Grafting CDs onto TiO_2 nanoparticles increases photocatalytic degradation efficiency by 2.7 times compared to pure TiO_2, while ensuring stability and controlled release of active ingredients, enhancing skin cancer prevention [58]. CDs act as structure-directing agents in TiO_2 self-assembly, improving crystallinity, porosity, and surface area—key factors for UV protection and photocatalytic performance [59, 60]. They enhance solubility and photostability of natural UV absorbers like ferulic acid, offering superior UVA protection compared to synthetic filters. Their biocompatibility and ability to boost TiO_2's UV and IR filtering capacity improve sunscreen formulations [61]. CDs enhance the efficacy and stability of sunscreens with zinc oxide, a common physical UV filter. In zinc oxide formulations, CDs prevent agglomeration of particles, ensuring better dispersion and transparency. CDs' ability to form inclusion complexes with UV filters like zinc oxide offers improved photostability, solubility, and reduced skin irritation [3].

11.6 FUTURE DIRECTIONS AND RESEARCH OPPORTUNITIES

Innovative CD inclusion complexes can be engineered to optimize release profiles of UV absorbers, ensuring long-term protection and mitigating risks of oxygen- and sunlight-induced degradation. Innovative techniques and materials combined with CD chemistry could create multifunctional sunscreens providing UV protection and other benefits, including enhanced skin hydration and antioxidant properties. This shift is tied to increased customer demand for safer and more effective skincare products. Using lower quantities of active drugs may be possible due to improved efficiency of UVR absorbers made possible by CDs, thus lowering undesirable effects. Leveraging advancements in CD formulations can improve solubility and stability of UV filters and antioxidants. The synergistic effect of nanotechnology and CDs offers research potential as this combination can enhance dispersion and skin penetration of active compounds, improving defences against UVR [31]. Research into safety profiles of these formulations must be continuously advanced to protect consumer safety. There is potential to improve encapsulation efficiency of UVR absorbers through advancements in CD chemistry, namely creating new derivatives with tailored characteristics. These advancements should result in safer and more effective sunscreen products, while meeting increased safety regulations.

11.7 CONCLUSION

CDs improve sunscreen formulations' performance by increasing solubility and stability of active components, critical for reaching high SPF levels. Their ability to create inclusion complexes with UV filters and antioxidants improves skin permeability and targeted distribution, enhancing sunscreens' efficiency in preventing UV-induced skin damage and cancer. CDs can slow degradation of sensitive chemicals, assuring long-term activity and stability. As consumer demand for effective and safe sun protection grows, future prospects for employing CDs in sunscreen formulations appear bright, potentially leading to novel solutions providing higher protection and better user experience.

REFERENCES

1. P. Kosmopoulos, "Ultraviolet Radiation Monitoring," in Panagiotis Kosmopoulos (ed.), *Planning and Management of Solar Power from Space*, United Kingdom, Elsevier, 2024, pp. 159–172. doi: 10.1016/B978-0-12-823390-0.00002-8.
2. A. Mishra, "Know your sunscreen," *Int J Multidiscipl Res*, vol. 6, no. 4, Jul. 2024, doi: 10.36948/ijfmr.2024.v06i04.24611.
3. L. Ferreira *et al.*, "Cyclodextrin-based dermatological formulations: Dermopharmaceutical and cosmetic applications," *Colloids Surf B Biointerfaces*, vol. 221, p. 113012, Jan. 2023, doi: 10.1016/j.colsurfb.2022.113012.
4. S. d'Agostino, A. Azzali, L. Casali, P. Taddei, and F. Grepioni, "Environmentally friendly sunscreens: Mechanochemical synthesis and characterization of β-CD inclusion complexes of Avobenzone and Octinoxate with improved photostability," *ACS Sustain Chem Eng*, vol. 8, no. 35, pp. 13215–13225, Sep. 2020, doi: 10.1021/acssuschemeng.0c02735.
5. A. D. Gholap *et al.*, "Drug delivery strategies for Avobenzone: A case study of photostabilization," *Pharmaceutics*, vol. 15, no. 3, p. 1008, Mar. 2023, doi: 10.3390/pharmaceutics15031008.
6. M. Mubeen *et al.*, "Elucidating the photoluminescence quenching in ensulizole: An artificial water soluble sunscreen," *J Fluoresc*, vol. 31, no. 4, pp. 1055–1063, Jul. 2021, doi: 10.1007/s10895-021-02736-x.

7. M. Sander, M. Sander, T. Burbidge, and J. Beecker, "The efficacy and safety of sunscreen use for the prevention of skin cancer," *Can Med Assoc J*, vol. 192, no. 50, pp. E1802–E1808, Dec. 2020, doi: 10.1503/cmaj.201085.
8. E. L. Holt *et al.*, "Insights into the photoprotection mechanism of the UV filter homosalate," *Phys Chem Chem Phys*, vol. 22, no. 27, pp. 15509–15519, 2020, doi: 10.1039/D0CP02610G.
9. N. d. N. Rodrigues, N. C. Cole-Filipiak, K. N. Blodgett, C. Abeysekera, T. S. Zwier, and V. G. Stavros, "Wavepacket insights into the photoprotection mechanism of the UV filter methyl anthranilate," *Nat Commun*, vol. 9, no. 1, p. 5188, Dec. 2018, doi: 10.1038/s41467-018-07681-1.
10. C. Peyrot, M. M. Mention, F. Brunissen, and F. Allais, "Sinapic acid esters: Octinoxate substitutes combining suitable UV protection and antioxidant activity," *Antioxidants*, vol. 9, no. 9, p. 782, Aug. 2020, doi: 10.3390/antiox9090782.
11. E. Venditti, F. Brugè, P. Astolfi, I. Kochevar, and E. Damiani, "Nitroxides and a nitroxide-based UV filter have the potential to photoprotect UVA-irradiated human skin fibroblasts against oxidative damage," *J Dermatol Sci*, vol. 63, no. 1, pp. 55–61, Jul. 2011, doi: 10.1016/j.jdermsci.2011.03.008.
12. M. M. Oliveira *et al.*, "Dihydrocaffeic acid prevents UVB-induced oxidative stress leading to the inhibition of apoptosis and MMP-1 expression via p38 signaling pathway," *Oxid Med Cell Longev*, vol. 2019, pp. 1–14, Jan. 2019, doi: 10.1155/2019/2419096.
13. L. A. Baker, M. D. Horbury, and V. G. Stavros, "Ultrafast photoprotective properties of the sunscreening agent octocrylene," *Opt Express*, vol. 24, no. 10, p. 10700, May 2016, doi: 10.1364/OE.24.010700.
14. X. Kang *et al.*, "A novel inclusion complex of oxybenzone with *C* -methylresorcin[4] arene deters skin permeation," *RSC Adv*, vol. 13, no. 37, pp. 25846–25852, 2023, doi: 10.1039/D3RA01890C.
15. X.-P. Chang, L. Yu, T.-S. Zhang, and G. Cui, "Quantum mechanics/molecular mechanics studies on the mechanistic photophysics of sunscreen oxybenzone in methanol solution," *Phys Chem Chem Phys*, vol. 24, no. 21, pp. 13293–13304, 2022, doi: 10.1039/D2CP01263D.
16. E. Fois, M. Oriani, and G. Tabacchi, "A post-HF approach to the sunscreen octyl methoxycinnamate," *J Chem Phys*, vol. 154, no. 14, Apr. 2021, doi: 10.1063/5.0046118.
17. S. J. Phelan-Dickinson, B. C. Palmer, Y. Chen, and L. A. DeLouise, "The UVR filter octinoxate modulates aryl hydrocarbon receptor signaling in keratinocytes via inhibition of CYP1A1 and CYP1B1," *Toxicol Sci*, vol. 177, no. 1, pp. 188–201, Sep. 2020, doi: 10.1093/toxsci/kfaa091.
18. M. C. Semones, C. M. Sharpless, A. A. MacKay, and Y.-P. Chin, "Photodegradation of UV filters oxybenzone and sulisobenzone in wastewater effluent and by dissolved organic matter," *Appl Geochem*, vol. 83, pp. 150–157, Aug. 2017, doi: 10.1016/j.apgeochem.2017.02.008.
19. M. Beghin, A. De Groote, and P. Kestemont, "Single and combined effects of titanium (TiO2) and zinc (ZnO) oxide nanoparticles in the rainbow trout gill cell line RTgill-W1," *Environ Sci Pollut Res*, vol. 31, no. 45, pp. 56523–56535, Sep. 2024, doi: 10.1007/s11356-024-34955-7.
20. J. Vera *et al.*, "Antioxidant activity as an indicator of the efficiency of plant extract-mediated synthesis of Zinc Oxide nanoparticles," *Antioxidants*, vol. 12, no. 4, p. 784, Mar. 2023, doi: 10.3390/antiox12040784.

21. S. L. Schneider and H. W. Lim, "A review of inorganic <scp>UV</scp> filters zinc oxide and titanium dioxide," *Photodermatol Photoimmunol Photomed*, vol. 35, no. 6, pp. 442–446, Nov. 2019, doi: 10.1111/phpp.12439.
22. G. Tsoucaris and N. Rysanek, "Cyclodextrins: Towards Large Assemblies for Efficient Slow Release Devices," in George Tsoucaris (ed.), *Current Challenges on Large Supramolecular Assemblies*, Dordrecht: Springer Netherlands, 1999, pp. 385–391. doi: 10.1007/978-94-011-5284-6_27.
23. M. Másson, T. Loftsson, and E. Stefánsson, "How Do Cyclodextrins Enhance Drug Permeability through Biological Membranes?" in *Proceedings of the Ninth International Symposium on Cyclodextrins*, Dordrecht: Springer Netherlands, 1999, pp. 363–366. doi: 10.1007/978-94-011-4681-4_86.
24. H. T. N. Le and H. K. Jeong, "Cyclodextrin-Graphite Oxide-Carbon nanotube composites for electrochemical supramolecular recognition," *Electrochim Acta*, vol. 232, pp. 7–12, Apr. 2017, doi: 10.1016/j.electacta.2017.02.119.
25. G. Wenz, "Recognition of Monomers and Polymers by Cyclodextrins,". In Gerhard Wenz (ed.), *Inclusion Polymers,* Berlin, Heidelberg, Springer Berlin Heidelberg, 2009, pp. 204–254. doi: 10.1007/12_2008_13.
26. X. Wang, J. Wu, X. Liu, X. Qiu, L. Cao, and Y. Ji, "Enhanced chiral recognition abilities of cyclodextrin covalent organic frameworks via Chiral/Achiral functional modification," *ACS Appl Mater Interfaces*, vol. 14, no. 22, pp. 25928–25936, Jun. 2022, doi: 10.1021/acsami.2c05572.
27. I. Tabushi, K. Yamamura, and Y. Kuroda, "Molecular Recognition by Cyclodextrin Hosts: Approach to New Functions," in Jerry L. Atwood (ed.), *Inclusion Phenomena and Molecular Recognition*, Boston, MA: Springer US, 1990, pp. 237–242. doi: 10.1007/978-1-4613-0603-0_24.
28. L. A. Felton, C. J. Wiley, and D. A. Godwin, "Influence of cyclodextrin complexation on the in vivo photoprotective effects of Oxybenzone," *Drug Dev Ind Pharm*, vol. 30, no. 1, pp. 95–102, Jan. 2004, doi: 10.1081/DDC-120027516.
29. S. Simeoni, S. Scalia, R. Tursilli, and H. Benson, "Influence of cyclodextrin complexation on the in vitro human skin penetration and retention of the sunscreen agent, Oxybenzone," *J Incl Phenom Macrocycl Chem*, vol. 54, no. 3–4, pp. 275–282, Apr. 2006, doi: 10.1007/s10847-005-9002-0.
30. E. Ricci-Júnior *et al.*, "Evaluation of octyl p-methoxycinnamate included in liposomes and cyclodextrins in anti-solar preparations: preparations, characterizations and in vitro penetration studies," *Int J Nanomedicine*, p. 3045, Jun. 2012, doi: 10.2147/IJN.S28550.
31. L. Dahabra, G. Broadberry, A. Le Gresley, M. Najlah, and M. Khoder, "Sunscreens containing cyclodextrin inclusion complexes for enhanced efficiency: A strategy for skin cancer prevention," *Molecules*, vol. 26, no. 6, p. 1698, Mar. 2021, doi: 10.3390/molecules26061698.
32. J. Shokri, D. Hasanzadeh, S. Ghanbarzadeh, M. Dizadji-Ilkhchi, and K. Adibkia, "The effect of Beta-Cyclodextrin on Percutaneous absorption of commonly used Eusolex® sunscreens," *Drug Res*, vol. 63, no. 11, pp. 591–596, Jul. 2013, doi: 10.1055/s-0033-1349089.
33. S. Scalia, G. Coppi, and V. Iannuccelli, "Microencapsulation of a cyclodextrin complex of the UV filter, butyl methoxydibenzoylmethane: In vivo skin penetration studies," *J Pharm Biomed Anal*, vol. 54, no. 2, pp. 345–350, Jan. 2011, doi: 10.1016/j.jpba.2010.09.018.

34. L. Yuan *et al.*, "Studies on the preparation and photostability of avobenzone and (2-hydroxy)propyl-β-cyclodextrin inclusion complex," *J Photochem Photobiol A Chem*, vol. 369, pp. 174–180, Jan. 2019, doi: 10.1016/j.jphotochem.2018.09.036.
35. G. L. N. Coelho *et al.*, "Preparation and evaluation of inclusion complexes of commercial sunscreens in cyclodextrins and Montmorillonites: Performance and substantivity studies," *Drug Dev Ind Pharm*, vol. 34, no. 5, pp. 536–546, Jan. 2008, doi: 10.1080/03639040701831769.
36. H. Dodziuk, "Rigidity versus flexibility. A review of experimental and theoretical studies pertaining to the cyclodextrin nonrigidity," *J Mol Struct*, vol. 614, no. 1–3, pp. 33–45, Sep. 2002, doi: 10.1016/S0022-2860(02)00236-3.
37. S. Scalia, A. Casolari, A. Iaconinoto, and S. Simeoni, "Comparative studies of the influence of cyclodextrins on the stability of the sunscreen agent, 2-ethylhexyl-p-methoxycinnamate," *J Pharm Biomed Anal*, vol. 30, no. 4, pp. 1181–1189, Nov. 2002, doi: 10.1016/S0731-7085(02)00433-8.
38. Z. Juvancz, J. Némethné-Katona, and R. Iványi, "Role of the Cyclodextins in Analytical Chemistry," In Imre J. Rudas, János Fodor & Janusz Kacprzyk (eds.), *Towards Intelligent Engineering and Information Technology,* Berlin, Heidelberg, Springer Berlin Heidelberg, 2009, pp. 703–718. doi: 10.1007/978-3-642-03737-5_50.
39. P. T. Loftsson and P. J. H. Olafsson, "Cyclodextrins: New drug delivery systems in dermatology," *Int J Dermatol*, vol. 37, no. 4, pp. 241–246, Apr. 1998, doi: 10.1046/j.1365-4362.1998.00369.x.
40. K. E. Burke, "Prevention and Treatment of Aging Skin with Topical Antioxidants," in Nava Dayan (ed.), *Skin Aging Handbook*, Norwich, NY, Elsevier, 2009, pp. 149–176. doi: 10.1016/B978-0-8155-1584-5.50012-0.
41. I. Kusumawati and G. Indrayanto, "Natural Antioxidants in Cosmetics," in Atta-ur-Rahman (ed.), *Studies in Natural Products Chemistry*, Netherlands, Elsevier, 2013, pp. 485–505. doi: 10.1016/B978-0-444-59603-1.00015-1.
42. O. Goral, G. Z. Zukowska, E. Zero, M. Siekierski, and A. Krzton-Maziopa, "Understanding avobenzone crystallization in sunscreen formulations: Role of metal Oxide-Driven nucleation and stabilization strategies," *Crystals (Basel)*, vol. 14, no. 7, p. 663, Jul. 2024, doi: 10.3390/cryst14070663.
43. D. K. Sahoo, P. Mohanty, H. S. Biswal, and K. H. Gowd, "Conformers influence on UV-absorbance of avobenzone," *J Photochem Photobiol A Chem*, vol. 453, p. 115671, Aug. 2024, doi: 10.1016/j.jphotochem.2024.115671.
44. T. Loftsson and M. Masson, "Cyclodextrins in topical drug formulations: theory and practice," *Int J Pharm*, vol. 225, no. 1–2, pp. 15–30, Aug. 2001, doi: 10.1016/S0378-5173(01)00761-X.
45. J. Vishwakarma, D. Takkella, S. Sharma, and K. Gavvala, "Modulation of excimer formation and photostability of avobenzone inside the nanocavities of cyclodextrins," *J Photochem Photobiol A Chem*, vol. 449, p. 115411, Apr. 2024, doi: 10.1016/j.jphotochem.2023.115411.
46. T. Mori, R. Tsuchiya, M. Doi, N. Nagatani, and T. Tanaka, "Solubilization of ultraviolet absorbers by cyclodextrin and their potential application in cosmetics," *J Incl Phenom Macrocycl Chem*, vol. 93, no. 1–2, pp. 91–96, Feb. 2019, doi: 10.1007/s10847-018-0846-5.
47. N. A. F. Al-Rawashdeh, K. S. Al-Sadeh, and M. B. Al-Bitar, "Inclusion complexes of sunscreen agents with ***β*** -Cyclodextrin: Spectroscopic and molecular modeling studies," *J Spectrosc*, vol. 2013, pp. 1–11, 2013, doi: 10.1155/2013/841409.

48. G. Nitulescu, D. Lupuliasa, I. Adam-Dima, and G. M. Nitulescu, "Ultraviolet filters for cosmetic applications," *Cosmetics*, vol. 10, no. 4, p. 101, Jul. 2023, doi: 10.3390/cosmetics10040101.
49. L. R. Gaspar and P. M. B. G. Maia Campos, "Evaluation of the photostability of different UV filter combinations in a sunscreen," *Int J Pharm*, vol. 307, no. 2, pp. 123–128, Jan. 2006, doi: 10.1016/j.ijpharm.2005.08.029.
50. M. Promkatkaew, P. Boonsri, S. Suramitr, T. Karpkird, P. Wolschann, and S. Hannongbua, "Stability improvement of UV-filter between methoxy cinnamic acid derivatives and cyclodextrins inclusion complexes based on DFT and TD-DFT investigations," *J Mol Graph Model*, vol. 125, p. 108619, Dec. 2023, doi: 10.1016/j.jmgm.2023.108619.
51. M. Balan-Porcărașu *et al.*, "Benzimidazolium bromide derivative inclusion complexes with native and modified beta-cyclodextrins," *Revue Roumaine de Chimie*, vol. 68, no. 3–4, pp. 119–125, Mar. 2024, doi: 10.33224/rrch.2023.68.3-4.01.
52. S. A. Eteer, "UV-Vis spectroscopic characterization of β-Cyclodextrin-Vanillin inclusion complex," *Mediterr J Chem*, vol. 12, no. 2, p. 175, Oct. 2022, doi: 10.13171/mjc02210111645eteer.
53. C. S. Stoicescu, A. D. Neacşu, C. D. Bădiceanu, and G. Munteanu, "Inclusion complexes of some thiourea derivatives in cyclodextrins," *J Incl Phenom Macrocycl Chem*, vol. 96, no. 3–4, pp. 275–283, Apr. 2020, doi: 10.1007/s10847-019-00968-w.
54. L. Sbârcea *et al.*, "Betulonic acid—cyclodextrins inclusion complexes," *J Therm Anal Calorim*, vol. 138, no. 4, pp. 2787–2797, Nov. 2019, doi: 10.1007/s10973-019-08359-6.
55. T. Karpkird, R. Khunsakorn, C. Noptheeranuphap, and S. Midpanon, "Inclusion complexes and photostability of UV filters and curcumin with beta-cyclodextrin polymers: Effect on cross-linkers," *J Incl Phenom Macrocycl Chem*, vol. 91, no. 1–2, pp. 37–45, Jun. 2018, doi: 10.1007/s10847-018-0796-y.
56. S. Deng, H. Liu, C. Qi, A. Yang, and Z. Li, "Study on preparation and inclusion behavior of inclusion complexes between β-cyclodextrin derivatives with benzophenone," *J Incl Phenom Macrocycl Chem*, vol. 90, no. 3–4, pp. 321–329, Apr. 2018, doi: 10.1007/s10847-018-0787-z.
57. M. I. Sancho, M. G. Russo, M. S. Moreno, E. Gasull, S. E. Blanco, and G. E. Narda, "Physicochemical characterization of 2-Hydroxybenzophenone with β-Cyclodextrin in solution and solid state," *J Phys Chem B*, vol. 119, no. 18, pp. 5918–5925, May 2015, doi: 10.1021/acs.jpcb.5b01742.
58. H. Liu *et al.*, "Cooperative catalysis between cyclodextrin and TiO2 nanoparticles," *Mater Lett*, vol. 365, p. 136453, Jun. 2024, doi: 10.1016/j.matlet.2024.136453.
59. S. Yoon and W. T. Nichols, "Cyclodextrin directed self-assembly of TiO2 nanoparticles," *Appl Surf Sci*, vol. 285, pp. 517–523, Nov. 2013, doi: 10.1016/j.apsusc.2013.08.086.
60. A. Lannoy *et al.*, "Photocatalysis of volatile organic compounds in water: Towards a deeper understanding of the role of cyclodextrins in the photodegradation of toluene over titanium dioxide," *J Colloid Interface Sci*, vol. 461, pp. 317–325, Jan. 2016, doi: 10.1016/j.jcis.2015.09.022.
61. R. Rajamohan *et al.*, "Formulation of ensulizole with Beta-Cyclodextrins for improved sunscreen activity and biocompatibility," *J Pharm Sci*, vol. 113, no. 6, pp. 1536–1545, Jun. 2024, doi: 10.1016/j.xphs.2023.12.019.

12 Cyclodextrin-Enhanced UVR Absorbers

Versatile Solutions for Fabrics, Polymers, and Beyond

Ayfer Koyuncu and Fatih Ciftci

12.1 INTRODUCTION

Ultraviolet radiation (UVR), a form of electromagnetic radiation, significantly impacts both organic and inorganic materials. Solar UVR poses risks to materials and human health, causing polymer degradation, dye fading, and increased skin cancer risk. To mitigate these effects, UVR absorbers are used in industries such as textiles, automotive, construction, and electronics to enhance product durability and safety [1–3].

UVR is categorized into UVA, UVB, and UVC, with varying wavelengths and energy levels. While UVC is absorbed by the Earth's atmosphere, UVA and UVB penetrate and degrade materials, altering their physical, chemical, and mechanical properties. This results in discolouration, brittleness, reduced strength, and shorter lifespans, posing challenges across industries. UVR absorbers are classified into organic, inorganic, and nanoparticle-based materials, each with distinct applications. UVR induces photochemical reactions, breaking chemical bonds and accelerating material ageing. Organic materials like plastics, rubber, textiles, and coatings are particularly vulnerable, experiencing fading, yellowing, tensile strength loss, and brittleness. Polymers may crack, and paints or coatings can peel or lose adhesion. In outdoor environments with constant UVR exposure, these effects are intensified, making protective measures essential to preserve material durability and appearance [1, 3, 4].

Cyclodextrins (CDs) are versatile tools for enhancing UVR absorbers' performance. These cyclic oligosaccharides, made of glucose units, form a truncated cone structure with a hydrophobic interior and hydrophilic exterior [5]. This structure enables CDs to form inclusion complexes with UVR absorbers, improving their solubility, stability, and dispersion in various matrices. Encapsulation within CDs' cavities protects UVR absorbers from photodegradation, enhances substrate compatibility, and extends their lifespan in UV protective coatings and treatments (Figure 12.1) [3, 6–8].

Organic UVR absorbers like benzotriazoles, benzophenones, and triazines are widely used for their strong UVA and UVB absorption, especially in clear coatings

DOI: 10.1201/9781003492757-12

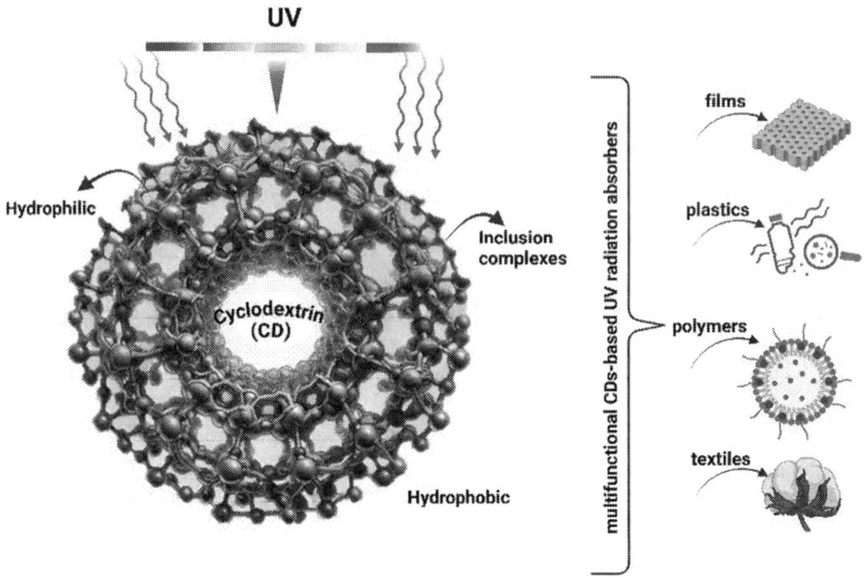

FIGURE 12.1 Cyclodextrin-enhanced UVR absorbers for diverse materials.

for glass and plastics where transparency and minimal material impact are crucial. However, they often face challenges like poor solubility, limited photostability, and migration from host materials. CDs address these issues by forming inclusion complexes that improve solubility, stability, uniform distribution, and sustained UV protection [3, 5, 9, 10].

Inorganic UVR absorbers like titanium dioxide and zinc oxide effectively block UVA, UVB, and UVC. Favoured for their durability, non-toxicity, and high refractive indices, they are used in sunscreens, paints, and coatings. However, their photocatalytic activity can degrade materials, and aggregation reduces effectiveness. CDs enhance dispersion and reduce photocatalytic activity by encapsulating particles, improving the performance and longevity of UV protective coatings [7, 9, 11, 12].

Nanoparticle-based UVR absorbers, like nanostructured metal oxides, offer superior UV protection due to their high surface area and tunable optical properties. They enable efficient UV absorption for advanced applications, such as multi-layered films and high-performance coatings. However, challenges like agglomeration, stability, and potential health risks must be addressed. CDs stabilize nanoparticles, prevent agglomeration, and improve dispersion in coating matrices, enhancing UV blocking efficiency while addressing safety concerns. Integrating CDs with UVR absorbers advances UV protection by improving stability, solubility, and distribution, enabling durable coatings for industries like textiles, automotive, and architecture, while preserving material properties [3, 4, 8].

In summary, combining cyclodextrins with UVR absorbers offers a promising way to overcome traditional UV protection limitations. CDs enhance the performance, stability, and safety of UV protective materials, enabling innovative solutions for diverse applications.

12.2 CYCLODEXTRINS AS HOSTS FOR UVR ABSORBERS

Cyclodextrins (CDs) are cyclic oligosaccharides of glucose units linked by α-1,4-glycosidic bonds [9, 13]. Their hydrophilic exterior and hydrophobic inner cavity make them ideal for encapsulating UVR absorbers. Encapsulation enhances the stability, solubility, and bioavailability of these molecules, effectively protecting materials and skin from UV damage. CDs improve the functionality of UVR absorbers and the formulation of consumer products, making them versatile and efficient hosts [1, 2, 7, 11–14].

12.3 CYCLODEXTRIN STRUCTURE AND PROPERTIES RELEVANT TO UVR FILTRATION

Cyclodextrins are cyclic oligosaccharides with a unique structure that enables effective encapsulation of UVR absorbers. Understanding their properties is key to optimizing UVR filtration and enhancing the performance, stability, and safety of UV protection products [2, 13]. The most common types cyclodextrins are (i) ***α-CD*** (six glucose units), (ii) ***β-CD*** (seven glucose units), and (iii) ***γ-CD*** (eight glucose units). The cavity size increases with the number of glucose units. For example, α-CD: Inner cavity diameter of ~4.7–5.3 Å; β-CD: Inner cavity diameter of ~6.0–6.5 Å; γ-CD: Inner cavity diameter of ~7.5–8.3 Å. Cavity size is key in determining which UVR absorbers can be effectively encapsulated. Cyclodextrins' structural flexibility allows them to accommodate guest molecules of various sizes and shapes, enhancing their versatility [6, 8, 9, 14–17].

CDs improve UVR absorption by enhancing the solubility, stability, and dispersion of UVR absorbers, while protecting them from degradation and adverse interactions [16, 18]. Their use in UVR filtration optimizes UV absorber performance in various applications. This leads to more effective, aesthetically pleasing UV protection products with reduced skin irritation and environmental impact [2, 12]. The reasons why cyclodextrins are preferred are as follows [16–20].

(i) **Inclusion Complex Formation**: Cyclodextrins have a hydrophobic inner cavity that encapsulates hydrophobic or poorly soluble UVR absorbers. The UVR absorber forms a non-covalent inclusion complex with cyclodextrins, stabilized by hydrophobic interactions, van der Waals forces, and hydrogen bonding. This encapsulation increases the concentration of UVR absorbers in water, enhancing their availability and effectiveness in UVR absorption.

(ii) **Enhanced Solubility and Dispersion**: Many UVR absorbers have poor water solubility, limiting their effectiveness in aqueous formulations like sunscreens and lotions. Cyclodextrins improve UVR absorber solubility, enhancing dispersion in the formulation and contact with the skin.

Enhanced solubility ensures UVR absorbers are evenly distributed on the skin, providing better coverage and UV protection.

(iii) **Increased Photostability**: UVR absorbers degrade with UV light exposure, reducing their effectiveness over time. Cyclodextrin encapsulation protects UVR absorbers from UVR, reducing photodegradation. CDs prevent interactions that could degrade UVR absorbers, preserving their integrity and function.

(iv) **Prevention of Aggregation**: UVR absorbers may aggregate in formulations, reducing their UV-absorbing capability. CDs maintain dispersion, preventing aggregation and ensuring maximum surface area for UV absorption.

(v) **Modulation of Release Profile**: CDs control the release of UVR absorbers, ensuring sustained release over time. This mechanism extends the duration of UV protection on the skin.

(vi) **Enhanced Compatibility with Other Formulation Components**: CDs enhance UVR absorber compatibility by reducing interactions that could affect stability or efficacy. Encapsulation minimizes adverse reactions, maintaining product stability and performance.

(vii) **Reduced Skin Penetration and Irritation**: CD encapsulation limits UVR absorber skin penetration, reducing systemic absorption and health risks. It also minimizes skin irritation and allergic reactions, making it suitable for sensitive skin.

(viii) **Improved Aesthetic Properties**: CDs improve sensory attributes, reducing greasiness and stickiness of UV filters. They enhance texture and spreadability, improving user compliance for better UV protection.

12.4 APPLICATIONS OF CD-ENCAPSULATED UVR ABSORBERS IN FABRICS, TEXTILES, AND POLYMER FILMS

Cyclodextrin-encapsulated UVR absorbers enhance UV protection in fabrics, textiles, and polymer films by improving stability, solubility, and retention. This extends the lifespan and effectiveness of materials, making them suitable for applications like protective clothing, agricultural films, and packaging. The approach provides superior UV protection while maintaining desirable properties for various industrial and consumer uses. Advantages of this method are given in Table 12.1 [2, 15, 21–25].

Cyclodextrin-based UVR absorption systems have shown versatility and effectiveness in enhancing UV protection across various industries. Studies highlight their use in personal care, textiles, coatings, packaging, and medical devices. For instance, Dahabra *et al.* (2021) [29] studied that β-cyclodextrins were used to stabilize organic UV filters like oxybenzone and avobenzone in sunscreens, improving solubility, stability, and skin compatibility while maintaining high SPF and broad-spectrum protection. This innovation led to more photostable, effective sunscreens with reduced skin irritation [24]. In another study on UV-protected clothing, γ-cyclodextrins encapsulated ZnO nanoparticles to increase the UV protection factor (UPF) while enhancing breathability and comfort. Cyclodextrin-functionalized TiO_2 and ZnO

TABLE 12.1
Applications in Fabrics, Textiles, and Polymer Films

Applications in Fabrics and Textiles	Applications in Polymer Films	Ref.
UV-Protective Clothing: CD-encapsulated UVR absorbers can be integrated into fibers during manufacturing or applied as a finishing treatment. This enhances the UV-blocking capabilities of textiles, making them ideal for UV-protective clothing and accessories. Encapsulation ensures UVR absorbers stay on the fabric after washing, improving UV protection durability.	**UV-Protective Coatings:** CD-encapsulated UVR absorbers can be incorporated into polymer coatings applied to surfaces, such as windows, car interiors, and displays to provide UV protection. These coatings help prevent UV-induced degradation of underlying materials and reduce the penetration of harmful UV rays, enhancing the durability and lifespan of the products.	[3] [16] [22] [24] [26] [27] [28]
Textile Finishes: CDs can be used as carriers for UVR absorbers in textile finishes, which can be applied through methods such as padding, coating, or spraying. This allows for the creation of fabrics with enhanced UV protection properties without altering the feel or appearance of the material. The use of CDs ensures that the UVR absorbers are more uniformly distributed and bound within the textile matrix, enhancing their effectiveness and longevity.	**Agricultural Films:** In agriculture, UV-protective films are used to cover greenhouses and crop fields to control light transmission and protect plants from harmful UVR. CD-encapsulated UVR absorbers can be embedded into these films to improve their UV-blocking performance, helping to create a controlled environment that promotes healthy plant growth. These films are more resistant to degradation from UV exposure, making them last longer and perform better in outdoor settings.	[3] [8] [16] [24] [27] [28]
Enhanced Fabric Performance: UVR absorbers in textiles help prevent fading, discolouration, and degradation of fibers caused by UV exposure. This is particularly valuable for outdoor fabrics, sportswear, and items that are frequently exposed to sunlight. CD-encapsulated UVR absorbers offer a solution that combines protection with minimal impact on fabric properties, such as breathability, softness, and comfort.	**Packaging Materials:** UV protection is critical in packaging materials, especially for products sensitive to light, such as pharmaceuticals, cosmetics, and food. Incorporating CD-encapsulated UVR absorbers into packaging films can help protect the contents from UV-induced deterioration, preserving product quality and extending shelf life. This approach can also enhance the aesthetic properties of packaging by preventing discolouration and maintaining the visual appeal of the packaging over time.	[12] [26] [27] [28]

nanoparticles were also used in automotive and architectural coatings to improve UV protection and thermal management, while maintaining transparency. For plastic packaging, α-cyclodextrins encapsulated ZnO nanoparticles to extend shelf life by preventing UV-induced degradation without compromising flexibility and clarity [30]. The study on cyclodextrin-functionalized nanoparticles in medical equipment enhanced UV protection, reducing material degradation [31]. These studies demonstrate the broad applications and benefits of cyclodextrin-enhanced UVR absorbers in diverse fields.

12.5 ORGANIC UVR ABSORBERS WITH CYCLODEXTRINS

Encapsulating organic UVR absorbers with cyclodextrins improves solubility, stability, and safety. This enhances UV protection while addressing concerns about skin irritation, absorption, and environmental impact, making it ideal for modern sunscreens and UV products [4, 8, 31]. Table 12.2 lists common absorbers and the benefits of encapsulation.

12.6 INORGANIC UVR ABSORBERS AND CYCLODEXTRIN COMPLEXES

Inorganic UVR absorbers like ZnO and TiO_2 offer strong UV protection but face issues like aggregation, poor dispersion, and aesthetic concerns. Cyclodextrins can form complexes with these absorbers, enhancing their performance. Common inorganic UVR absorbers with cyclodextrins are listed in Table 12.3 [2, 9, 24].

Cyclodextrin complexes with inorganic UVR absorbers like zinc oxide and titanium dioxide improve dispersion, stability, and cosmetic appeal. By addressing issues like aggregation, photocatalytic activity, and whitening, cyclodextrins enhance performance and safety in sunscreens, skincare, textiles, and polymer films. This approach broadens the applications of inorganic UVR absorbers in effective, consumer-friendly products [3, 9, 40].

12.6.1 Integration into Glass and Plastic Materials for Enhanced UV Protection

Integrating cyclodextrin-modified inorganic UVR absorbers into glass and plastic substrates enhances UV protection while maintaining transparency and aesthetics. This improves dispersion, stability, and performance, making it suitable for automotive, architectural glass, eyewear, packaging, and agricultural films. Benefits include extended material life, improved safety, and environmental performance. Table 12.4 lists methods, benefits, and applications [2, 41, 42].

TABLE 12.2
Common Organic UVR Absorbers with Cyclodextrins

Organic UVR Absorbers	Function	Challenges	Cyclodextrin Encapsulation	Ref.
Avobenzone (Butyl Methoxydibenzoylmethane)	Absorbs UVA radiation	Prone to photodegradation when exposed to sunlight, reducing its efficacy.	Enhances photostability and solubility, reducing degradation and extending the duration of UVA protection.	[32]
Octinoxate (Ethylhexyl Methoxycinnamate)	Absorbs UVB radiation	Can degrade under UV exposure and has limited water solubility.	Improves solubility, stability, and UV absorption efficiency, enhancing the overall effectiveness of UVB protection.	[33]
Oxybenzone (Benzophenone-3)	Absorbs UVA and UVB radiation	Associated with skin irritation and potential hormonal effects.	Reduces skin penetration and potential irritation, while also enhancing stability and reducing environmental impact.	[34]
Homosalate	Absorbs UVB radiation	Limited water solubility and potential for degradation.	Improves solubility and stability, offering better UVB protection with reduced risk of skin irritation.	[32, 35]
Octocrylene	Absorbs UVB and short-wave UVA radiation	Can cause skin irritation and has potential allergenic effects.	Reduces direct skin exposure, improving safety and minimizing adverse effects while maintaining UV protection efficacy.	[34]
Tinosorb S (Bis-Ethylhexyloxyphenol Methoxyphenyl Triazine)	Broad-spectrum absorber of UVA and UVB radiation	Expensive and has solubility issues.	Enhances solubility, broadening its application in aqueous formulations and improving overall stability.	[35]
Ensulizole (Phenylbenzimidazole Sulfonic Acid)	Primarily absorbs UVB radiation	Water-soluble but prone to photodegradation.	Provides enhanced photostability, helping maintain UV absorption capacity over time.	[36]

TABLE 12.3
Common Inorganic UVR Absorbers with Cyclodextrins

Inorganic UVR Absorbers	Zinc Oxide (ZnO)	Titanium Dioxide (TiO_2)	Ref.
Function	Broad-spectrum UVR absorber that provides both UVA and UVB protection.	Primarily absorbs UVB and short-wave UVA radiation.	
Challenges	Can form aggregates, leading to a whitening effect on the skin, and has limited dispersion in some formulations.	Prone to aggregation, can cause a whitening effect, and may generate reactive ROS under UV exposure.	[37] [38] [39]
Properties	Known for its photostability, antimicrobial properties, and low irritation potential, making it suitable for sensitive skin.	Offers excellent photostability and is widely used in sunscreens, coatings, and plastics.	

TABLE 12.4
Methods of Integration into Glass and Plastic Materials

	Ref.
Surface Coatings and Films: CD-enhanced inorganic UVR absorbers can be applied as coatings or films on the surface of glass and plastic materials. These coatings are typically applied using techniques such as dip-coating, spray-coating, or spin-coating, followed by curing processes to ensure strong adhesion and durability. The coatings act as a barrier to UVR, preventing it from penetrating the material and reaching the interior or the objects behind the glass or plastic.	
Bulk Incorporation: Inorganic UVR absorbers modified with CDs can be incorporated directly into the bulk of plastics during the manufacturing process, such as extrusion or injection moulding. This method distributes the UVR absorbers throughout the entire material, providing uniform protection and avoiding degradation issues that can occur with surface-only treatments. CDs help maintain the dispersion and stability of UVR absorbers within the plastic matrix, ensuring consistent UV protection.	[3] [40] [41] [42]
Sol-Gel Processes for Glass: In glass applications, CD-enhanced UVR absorbers can be integrated using sol-gel processes, which involve the formation of a gel from a liquid precursor that contains the UVR absorbers. The sol-gel method allows for the incorporation of UVR absorbers into the glass matrix at a molecular level, providing effective UV blocking without compromising the transparency or mechanical properties of the glass. This method is particularly useful for producing UV-protective coatings on architectural or automotive glass.	[43] [44]
Lamination: Laminating glass or plastic with a layer that contains CD-modified UVR absorbers is another effective method. Laminated glass, often used in car windshields and building windows, can incorporate UVR absorbers between layers of glass or plastic, providing enhanced protection while maintaining the structural integrity of the material. This approach allows for the incorporation of a high concentration of UVR absorbers without affecting the surface characteristics of the outer layers.	

TABLE 12.5
Mechanisms of Cyclodextrin Interaction with Nanoparticle-Based UVR Absorbers

	Ref.
Surface Modification and Encapsulation: CDs modify nanoparticle surfaces via non-covalent interactions (hydrogen bonding, van der Waals forces) or covalent attachment to functional groups. This improves nanoparticle hydrophilicity or hydrophobicity, enhances compatibility with formulations, and reduces aggregation. CDs can also encapsulate small molecules or surface modifiers, stabilizing nanoparticles and improving dispersion.	
Prevention of Aggregation: Nanoparticles often aggregate due to high surface energy, reducing UVR absorption efficiency. CDs stabilize nanoparticles by forming a protective layer, creating steric or electrostatic barriers that prevent clumping. This ensures proper dispersion and preserves their nanoscale properties for effective UV absorption and scattering.	
Reduction of Photocatalytic Activity: Metal oxide nanoparticles like TiO_2 can generate ROS under UV exposure, causing oxidative damage. CDs reduce this photocatalytic activity by encapsulating nanoparticles or surface modifiers, restricting their interaction with oxygen and water.	[3] [4] [26] [45]
Enhanced Solubility and Dispersibility: CDs enhance the solubility and dispersibility of nanoparticle UVR absorbers in various solvents, ensuring versatility for formulations like sunscreens, where uniform dispersion is vital for consistent UV protection.	
Controlled Release: CDs enable controlled release of UVR absorbers, ensuring sustained UV protection for prolonged exposure, ideal for outdoor use and long-lasting textiles or coatings.	

12.7 NANOPARTICLE-BASED UVR ABSORBERS WITH CYCLODEXTRINS

Using cyclodextrins with nanoparticle-based UVR absorbers brings a range of benefits, combining the unique properties of cyclodextrins with the advanced capabilities of nanoparticles. CD-modified nanoparticle UVR absorbers offer enhanced UV protection, stability, and material compatibility, making them suitable for a broad array of applications across various industries [2, 3]. Cyclodextrin-modified nanoparticle UVR absorbers, such as zinc oxide and titanium dioxide, provide superior UV protection. However, challenges like aggregation, photocatalytic activity, and poor dispersion affect their effectiveness. CDs help improve the performance and stability of these nanoparticles. Table 12.5 outlines mechanisms of cyclodextrin interaction with nanoparticle-based UVR absorbers [2, 3, 7].

12.8 COMPARISON OF ORGANIC, INORGANIC, AND NANOPARTICLE UVR ABSORBERS WITH CDS

Cyclodextrins enhance the performance and safety of organic, inorganic, and nanoparticle UVR absorbers, making them versatile for various applications. CD organic absorbers are ideal for personal care due to their solubility and low irritation

TABLE 12.6
A Comparative Analysis of These UVR Absorbers with Cyclodextrin

Absorbers	Characteristics	Advantages	Challenges	Applications	Ref.
Organic UVR Absorbers with CDs	• **Common Types**: Benzophenones, cinnamates, avobenzone, and p-aminobenzoic acid (PABA). • **Mechanism**: Organic absorbers function by absorbing UVR and converting it into less harmful forms of energy, such as heat. • **Cyclodextrin Role**: CDs enhance the solubility, stability, and photostability of organic absorbers, often encapsulating the UVR absorbers within their hydrophobic cavities.	• **Improved Solubility**: CDs significantly increase the aqueous solubility of poorly soluble organic UVR absorbers, enhancing their effectiveness in water-based formulations. • **Enhanced Stability**: CDs protect organic UVR absorbers from photodegradation by shielding them from direct UV exposure, prolonging their effectiveness. • **Versatility**: Compatible with various formulations, including sunscreens, cosmetics, and personal care products. • **Reduced Irritation**: Encapsulation in CDs can reduce the skin irritation potential of some organic UVR absorbers.	• **Lower UV Blocking Efficiency**: Compared to inorganic and nanoparticle absorbers, organic UVR absorbers generally provide less efficient UV blocking, particularly against UVA rays. • **Potential for Skin Penetration**: Some organic UVR absorbers can penetrate the skin, raising concerns about long-term safety. • **Environmental Concerns**: Certain organic UVR absorbers can have adverse environmental effects, such as contributing to coral bleaching.	Commonly used in sunscreens, moisturizers, and other personal care products where high solubility and skin compatibility are required.	[3] [4] [5] [10] [12]

Inorganic UVR Absorbers with CDs	• **Common Types**: Titanium dioxide (TiO_2), zinc oxide (ZnO), and cerium oxide (CeO_2). • **Mechanism**: Inorganic absorbers block UVR through reflection, scattering, and absorption due to their high refractive index and wide bandgap. • **Cyclodextrin Role**: CDs can reduce the photocatalytic activity of inorganic particles, enhance their dispersibility in formulations, and prevent aggregation.	• **Broad-Spectrum Protection**: Inorganic UVR absorbers provide excellent broad-spectrum UV protection, covering both UVA and UVB regions. • **Photostability**: They are inherently photostable and do not degrade upon UV exposure, maintaining their protective properties over time. • **Reduced Photocatalytic Activity**: CDs can mitigate the photocatalytic effects of metal oxides, reducing the generation of reactive oxygen species (ROS) that can cause skin damage. • **Non-Irritating**: Generally regarded as safe and non-irritating for the skin, making them suitable for sensitive skin and children's formulations.	• **Aesthetic Issues**: Inorganic absorbers can leave a white cast on the skin, especially at higher concentrations. CDs help improve dispersibility but may not completely eliminate this issue. • **Aggregation**: In the absence of proper stabilization, inorganic particles tend to aggregate, reducing their effectiveness. CDs aid in dispersion but may require additional stabilizers. • **Formulation Challenges**: Integrating inorganic absorbers into certain formulations can be challenging due to their particulate nature and impact on texture and feel.	Widely used in sunscreens, daily wear cosmetics, and protective coatings for outdoor materials and surfaces where high UVR blocking is needed without compromising safety.	[4] [5] [8] [45]

(continued)

TABLE 12.6 (Continued)
A Comparative Analysis of These UVR Absorbers with Cyclodextrin

Absorbers	Characteristics	Advantages	Challenges	Applications	Ref.
Nanoparticle UVR Absorbers with CDs	• **Common Types**: Nanoparticles of TiO_2, ZnO, fullerene derivatives, quantum dots, and silica-coated UV absorbers. • **Mechanism**: Nanoparticles provide UV protection through absorption, scattering, and reflection at the nanoscale, offering enhanced UVR absorption efficiency. **Cyclodextrin Role**: CDs help to stabilize nanoparticles, prevent aggregation, enhance dispersibility, and functionalize the surface for specific applications.	• **High UV Blocking Efficiency**: Nanoparticles have a larger surface area and can be engineered for optimized UV absorption, providing superior protection compared to bulk inorganic particles. • **Reduced White Cast**: Due to their nanoscale size, these particles are less visible on the skin, overcoming aesthetic limitations associated with larger inorganic particles. • **Enhanced Functionalization**: CDs allow for easy surface modification of nanoparticles, enabling the addition of functional groups for targeted applications, such as controlled release of UVR absorbers. **Environmental Safety**: Cyclodextrin-functionalized nanoparticles are less likely to leach and pose reduced environmental risks compared to conventional nanoparticles.	• **Regulatory and Safety Concerns**: The use of nanoparticles raises concerns about safety and potential toxicity, especially regarding their ability to penetrate the skin or enter the environment. • **Cost and Scalability**: Synthesis and functionalization of nanoparticles with CDs can be more complex and costly compared to traditional organic and inorganic UVR absorbers. **Potential for Photocatalytic Activity**: Similar to inorganic absorbers, certain nanoparticles can exhibit photocatalytic properties, necessitating careful design and stabilization to avoid ROS generation.	Used in high-performance sunscreens, advanced coatings, and films for automotive, architectural, and agricultural purposes, where enhanced UV protection and functionalization are desired.	[3] [4] [10] [12]

potential. CD inorganic absorbers offer broad-spectrum protection with minimal safety concerns, while CD nanoparticle absorbers provide superior protection and functionalization for advanced applications. When comparing these types, factors like stability, resolution, security, and application must be considered. Table 12.6 provides a comparative analysis of UVR absorbers with cyclodextrins [2–4, 8, 10, 46, 47].

12.9 RESULTS AND DISCUSSIONS

While cyclodextrin-enhanced UV protective coatings offer many benefits, there are several challenges that need to be addressed to fully leverage their potential in commercial applications [5, 7, 10, 29, 31, 46].

For example, achieving uniform application of CD-UVR absorber complexes on substrates is challenging, particularly for large-scale production. Variations in coating thickness can impact UV protection, and strong adhesion to diverse substrates (e.g., glass, polymers) is essential for durability under mechanical stress, UV exposure, and thermal cycling.

- High-purity CDs and specialized UVR absorbers increase synthesis costs. Scaling up production while maintaining quality poses challenges, requiring a balance between cost and benefits for industries like packaging and textiles.
- Despite enhancing UVR absorber stability, coatings must maintain performance under long-term UV exposure and variable environmental conditions, resisting degradation from contamination, moisture, and chemicals.
- Compatibility with diverse substrates, including glass, polymers, and composites, necessitates tailored formulations and application methods for optimal adhesion and performance.
- Maintaining optical clarity, colour, and mechanical properties is critical, particularly for clear glass and polymer applications.
- Coatings must meet regulatory safety standards for human contact or food packaging, with consideration for environmental impact, toxicity, and biodegradability.

Despite these challenges, there are numerous opportunities for the advancement of CD-based UV protective coatings, driven by ongoing research and innovation [3, 10, 44, 42, 46]. For example, as follows:

- Modified CDs with specific functional groups enhance encapsulation, stability, and compatibility of UVR absorbers, enabling tailored coatings with added functionalities like antireflective, antifog, or antimicrobial properties for broader applications.
- Combining CDs with nanoparticles, leveraging high surface area and tunable optical properties, enhances UV blocking and stability. Stimuli-responsive materials, adapting to UV exposure, enable smart coatings for varying conditions.

- Biodegradable CDs and non-toxic UVR absorbers support eco-friendly products. Bio-based CDs and greener synthesis reduce environmental impact, while removable or recyclable coatings improve lifecycle management.
- Coatings resistant to extreme conditions, like high UV intensity or marine environments, expand applications to aerospace, military, infrastructure, automotive, and electronics. Long-lasting UV protection under harsh conditions is essential.
- Growing demand for portable and outdoor electronics requires UV coatings to protect screens and components. Biocompatible, antimicrobial UV coatings suit medical devices and UV-exposed equipment.
- Collaboration between academia, industry, and regulators accelerates CD-based coating development, addressing scaling, cost, and compliance. Promoting health and safety benefits can boost consumer demand and market growth.

12.10 CONCLUSION

CD-based UV protective coatings enhance UV protection for materials like glass, polymers, and textiles. By encapsulating UVR absorbers, cyclodextrins improve solubility, stability, and photostability, ensuring durable and efficient coatings. Their inclusion complexes maintain substrate optical clarity and mechanical integrity, enabling diverse industrial applications.

This chapter discusses advancements through cyclodextrin integration, including better stabilization of organic UV absorbers, improved dispersion of inorganic absorbers, and enhanced nanoparticle system functionality. These innovations tackle challenges like material degradation and limited durability, ensuring long-term UV protection. However, scalability, economic viability, and environmental concerns need further investigation.

Future research should focus on integrating nanoparticles and smart materials for next-gen coatings with enhanced UV protection, self-healing, and environmental responsiveness. Green chemistry approaches, like biodegradable cyclodextrins and eco-friendly UV absorbers, should be prioritized for sustainability. Exploring coatings for extreme conditions could expand applications in aerospace, automotive, and outdoor industries. Collaboration between academia, industry, and regulators is crucial to overcome technical, economic, and regulatory challenges, enabling cyclodextrin-based UV coatings to offer sustainable, high-performance solutions across industries.

REFERENCES

1. El-Tahlawy K., El-Nagar K., Elhendawy A. G. (2007) 'Cyclodextrin-4 Hydroxy benzophenone inclusion complex for UV protective cotton fabric' *Journal of the Textile Institute*, Vol 98, pp 453–462. https://doi.org/10.1080/00405000701556327
2. Carvalho P. G., Naveira R. I., Granone L. I., Mendive C. B., Massa A. E., Churio M. S. (2023) 'A comparative review of natural and synthetic UV filters: Gadusol and benzophenone-3 as representative examples' *Environmental Advances*, Vol 13, p 100404. https://doi.org/10.1016/j.envadv.2023.100404

3. Morabito K., Shapley N. C., Steeley K. G., Tripathi A. (2011) 'Review of sunscreen and the emergence of non-conventional absorbers and their applications in ultraviolet protection' *International Journal of Cosmetic Science*, Vol 33, pp 385–390. https://doi:10.1111/j.1468-2494.2011.00654.x
4. Hassabo A. G., Ragab M. (2022) 'Ultarvielote protection of cellulosic fabric' *Journal of Textiles, Coloration and Polymer Science*, Vol 19, pp 51–61. DOI: 10.21608/jtcps.2021.105105.1092
5. Jesus A., Sousa E., Cruz M. T., Cidade H., Lobo J. M. S., Almeida I. F. (2022) 'UV filters: Challenges and prospects' *Pharmaceuticals*, Vol 15, p 263. https://doi.org/10.3390/ph15030263
6. Celebioglu A., Uyar T. (2012) 'Electrospinning of nanofibers from non-polymeric systems: Polymer-free nanofibers from cyclodextrin derivatives' *Nanoscale*, Vol 4, pp 621–631. https://doi.org/10.1039/C1NR11364J
7. Landy D., Mallard I., Ponchel A., Monflier E., Fourmentin S. (2012) 'Cyclodextrins for Remediation Technologies'. In Eric Lichtfouse, Jan Schwarzbauer & Didier Robert (eds.), *Environmental Chemistry for a Sustainable World: Volume 1: Nanotechnology and Health Risk*, Springer Dordrecht, pp 47–81. DOI 10.1007/978-94-007-2442-6
8. Cova T. F., Murtinho D., Pais A. A., Valente A. J. (2018) 'Combining cellulose and cyclodextrins: Fascinating designs for materials and pharmaceutics' *Frontiers in Chemistry*, Vol 6, p 271. https://doi.org/10.3389/fchem.2018.00271.
9. Kali G., Haddadzadegan S., Bernkop-Schnürch A. (2024) 'Cyclodextrins and derivatives in drug delivery: New developments, relevant clinical trials, and advanced products' *Carbohydrate Polymers*, Vol 324, p 121500. https://doi.org/10.1016/j.carbpol.2023.121500
10. Santos B. A., da Silva A. C., Bello M. L., Goncalves A. S., Gouvea T. A., Rodrigues R. F., Rodrigues C. R. (2018) 'Molecular modeling for the investigation of UV absorbers for sunscreens: Triazine and benzotriazole derivatives' *Journal of Photochemistry and Photobiology A: Chemistry*, Vol 356, pp 219–229. https://doi.org/10.1016/j.jphotochem.2017.12.036
11. Manaia E. B., Kaminski R. C. K., Corrêa M. A., Chiavacci L. A. (2013) 'Inorganic UV filters' *Brazilian Journal of Pharmaceutical Sciences*, Vol 49, pp 201–209. https://doi.org/10.1590/S1984-82502013000200002
12. Abou Elmaaty T. M., Mandour B. A. (2018) 'ZnO and TiO2 nanoparticles as textile protecting agents against UV radiation: A review' *Asian Journal of Chemical Science*, Vol 4, pp 1–14. DOI: 10.9734/AJOCS/2018/40329
13. Loftsson T., Brewster M. E. (1996) 'Pharmaceutical applications of cyclodextrins. 1. Drug solubilization and stabilization' *Journal of Pharmaceutical Sciences*, Vol 85, pp 1017–1025. https://doi.org/10.1021/js950534b
14. Alrbaihat M. R. (2023) 'Mechanochemical synthesis of Host-Guest inclusion complexes of Cyclodextrin: A review' *Biomedicine and Chemical Sciences*, Vol 2, pp 102–108. https://doi.org/10.48112/bcs.v2i2.287
15. Poulson B. G., Alsulami Q. A., Sharfalddin A., El Agammy E. F., Mouffouk F., Emwas A. H., Jaremko M. (2021) 'Cyclodextrins: Structural, chemical, and physical properties, and applications' *Polysaccharides*, Vol 3, pp 1–31. https://doi.org/10.3390/polysaccharides3010001
16. Bezerra F. M., Lis M. J., Firmino H. B., Dias da Silva J. G., Curto Valle R. D. C. S., Borges Valle J. A., Tessaro A. L. (2020) 'The role of β-cyclodextrin in the textile industry' *Molecules*, Vol 25, p 3624. https://doi.org/10.3390/molecules25163624

17. Kumari P., Singh P., Singhal A. (2020) 'Cyclodextrin-based nanostructured materials for sustainable water remediation applications' *Environmental Science and Pollution Research*, Vol 27, pp 32432–32448. https://doi.org/10.1007/s11356-020-09519-0
18. Krishnamoorthy S., Allabasha N., Mani M. K., Sarkar A. K. (2022) 'Michler's Ketone: β-Cyclodextrin Host–Guest inclusion complex for enhancing the ultraviolet protection factor of Poplin Cotton fabric' *Photochemistry and Photobiology*, Vol 98, pp 1284–1292. https://doi.org/10.1111/php.13619
19. Eteer S. A. (2022) 'UV-Vis spectroscopic characterization of β-Cyclodextrin-Vanillin inclusion complex' *Mediterr. J. Chem.*, Vol 12, pp 10–13171. DOI:10.13171/mjc02210111645eteer
20. Kou X., Zhang Y., Su D., Wang H., Huang X., Niu Y., Meng Q. (2023) 'Study on host-guest interaction of aroma compounds/γ-cyclodextrin inclusion complexes' *LWT*, Vol 178, 114589. https://doi.org/10.1016/j.lwt.2023.114589
21. Zhang J., Zhang X., Wang R., Wang W., Zhao H., Yang S., Pan Y. T. (2024) 'Cyclodextrin-based host-guest hierarchical fire retardants: Synthesis and novel strategy to endow polylactic acid fire retardancy and UV resistance' *Carbohydrate Polymers*, p 122313. https://doi.org/10.1016/j.carbpol.2024.122313
22. Martel B., Morcellet M., Ruffin D., Vinet F., Weltrowski L. (2002) 'Capture and controlled release of fragrances by CD finished textiles' *Journal of Inclusion Phenomena and Macrocyclic Chemistry*, Vol 44, pp 439–442.
23. Dastjerdi R., Montazer M. (2010) 'A review on the application of inorganic nano-structured materials in the modification of textiles: Focus on anti-microbial properties' *Colloids and Surfaces B: Biointerfaces*, Vol 79, pp 5–18. https://doi.org/10.1016/j.colsurfb.2010.03.029
24. Khan A., Nazir A., Rehman A., Naveed M., Ashraf M., Iqbal K., Maqsood H. S. (2020) 'A review of UV radiation protection on humans by textiles and clothing' *International Journal of Clothing Science and Technology*, Vol 32, pp 869–890. https://doi:10.1108/ijcst-10-2019-0153
25. Zille A., Almeida L., Amorim T., Carneiro N., Esteves M. F., Silva C. J., Souto A. P. (2014) 'Application of nanotechnology in antimicrobial finishing of biomedical textiles' *Materials Research Express*, Vol 1, p 032003. DOI: 10.1088/2053-1591/1/3/032003
26. Abdulaziz F., Salah D. (2021) 'Gold nanoparticles incorporated with cyclodextrins and its applications' *Journal of Biomaterials and Nanobiotechnology*, Vol 12, pp 79–97. DOI: 10.4236/jbnb.2021.124007
27. El-Sayed E., Othman A. H., Hassabo A. G. (2021) 'Cyclodextrin usage in textile industry' *Journal of Textiles, Coloration and Polymer Science*, Vol 18, pp 111–119. DOI: 10.21608/jtcps.2021.76238.1059
28. Sfameni S., Hadhri M., Rando G., Drommi D., Rosace G., Trovato V., Plutino, M. R. (2023) 'Inorganic finishing for textile fabrics: Recent advances in wear-resistant, UV protection and antimicrobial treatments' *Inorganics*, Vol 11, pp 19. https://doi.org/10.3390/inorganics11010019
29. Dahabra L., Broadberry G., Le Gresley A., Najlah M., Khoder M. (2021) 'Sunscreens containing cyclodextrin inclusion complexes for enhanced efficiency: A strategy for skin cancer prevention' *Molecules*, Vol 26, pp 1698. https://doi.org/10.3390/molecules26061698
30. Velázquez-Contreras F., Zamora-Ledezma C., López-González I., Meseguer-Olmo L., Núñez-Delicado E., Gabaldón J. A. (2021) 'Cyclodextrins in polymer-based active food packaging: A fresh look at nontoxic, biodegradable, and sustainable technology trends' *Polymers*, Vol 14, pp 104. https://doi.org/10.3390/polym14010104

31. Visakh P. M. (2022) 'Nanomaterials and Nanotechnology in Medicine: Medical Applications: Challenges and Opportunities'. In P.M. Visakh (ed.), *Nanomaterials and Nanotechnology in Medicine*, John Wiley & Sons Ltd., pp 1–25. https://doi.org/10.1002/9781119558026.ch1
32. Gholap A. D., Sayyad S. F., Hatvate N. T., Dhumal V. V., Pardeshi S. R., Chavda V. P., Vora L. K. (2023) 'Drug delivery strategies for avobenzone: A case study of photostabilization' *Pharmaceutics*, Vol 15, p 1008. https://doi.org/10.3390/pharmaceutics15031008
33. Lorigo M., Quintaneiro C., Breitenfeld L., Cairrao E. (2024) 'Exposure to UV-B filter octylmethoxycinnamate and human health effects: Focus on endocrine disruptor actions' *Chemosphere*, Vol 358, p 142218. https://doi.org/10.1016/j.chemosphere.2024.142218
34. Messias M. A., Ferreira S. M., Tavares L., Santos L. (2023) 'A comparative study between onion peel extracts, free and complexed with β-Cyclodextrin, as a natural UV filter to cosmetic formulations' *International Journal of Molecular Sciences*, Vol 24, p 15854. https://doi.org/10.3390/ijms242115854
35. Ferreira L., Mascarenhas-Melo F., Rabaça S., Mathur A., Sharma A., Giram P. S., Paiva-Santos A. C. (2023) 'Cyclodextrin-based dermatological formulations: Dermopharmaceutical and cosmetic applications' *Colloids and Surfaces B: Biointerfaces*, Vol 221, p 113012. https://doi.org/10.1016/j.colsurfb.2022.113012
36. Rajamohan R., Kamaraj E., Govindasamy C., Subramanian S. A., Kim S. J., Krishnan M. M., Lee Y. R. (2024) 'Formulation of ensulizole with Beta-Cyclodextrins for improved sunscreen activity and biocompatibility' *Journal of Pharmaceutical Sciences*, Vol 113, pp 1536–1545. https://doi.org/10.1016/j.xphs.2023.12.019
37. Gholap A. D., Pardeshi S. R., Hatvate N. T., Dhorkule N., Sayyad S. F., Faiyazuddin M., Khalid M. (2024) 'Environmental implications and nanotechnological advances in Octocrylene-Enriched sunscreen formulations: A comprehensive review' *Chemosphere*, p 142235. https://doi.org/10.1016/j.chemosphere.2024.142235
38. Chavda V. P., Acharya D., Hala V., Daware S., Vora L. K. (2023) 'Sunscreens: A comprehensive review with the application of nanotechnology' *Journal of Drug Delivery Science and Technology*, Vol 86, p 104720. https://doi.org/10.1016/j.jddst.2023.104720
39. Rajamohan R., Muthuraja P., Govindasamy C., Subramanian S. A., Kim S. J., Krishnan M. M., Lee Y. R. (2023) 'Enhanced photostability and biocompatibility of sunscreen formulation of 2-phenylbenzimidazole-5-sulfonic acid with methyl-beta-cyclodextrin' *Journal of Molecular Liquids*, Vol 390, p 123013. https://doi.org/10.1016/j.molliq.2023.123013
40. Fenyvesi É., Puskás I., Szente L. (2021) 'Encapsulation for Packaging' *Functionality of Cyclodextrins in Encapsulation for Food Applications*, pp 187–222. https://doi.org/10.1007/978-3-030-80056-7_11
41. Shi C., Zhou A., Fang D., Lu T., Wang J., Song Y., Li W. (2022) 'Oregano essential oil/β-cyclodextrin inclusion compound polylactic acid/polycaprolactone electrospun nanofibers for active food packaging' *Chemical Engineering Journal*, Vol 445, pp 136746. https://doi.org/10.1016/j.cej.2022.136746
42. Yang J., Wiley C. J., Godwin D. A., Felton L. A. (2008) 'Influence of hydroxypropyl-β-cyclodextrin on transdermal penetration and photostability of avobenzone' *European Journal of Pharmaceutics and Biopharmaceutics*, Vol 69, pp 605–612. https://doi.org/10.1016/j.ejpb.2007.12.015
43. Prorokova N. P., Odintsova O. I., Rumyantseva V. E., Rumyantsev E. V., Konovalova V. S. (2023) 'Giving improved and new properties to fibrous materials by surface modification' *Coatings*, Vol 13, pp 139. https://doi.org/10.3390/coatings13010139

44. Giannossa L. C., Longano D., Ditaranto N., Nitti M. A., Paladini F., Pollini M., Cioffi N. (2013) 'Metal nanoantimicrobials for textile applications' *Nanotechnology Reviews*, Vol 2, pp 307–331. https://doi.org/10.1515/ntrev-2013-0004
45. Crini G., Fourmentin S., Fenyvesi É., Torri G., Fourmentin M., Morin-Crini N. (2018) 'Cyclodextrins, from molecules to applications' *Environmental Chemistry Letters*, Vol 16, pp 1361–1375. https://doi.org/10.1007/s10311-018-0763-2
46. Cabrales L., Abidi N., Hammond A., Hamood A. (2012) 'Cotton fabric functionalization with cyclodextrins' *Surfaces*, Vol 6, ISSN: 2028-2508.

13 Impact of Cyclodextrins on the Ultraviolet/ Sun Protection Factors (UVPF/SPF) of UVR Absorbers

Subhadeep Saha, Tanmoy Dutta, Sudipta K. Kundu, Ekramul Kabir, and Abdul Ashik Khan

13.1 INTRODUCTION

Skin cancer, the most frequent type of cancer around the world, commonly known as non-melanoma skin cancer (NMSC) or melanoma [1]. Both melanoma and non-melanoma skin cancers are linked to ultraviolet radiation (UVR), and using sunscreens and UV-protective textiles has been advised as the most effective way to avoid these diseases [2]. UVR is a component of the electromagnetic spectrum, with wavelengths ranging from visible light to gamma radiation. UV energy is classified into UVA, UVB, and UVC components based on electrophysical parameters, with UVC photons having the shortest wavelengths (100–280 nm) and the highest energy, UVA having the longest (320–400 nm) but the least energetic photons, and UVB falling in between [3]. UVB exposure can directly damage DNA, resulting in skin cancer. On the other hand, while having the most energy, UVC radiation is prevented by the atmospheric ozone layer and hence has no adverse effect on skin [4]. Different problems arise due to direct contact of UVB and UVA with the skin.

Sunscreens are topical preparations that protect the skin from ultraviolet radiation (UV). Sunscreens contain UV filters, which absorb, scatter, and/or reflect UV radiation [5]. Natural antioxidants are added to sunscreens to prevent the creation of free radicals and repair DNA damage, providing a second layer of UV protection [6]. Antioxidants can assist these compositions stay stable during the production process and after application to the skin. However, both UV filters and antioxidants degrade when exposed to sunshine and oxygen [7].

Sunscreens and UV-protective textiles are critical in mitigating these risks. The effectiveness of these protective measures is often quantified by the sun protection factor (SPF) and ultraviolet protection factor (UVPF) [8]. SPF and UVPF factors

DOI: 10.1201/9781003492757-13

mainly reflect the minimizing capability of the risk factor from UVR. Recent research has focused on the enhancement of these factors through various means, one of which is the inclusion of cyclodextrins (CDs) in formulations.

Cyclodextrins (CDs) are cyclic oligosaccharides known for their ability to form inclusion complexes with various molecules, including UVR absorbers [9]. CDs have been studied for their potential to enhance the performance of UVR absorbers by improving their solubility and stability. This review critically examines the role of CDs in enhancing the UVPF/SPF of UVR absorbers, focusing on the chemical interactions, the photostability of the complexes, and the practical applications in sunscreens and UV-protective textiles.

13.2 CHEMISTRY OF CYCLODEXTRINS

Cyclodextrins (CDs) are a family of macrocyclic oligosaccharides composed of 6–12 D (+)-glucopyranose monomers in a chair conformation linked by α-1,4-glycosidic bonds (Figure 13.1). It is extracted from the degradation of starch by

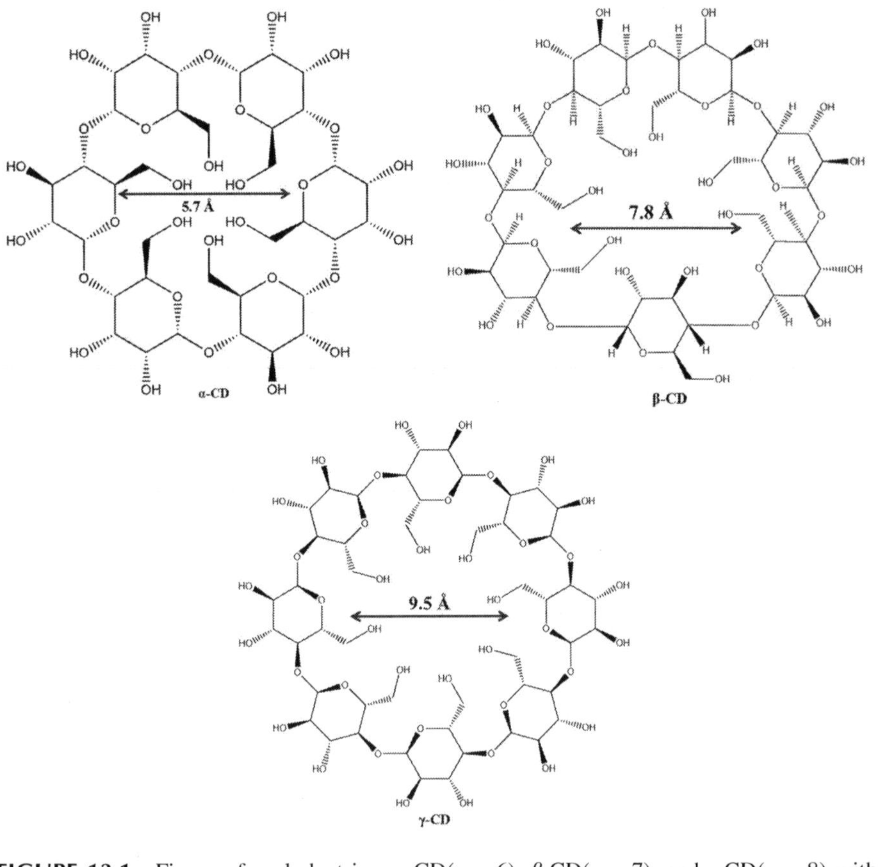

FIGURE 13.1 Figure of cyclodextrins: α-CD(n = 6), β-CD(n = 7), and γ-CD(n = 8) with diameter [13].

glucosyltransferase [10]. The most common types are α-, β-, and γ-cyclodextrins, which differ in the number of glucose units (six, seven, and eight, respectively). The number of glucose units in CDs determine the cavity size and diameter, i.e., 0.57 nm for α-CD, 0.78 nm for β-CD and 0.95 nm for γ-CD. With eight glucose units, γ-CD has the biggest cavity size and can accommodate significant antioxidants [11]. However, greater cavities may not necessarily equal higher inclusion complex stability, as large cavities may allow for the easier escape of guest molecules. Although β-CD has low solubility in water (25°C, 18.5 mg mL^{-1}), highly hemolytic and highly nephrotoxic, β-CD is frequently employed due to its medium cavity size, which allows for robust fitting of drugs with molecular weights ranging from 200 to 800 Daltons, as well as low production costs [12]. Therefore, scientists need to boost the solubility of β-CD by producing several CD derivatives such as methylated-β-CD, hydroxypropyl-β-CD, sulfobutylether-β-cyclodextrin and acetylated-β-CD.

The unique truncated-cone-shaped structure of CDs creates a hydrophobic inner cavity and a hydrophilic outer surface, allowing a variety of hydrophobic molecules (such as natural antioxidants and anticancer drugs) of suitable size and shape to form inclusion complexes and improving molecular stability and solubility [14]. Several factors contribute to the successful formation of an inclusion complex. Non-covalent forces including hydrogen bonding, hydrophobic interactions, and van der Waals forces, play a crucial role in retaining the guest molecule inside the cavity [15].

The ability of CDs to encapsulate guest molecules within their hydrophobic cavity is a key property that has been exploited in many fields, including pharmaceuticals, food, and cosmetics [16]. In the context of UV protection, CDs can encapsulate UVR absorbers and antioxidant, potentially altering their solubility, stability, and efficacy. The absence of lowering glycosidic units in CD structures contributes significantly to their chemical stability. The neighboring C-2 and C-3 hydroxyl groups of the glucopyranose units generate hydrogen bonds, which increase the chemical stability of CD structures and boost water solubility [17]. This complexation phenomenon can alter the guest's physicochemical and chemical properties, such as boosting its oxidant and light stability and apparent aqueous solubility [18].

There are different methods involved for the synthesis of inclusion complexes.

Co-precipitation: it involves the simultaneous formation of solid particles and the precipitation of a soluble compound from a solution.

Freeze drying: it is a process that removes water from a frozen substance by placing it under a vacuum.

Ball milling: it is a mechanical process that uses grinding balls to crush materials into fine particles.

Kneading: it is a technique that involves stretching and folding of dough to strengthen gluten strands.

In comparison, the kneading process demonstrates decreased effectiveness in entrapment capacities. As a result, co-precipitation is increasingly widely used in producing a stable inclusion complex of drugs [18].

The generated inclusion complex was then characterized using Fourier transform infrared spectroscopy (FT-IR), ultraviolet-visible spectroscopy (UV-vis), ^{1}H nuclear magnetic resonance (^{1}H NMR), X-ray diffraction (XRD), differential scanning calorimetry (DSC), and scanning electron microscopy (SEM) [19].

13.3 UV RADIATIONS AND UV/SUN PROTECTION FACTORS (UVPF/SPF)

Human skin is composed of three primary layers: the epidermis, the dermis, and the hypodermis. Epidermis is the outermost layer that provides a waterproof barrier and contributes to skin tone. Dermis is the middle layer that contains connective tissue, hair follicles, blood vessels, lymphatic vessels, and sweat glands and hypodermis is the deepest layer made of fat and connective tissue. There are three types of UV radiation: UVA, UVB, and UVC. The majority of UV energy that reaches Earth is UVA, or long wave UV radiation, which has a wavelength of 320–400 nm. Due to its ability to reach the skin's dermis layer, UVA is the primary cause of skin ageing, wrinkles and photo ageing, skin cancer, and damage to the eyes. UVB, or mid-strength UV light, has a 280–320 nm wavelength. Most of it is absorbed by the atmosphere, and when it reaches the skin, it can only enter the epidermal layer, typically burning the skin's outermost layers. Since UVC is the weakest UV wavelength (100–280 nm), it does not impact the biota since the ozone layer absorbs it and does not reach Earth (Figure 13.2).

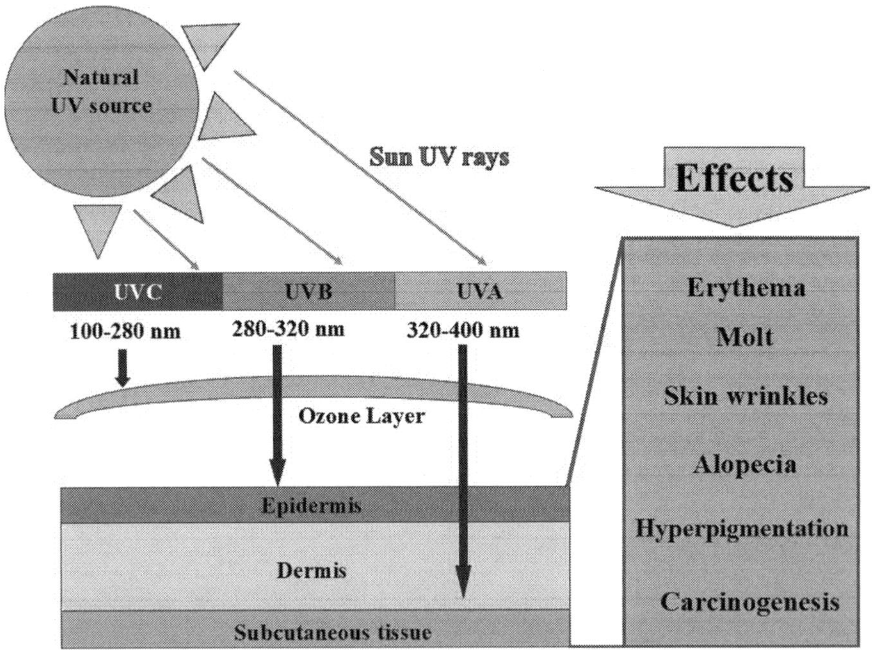

FIGURE 13.2 Effect of UV rays on skin [20].

The degree of UV protection provided by fabric or sunscreen is measured using ultraviolet protection factor, or UVPF. The primary distinction between UVPF and SPF values lies in their determination methods. UVPF values are derived from instrumental measurements, while SPF values are determined through human testing for sunscreens [21]. The formula for calculating UV protection factor (UVPF) is the ratio of the average effective UV irradiance on exposed skin to the average UV irradiance on skin covered by the test fabric. A higher UVPF value indicates that a person can stay in the sun longer before the skin beneath until the region of skin beneath the fabric turns redden [22].

Fabrics with a UVPF value in the range 15–24 were classified as having "Good UV Protection" when the UVPF values were between 25 and 39, fabrics were classified as having "Very Good UV Protection," and "Excellent UV Protection" classification was used when the UVPF was 40 or greater [23]. In no event was a fabric assigned a UVPF rating greater than 50.

13.4 CYCLODEXTRINS AND UVR ABSORBERS

Cyclodextrins are really having interest in the formulation of different highly effective UV radiation absorbing molecules with various applications. Felton *et al.* showed the photoprotective effects of a model ultraviolet absorber, oxybenzone, with hydroxypropyl-β-cyclodextrin (HPCD) [24]. Oxybenzone is an organic compound belonging to the class of aromatic ketones known as benzophenones. It takes the form of pale-yellow crystals that are readily soluble in most organic solvents. It is widely used in sunscreen formulations, plastics, toys, furniture finishes, and other products to limit UV degradation (Figure 13.3). They also compared these novel sunscreens to a commercial SPF30 sunscreen by applying these on the skin of SKH-1 hairless mice (Table 13.1) [24]. The number of sunburn cells (SBCs) was measured 24 hours post exposure. They showed in their results that all oxybenzone-containing formulations decreased the number of SBCs formed, diminished swelling, and reduced the physical damage to the skin structure, in comparison to control (Figure 13.4). The 20% HPCD formulation exhibited more substantial photo-protection at UV exposures as evidenced by the formation of fewer SBCs. The 5% HPCD formulation also provided substantial protection. These studies demonstrate that inclusion of HPCD in sunscreen formulations may enhance the in vivo photoprotective effects of the UV absorbers.

The solubilization of ultraviolet absorbers by cyclodextrin and their potential application in cosmetics were shown by Mori et al. [25]. Protection of our skin from ultraviolet (UV) rays is very important to keep the skin healthy and beautiful. UV

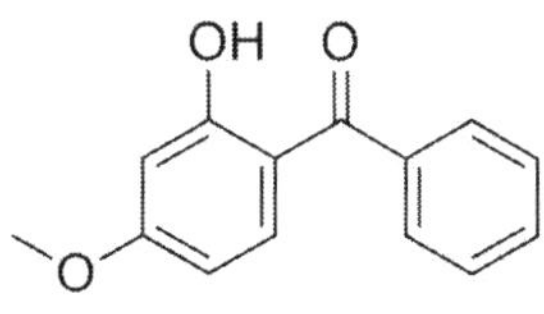

FIGURE 13.3 Molecular structure of oxybenzone.

TABLE 13.1
Comparison of photoprotective Effects of a Commercial SPF 30 Sunscreen and the HPCD-Containing Oxybenzone Formulations at Two MEDs

Formulation	Sunburn Cells / Per cm
Control	35.50 ± 5.32
0% HPCD	15.75 ± 4.86
5% HPCD	14.00 ± 3.92
10% HPCD	14.00 ± 2.16
20% HPCD	6.75 ± 2.63
SPF 30 Sunscreen[a]	4.75 ± 2.75

Data reported as the ± standard deviation.
[a] No-Ad® Ultra Sun Block lotion [24].

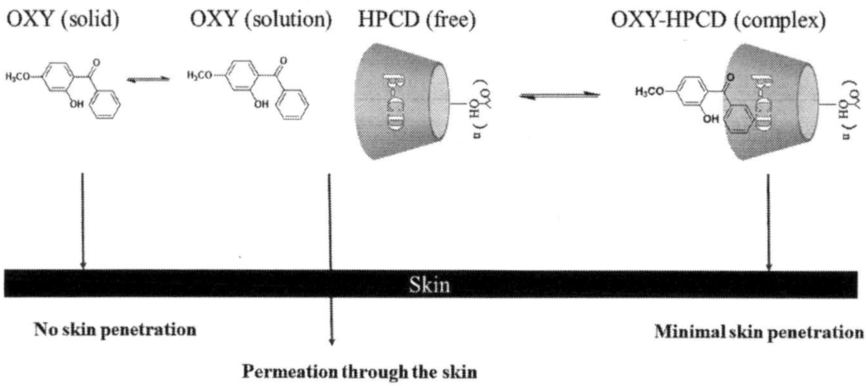

FIGURE 13.4 Schematic representation of the complexation and transdermal permeation of the oxybenzone-HPCD systems [24].

radiation is subdivided into UVA, UVB, and UVC. UVC (wavelength 200–280 nm) is absorbed by the ozone layer and does not reach the Earth. But UVA (wavelength 320–395 nm) makes about 95% of the UV radiation that reaches the Earth and penetrates into the dermis layer of the skin and damages it. UVB (wavelength 280–320 nm) most of which is absorbed by the atmosphere, and while reaching the skin, it can only penetrate into the epidermis layer and usually burns the superficial layers of skin. Because of the relatively short wavelength of UV radiation, it stimulates our skin strongly and therefore, even a short irradiation time may cause sunburns. They showed that cyclodextrin can make inclusion complex with the UV absorber in aqueous solution. It was established by a phase solubility test and Job plot that hydroxypropylated-γ-cyclodextrin (HP-γ-CD) made an inclusion complex with ferulic acid in a stoichiometry of 1:2. Ferulic acid is a hydroxycinnamic acid derivative and a phenolic compound. The name is derived from the genus *Ferula*, referring to the giant fennel. Classified as a phenolic phytochemical, ferulic acid is an amber coloured solid. In cosmetic applications, the formulation containing the ferulic acid/HP-γ-CD inclusion complex showed a higher sun-cut effect than that containing octylmethoxycinnamate (OMC) (Figure 13.5). Octyl methoxycinnamate is an organic compound that is an ingredient in some sunscreens and lip balms. It is an ester formed from methoxycinnamic acid and 2-ethylhexanol. It is a liquid that is insoluble in water. Thus, the ferulic acid/HP-γ-CD inclusion complex showed extra ordinarily high protection values for UV, which may be useful for preparing effective and safe sunscreen formulations. The data was reported for trans-ferulic acid (*t*FA), rutin, alloxazine, OMC, etc., encapsulated into α-CD, methylated-β-cyclodextrin (Me-β-CD), hydroxypropylated-β-cyclodextrin (HP-β-CD), γ-CD, and hydroxylpropylated-γ-cyclodextrin (HP-γ-CD) (Figure 13.6). Rutin is the glycoside combining the flavonol quercetin and the disaccharide rutinose. It is a flavonoid glycoside found in a wide variety of plants, including citrus.

It is observed that upon complexation with the CDs, a bathochromic shift of *t*FA and a hypsochromic shift of rutin were detected, whereas no shifts were observed in the spectra of alloxazine. The absorbance maximum of *t*FA was slightly shifted, from 312 to 316 nm, upon formation of an inclusion complex. Some researchers have reported that these shifts in the UV–Vis spectra are induced by interactions between the guest compound and host CD [26]. Furthermore, because of this bathochromic shift of the *t*FA/CD inclusion complexes, the UV absorption abilities between 320 and 370 nm were greatly increased. This property might be useful for preparing

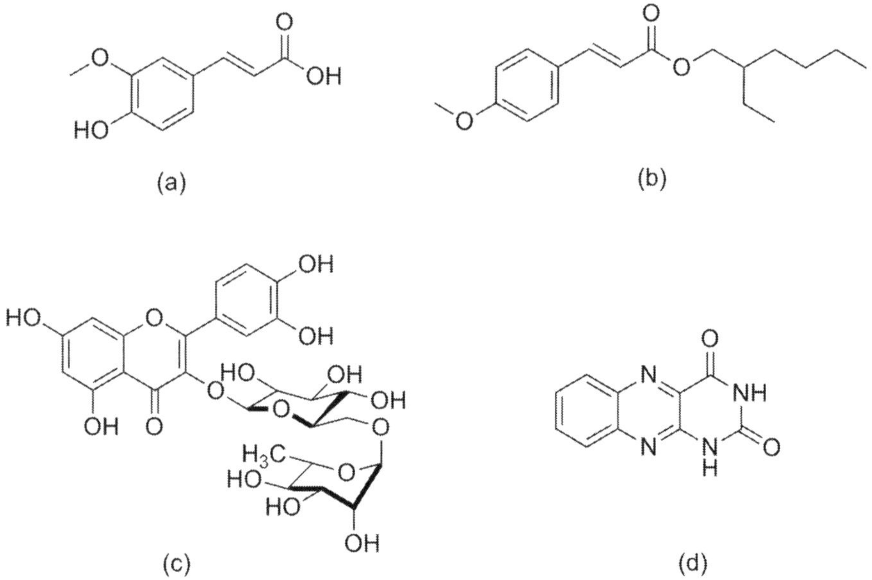

FIGURE 13.5 Molecular structure of (a) trans-ferulic acid, (b) octylmethoxycinnamate, (c) rutin, and (d) alloxazine.

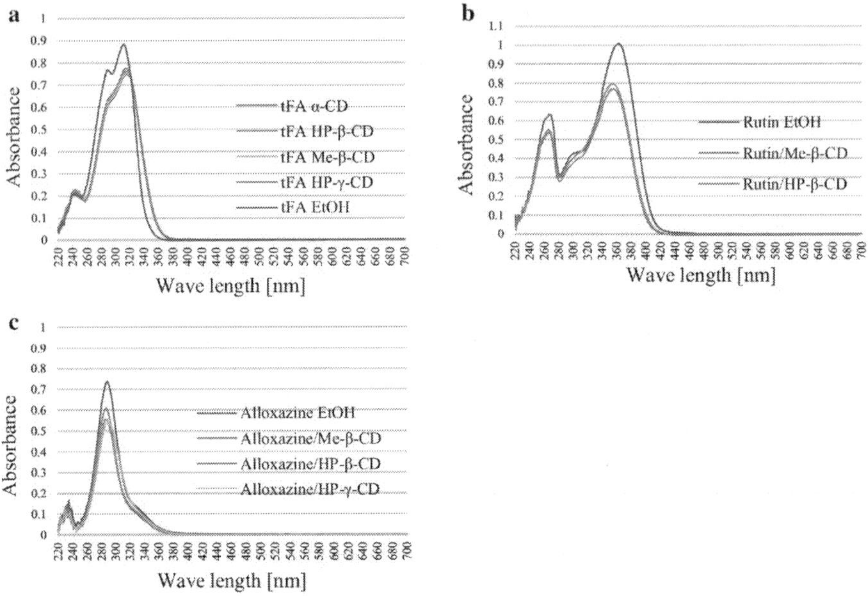

FIGURE 13.6 UV–Vis absorption spectra of *t*FA (a), rutin (b), and alloxazine (c) upon complexation with CDs. Solutions of each sample (0.05 mM) were scanned at 25 °C to obtain these spectra [25].

effective cosmetic sunscreens, because UV rays between 320 and 400 nm (UVA) cause wrinkling and ageing.

Phase-solubility analysis of *t*FA/CD at 25 °C using α-CD, Me-β-CD, HP-β-CD, and HP-γ-CD showed the phase-solubility of these *t*FA/CDs inclusion complexes (Figure 13.7). All diagrams show linear increases in *t*FA solubility with increasing CD concentrations.

The influence of cyclodextrins on in vitro human skin absorption of sunscreen butyl-methoxydibenzoylmethane (BM-DBM) was demonstrated by Simeoni *et al.* [27] (Figure 13.8). BM-DBM is an organic molecule and an oil-soluble ingredient used in sunscreen products to absorb the full spectrum of UVA rays. They showed the effects of hydroxypropyl-β-cyclodextrin (HP-β-CD) and sulfobutylether-β-CD (SBE7-β-CD) on in vitro human skin penetration and retention of the sunscreen agent butyl-methoxydibenzoylmethane. The interaction between the UV filter and the cyclodextrins was studied in water by phase-solubility analysis. Solid complexes were prepared by the co-evaporation method and characterized by ^{1}H NMR spectroscopy, thermal analysis and powder X-ray diffraction. Solutions containing free BM-DBM and complexed with cyclodextrins were applied to excised human skin in Franz diffusion cells.

As much as 14.10–16.78% of the applied dose of BM-DBM penetrated within the skin tissue. No sunscreen was detected in the dermis and in the receiver phase.

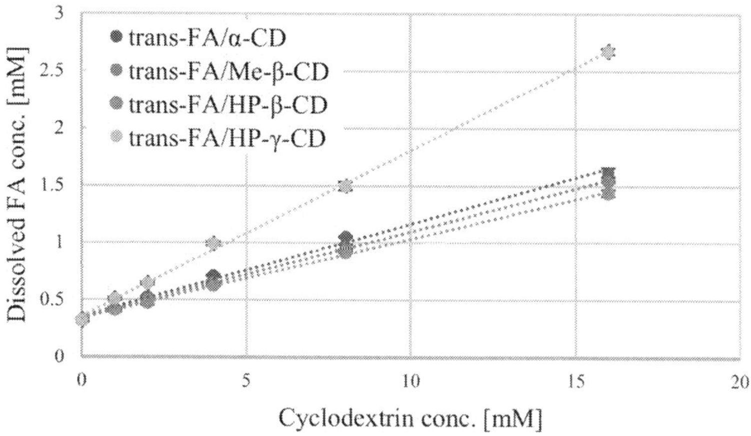

FIGURE 13.7 Phase-solubility diagrams of the inclusion complexes between *t*FA and various CDs [25].

Reproduced with permission from Springer Nature.

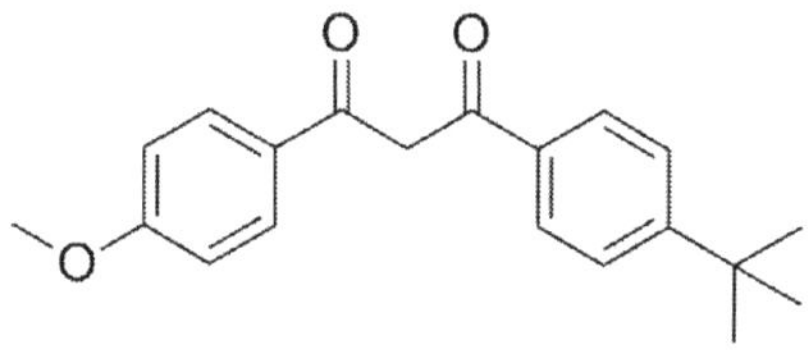

FIGURE 13.8 Molecular structure of BM-DBM.

The greater proportion (84.6–95.5%) of the absorbed UV filter was localized in the stratum corneum with no significant differences between uncomplexed and complexed BM-DBM. Notable levels (2.29% of the applied dose) of the sunscreen agent accumulated in the epidermis from the preparation containing free BM-DBM. The epidermal concentration of the UV filter was markedly reduced by complexation with SBE7-β-CD, whereas HP-β-CD had no effect. The decreased BM-DBM retention in the epidermal region achieved by SBE7-β-CD limits direct contact of the sunscreen and of its reactive photolytic products with the skin viable tissues (Table 13.2).

The incorporation of the sunscreen agent, octyl methoxycinnamate in a cellulosic fabric grafted with β-cyclodextrin was demonstrated by Santo Scalia *et al.* (Figure 13.9) [28], where they showed that the sunscreen agent, octyl methoxycinnamate was encapsulated into cyclodextrin cavities covalently bound to cloth fibres. A cellulosic fabric, Tencel, was modified by grafting β-cyclodextrin molecules using monochlorotriazinyl-β-cyclodextrin (β-CDMCT). Both treated and untreated textiles were soaked in water-methanol mixtures with 2% (v/v) sunscreen and then subjected to multiple washing cycles. The fabrics were analyzed using UV spectrophotometry, and the amount of octyl methoxycinnamate trapped in the Tencel

TABLE 13.2
In Vitro Distribution of BM-DBM in Human Skin After 6 h from Topical Application of Free or Complexed Sunscreen Agent in Ethanol–Propylene Glycol–H_2O (20:30:50, v/v) Solutions

	Percent of Applied Dose Permeated[a]		
Sample	**Stratum Corneum**	**Epidermis**	**Dermis**
Free BM-DBM	14.49 ± 5.69	2.29 ± 1.10	0
HP-β-CD complex	13.01 ± 3.34	2.37 ± 1.23	0
SBE7-β-CD complex	14.03 ± 2.63	0.66 ± 0.52[b]	0
SBE7-β-CD complex + 1% SBE7-β-CD	13.14 ± 3.82	0.96 ± 0.63[b]	0

[a] Each value is the mean ± S.D. of 10 determinations.
[b] Indicates $P < 0.05$ compared to free BM-DBM.
Reproduced with permission from Elsevier.

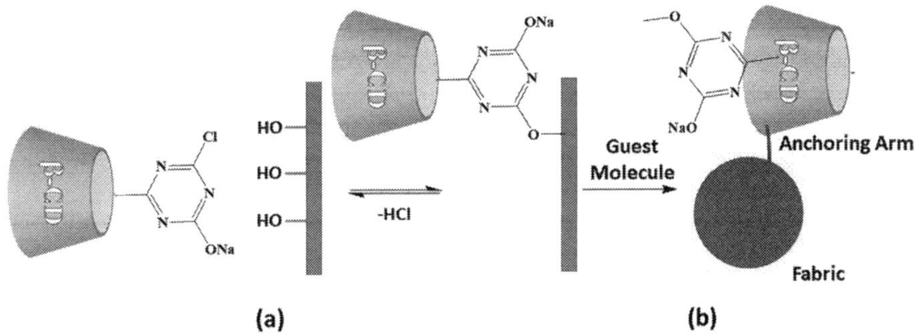

FIGURE 13.9 (a) Chemical grafting of monochlorotriazinyl-β-cyclodextrin onto a cellulosic fibre. (b) Scheme of a host–guest inclusion complex grafted on the textile surface [28].

was measured via high-performance liquid chromatography. The results showed that the β-CDMCT-functionalized fabric retained significantly more sunscreen (0.0203%, w/w) compared to the unmodified version (0.0025%, w/w) (Table 13.3). Additionally, spectrophotometric evaluation of UV transmission through the fabric samples using the Transpore™ test revealed a notable enhancement in the in vitro sun protection factor, with a 3.2-fold increase in the β-CDMCT-grafted textile. Consequently, even after repeated washings, the β-CD-finished fabric demonstrated superior retention of the sunscreen agent and improved photoprotective properties compared to the untreated textile.

In this study, the researchers showed that treating textiles with β-CDMCT enhances the absorption of OMC by the fabric, improving its UV protection. Furthermore, the covalent attachment of β-CD to the textile fibers increases the sunscreen's resistance to washing, extending the fabric's UV protective capabilities. This chemical grafting

TABLE 13.3
Sun Protection Factor for Two Commercial Sunscreen Lotions (Melascreen™ and Ayur™), with and Without Inclusion Complex of Dinitrocompounds [29]

Wavelength (nm)	Sunscreen Lotion		Without Inclusion	Inclusion Complex
	Melascreen™	Ayur™		
UVA (320–400 nm)				
400	16.1968	21.4739	13.7662	17.7958
395	16.7161	23.7274	14.2878	18.3136
390	17.0610	25.9198	14.6352	18.6569
385	17.3237	28.0100	14.9005	18.9179
380	17.5486	29.8223	15.1283	19.1408
375	17.7155	30.0899	15.2980	19.3059
370	17.9412	30.6750	15.5265	19.5298
365	18.1449	32.3590	15.7323	19.7321
360	18.2202	31.3761	15.8201	19.7992
355	18.3956	31.1677	16.0057	19.9679
350	18.6913	31.2255	16.3037	20.2620
345	18.9574	31.0623	16.5736	20.5257
340	18.0783	31.0855	16.6988	20.6437
335	18.2339	31.0020	16.8595	20.7961
330	18.4526	31.9259	17.0834	21.0113
325	18.5335	31.5455	17.1697	21.0886
320	18.5749	31.7496	17.2167	21.1263
315	18.5989	30.6587	17.2468	21.1463
UVB (280–320 nm)				
310	18.5971	30.5525	17.2526	21.1396
305	18.6796	32.5121	17.3386	21.2198
300	18.6195	31.4403	17.2836	21.1563
295	18.7374	31.3045	17.4111	21.2678
290	18.7166	31.1653	17.4007	21.2402
SPF	18.1	30.1	16.2	20.16
LabeledSPF	18	30	–	–

Reproduced with permission from Elsevier.

of cyclodextrins onto cotton fibers is an effective strategy for creating sun-protective clothing.

13.5 MECHANISMS OF CYCLODEXTRIN INTERACTION WITH UVR ABSORBERS

The interaction between CDs and UVR absorbers typically involves non-covalent forces such as van der Waals interactions, hydrogen bonding, and hydrophobic effects. The formation of inclusion complexes is influenced by the size and shape compatibility between the CD cavity and the UVR absorber molecule.

(i) **Formation of Inclusion Complexes:** The primary mechanism by which CDs enhance the performance of UVR absorbers is through the formation of inclusion complexes. These complexes can protect the UVR absorbers from photodegradation by isolating them from the environment.

(ii) **Enhancement of UVR absorption by encapsulation:** The absorption capacity of the UVR absorbers is enhanced by the changes in their molar extinction coefficient when they are incorporated into the cavity of CD molecule, which is owing to the changes in the polarity of the environment of the chromophore of the UVR absorbers when it goes from the polar aqueous environment to the apolar cavity of CD [29].

(iii) **Synergistic Effects:** The encapsulation of UVR absorbers by CDs can also lead to synergistic effects where the protective properties of the absorber are enhanced due to the stabilized microenvironment provided by the CD cavity.

(iv) **Impact on Photostability:** Studies have shown that the photostability of UVR absorbers like oxybenzone and avobenzone is significantly improved when complexed with CDs. This stability translates to higher SPF/UVPF values as the active ingredients remain effective for longer periods under UV exposure.

13.6 EXPERIMENTAL EVIDENCE

(i) Stabilization of Avobenzone with (2-hydroxy)propyl-β-cyclodextrin

Yuan *et al.* demonstrated the photostabilizing effect of (2-hydroxy)propyl-β-cyclodextrin (2-HP-β-CD) on avobenzone, a common UVA filter [30]. The solubility, thermal stability, and photostability of avobenzone have been increased due to the formation of avobenzone and 2-HP-β-CD inclusion complexes. The study used UV-visible spectrophotometry to assess the degradation rate of avobenzone with and without cyclodextrin under UV irradiation. The inclusion complex of avobenzone and 2-HP-β-CD exhibited significantly reduced photodegradation compared to avobenzone alone. The formation of avobenzone-β-cyclodextrin complexes improved photostability, highlighting the role of cyclodextrins as effective stabilizers of this UVR absorber.

(ii) Photostabilization of octyl methoxycinnamate (OMC) by β-cyclodextrin

Researchers studied the impact of β-cyclodextrin on the photostability of octyl methoxycinnamate (OMC), a UVB filter, under UV exposure [31]. The experiments involved irradiating OMC solutions with and without cyclodextrin under simulated sunlight, followed by analysis using HPLC and UV spectrophotometry. OMC, when complexed with β-cyclodextrin, showed a much lower degradation rate compared to its un-complexed counterpart. The complex retained 70% of its original absorbance after 2 hours of UV irradiation, whereas the free OMC degraded by more than 50% over the same period. The study provided direct evidence that cyclodextrins can stabilize

UV filters like OMC by reducing their exposure to harmful UV radiation, preventing their breakdown.

(iii) Stabilization of oxybenzone by cyclodextrins

The oxybenzone, a widely used UVR absorber, is nicely stabilized by using cyclodextrins [24]. The interaction between oxybenzone and hydrophilic α-, β- and γ-cyclodextrin derivatives was studied in water by phase-solubility analysis. Among the available cyclodextrins, hydroxypropyl-β-cyclodextrin (HP-β-CD) and especially sulfobutylether-β-cyclodextrin (SBE-β-CD) had the greatest solubilizing activity. When oxybenzone is encapsulated into the SBE-β-CD cavity, a more localized distribution of oxybenzone on the skin surface is attained, which should enhance the sunscreen's photoprotective power [32]. The photostability of oxybenzone increased significantly, with the cyclodextrin complex maintaining over 85% of its absorbance after prolonged UV exposure, compared to a sharp decline in the absorbance of the un-complexed oxybenzone.

(iv) Photostability of padimate O with β-cyclodextrin

Researchers evaluated the photostability of padimate O (PMO), a UVB absorbing agent which was enhanced when it was complexed with α- and β-cyclodextrin in both aqueous environment and solid state (Figure 13.10). UV-Vis spectroscopy, FTIR, mass spectra, and powder X-ray diffraction studies suggest that α- and β-cyclodextrin can be used as good sunscreen agent carriers and photo-stabilizer additives. It was revealed that β-cyclodextrin formed a more firmly inclusion complex than α-cyclodextrin with PMO. β-cyclodextrin effectively stabilizes padimate O, preventing its degradation under UV exposure, thus extending its UV blocking capacity [33].

(v) Stabilization of phenylbenzimidazole sulfonic acid by β-cyclodextrin

The photostabilization of phenylbenzimidazole sulfonic acid (PBSA), a UVA and UVB absorber, was investigated in the presence of hydrophilic α-, β-, and γ-cyclodextrin derivatives (Figure 13.11). The complexation of the sunscreen agent with hydroxypropyl-β-cyclodextrin was confirmed by NMR spectroscopy. PBSA is prone to photodegradation, which was remarkedly reduced by

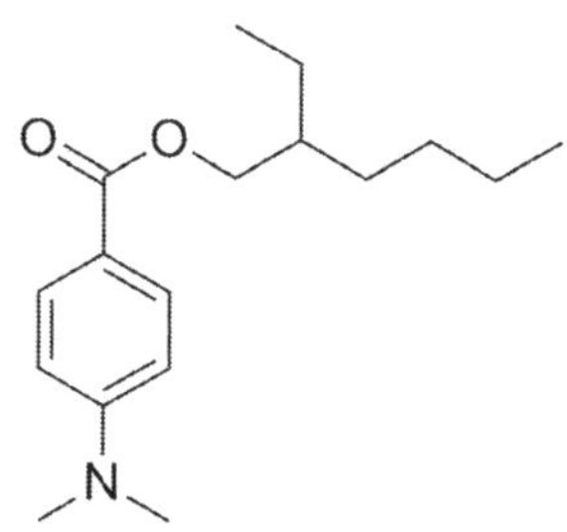

FIGURE 13.10 Molecular structure of padimate O.

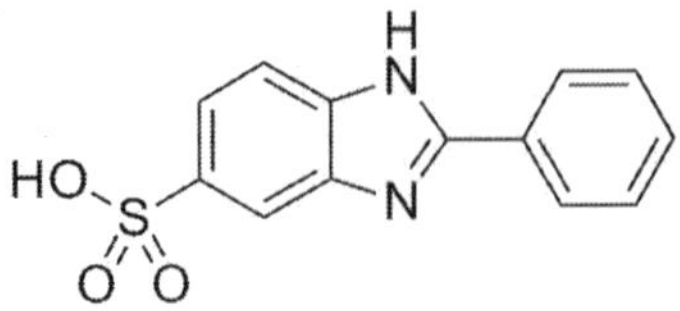

FIGURE 13.11 Molecular structure of phenylbenzimidazole sulfonic acid.

complexation with hydroxypropyl-β-cyclodextrin. The researchers observed that the extent of degradation was 3.9% for complex compared to 9.1% for un-complexed PBSA. The inclusion complex with β-cyclodextrin effectively enhanced the stability of PBSA, making it a more reliable UV filter in sun protection products [34].

These experimental studies provide strong evidence that cyclodextrins act as effective stabilizers of various UVR absorbers, enhancing their photostability, reducing degradation, and improving their overall performance in formulations. Through complexation, cyclodextrins shield these molecules from UV-induced damage, making them more durable and reliable in sunscreens, cosmetics, and other UV-protection applications.

13.7 PRACTICAL APPLICATIONS

(i) **Cosmetic Industry**

- Cosmetic ingredients are hydrophobic; however, these substances can produce inclusion compounds with cyclodextrins (CDs) to achieve enhanced hydrophilicity. In many self-care and cosmetic products, triclosan serves as a sanitizer and surface antiseptic. It dissolves in organic liquid and is minimally soluble in alkaline solutions but is almost water-resistant. The complex of CD with triclosan is fully soluble in water, producing a clear solution. CDs are also utilized in antiperspirant sticks and can create complexes with moisture odours [35]. Cyclodextrins are extensively employed to encapsulate volatile fragrance agents that tend to evaporate rapidly, thereby diminishing the scent's duration. By confining these molecules within their hydrophobic cavity, CDs safeguard them from volatilization and degradation. This leads to a more sustained and consistent release of the fragrance over time. Fenyvesi *et al.* (2023) examined how cyclodextrins, especially β-cyclodextrin, are employed in fragrance encapsulation to boost their stability and extend their release in cosmetic products. This research emphasized the growing trend of incorporating CDs in premium perfumes and personal care items [36].
- Cyclodextrins can boost the efficacy of active components in skincare formulations by enhancing their solubility and stability. For example, antioxidants such as vitamin E and retinoids, which are susceptible to oxidation and breakdown, can be efficiently stabilized with CDs. This not only maintains their effectiveness but also prolongs the product's shelf life.

This guarantees that the active substances penetrate deeper skin layers, improving their therapeutic benefits [37].

- Cyclodextrins are essential for stabilizing UV filters found in sunscreens, which are often susceptible to photodegradation. By forming complexes with these filters, cyclodextrins shield them from deterioration when exposed to sunlight, thereby preserving the sunscreen's protective function. Certain UV filters may induce skin irritation, particularly in those with sensitive skin. Encapsulation within cyclodextrins can minimize direct contact between the skin and these potentially irritating substances, thereby reducing the likelihood of adverse reactions [38].
- Sunscreen agents, particularly those that are hydrophobic, such as avobenzone, oxybenzone, and octinoxate, often exhibit poor water solubility, which limits their effectiveness in aqueous-based formulations and can lead to uneven distribution on the skin. Cyclodextrins, with their unique molecular structure featuring a hydrophobic cavity and hydrophilic exterior, address this issue by forming inclusion complexes with these molecules. This encapsulation process enhances the water solubility of the sunscreen agents, enabling their incorporation into more versatile and user-friendly formulations. Improved solubility ensures better bioavailability of the active ingredients, facilitating even application on the skin and increasing the overall efficacy of the product. Moreover, this capability of cyclodextrins contributes to the development of lightweight and non-greasy sunscreens, which are increasingly preferred by consumers.

(ii) Drug Delivery Systems

Cyclodextrins (CDs) have transformed drug delivery systems due to their distinctive capacity to create inclusion complexes with a variety of drug molecules. These complexes can greatly enhance the pharmacokinetic and pharmacodynamic characteristics of drugs, rendering cyclodextrins indispensable in the pharmaceutical sector. Numerous drugs experience poor water solubility, which restricts their bioavailability and therapeutic effectiveness. Cyclodextrins can encapsulate these hydrophobic drugs within their hydrophobic cavity, thereby effectively enhancing their solubility in aqueous environments. This results in better absorption in the gastrointestinal tract, leading to improved bioavailability. Cyclodextrins not only boost their solubility but also protect them from degradation in the biological environment, ensuring a higher proportion of the drug reaches the target site [39].

13.8 FUTURE DIRECTIONS

Ongoing studies concentrate to put some lights on the innovative CD derivatives for precise drug delivery, gene therapy, and customized medicine. More and more integration of CDs with nanotechnology are required to achieve an effective and accurate delivery mechanism in the near future. CDs could be instrumental in developing personalized medicines, where drug release profiles are customized to meet individual patient needs. For the future scientists, CDs can be promising candidates in emerging areas, such as nanotechnology, sustainable chemistry, and personalized medicine to propel substantial progress, making them increasingly essential to future technological and scientific advancements.

13.9 CONCLUSION

Cyclodextrins have a significant impact on the SPF and UVPF of UVR absorbers due to their ability to form stable inclusion complexes, enhance photostability, and improve the overall efficacy of UV protection products. The ultraviolet (Sun) defence provided by UV radiation absorbers can be enhanced and increased by encapsulating them within the cavity of cyclodextrins. The enhancement in UPF/SPF values is due to the formation of inclusion complexes between cyclodextrins and UVR absorbers, which can boost the solubility, stability, and skin penetration of the UVR absorbers. A few researchers also suggested the limitations of CDs regarding its interactions with cholesterol and formation of a nephrotoxic complex, and limited penetration and bioavailability in deeper skin layer. Hence, a continued research and innovation are required in this field to lead to more effective and long-lasting sun protection solutions.

REFERENCES

1. Zambrano-Román M, Padilla-Gutiérrez JR, Valle Y, et al. Non-melanoma skin cancer: A genetic update and future perspectives. *Cancers*. 2022;14(10):2371.
2. Saes da Silva E, Tavares R, da Silva Paulitsch F, et al. Use of sunscreen and risk of melanoma and non-melanoma skin cancer: A systematic review and meta-analysis. *European Journal of Dermatology*. 2018;28:186–201.
3. Orazio J, Jarrett S, et al. UV radiation and the skin. *International Journal of Molecular Sciences*. 2013;14(6):12222–12248.
4. Mohania D, Chandel S, Kumar P, et al. Ultraviolet radiations: Skin defense-damage mechanism. In Shamim I. Ahmad (ed.), *Ultraviolet Light in Human Health, Diseases and Environment*. Switzerland, Springer, 2017:71–87.
5. Moloney FJ, Collins S, Murphy GM. Sunscreens: Safety, efficacy and appropriate use. *American Journal of Clinical Dermatology*. 2002;3:185–191.
6. Dunaway S, Odin R, Zhou L, et al. Natural antioxidants: Multiple mechanisms to protect skin from solar radiation. *Frontiers in Pharmacology*. 2018;9:392.
7. Lorigo M, Cairrao E. Antioxidants as stabilizers of UV filters: An example for the UV-B filter octylmethoxycinnamate. *Biomedical Dermatology*. 2019;3(1):1–9.
8. Osterwalder U, Herzog B. Sun protection factors: World wide confusion. *British Journal of Dermatology*. 2009;161(s3):13–24.
9. Martin J, Díaz-Montaña EJ, Asuero AG. Cyclodextrins: Past and present. *Cyclodextrin: A Versatile Ingredient*. 2018;1:1-332.
10. Crini G. A history of cyclodextrins. *Chemical Reviews*. 2014;114(21):10940–10975.
11. Sandilya AA, Natarajan U, Priya MH. Molecular view into the cyclodextrin cavity: Structure and hydration. *ACS Omega*. 2020;5(40):25655–25667.
12. Ferreira L, Campos J, Veiga F, et al. Cyclodextrin-based delivery systems in parenteral formulations: A critical update review. *European Journal of Pharmaceutics and Biopharmaceutics*. 2022;178:35–52.

13. Li Z, Chen S, Gu Z, et al. Alpha-cyclodextrin: Enzymatic production and food applications. *Trends in Food Science & Technology*. 2014;35(2):151–160.
14. Lai W-F, Rogach AL, Wong W-T. Chemistry and engineering of cyclodextrins for molecular imaging. *Chemical Society Reviews*. 2017;46(20):6379–6419.
15. Chen G, Jiang M. Cyclodextrin-based inclusion complexation bridging supramolecular chemistry and macromolecular self-assembly. *Chemical Society Reviews*. 2011;40(5):2254–2266.
16. Zarzycki PK, Głód BK. Cyclodextrins-based nanocomplexes for encapsulation of bioactive compounds in food, cosmetics, and pharmaceutical products: Principles of supramolecular complexes formation, their influence on the antioxidative properties of target chemicals, and recent advances in selected industrial applications. *Encapsulations*. 2016;2:717–767.
17. Jin Z-Y. *Cyclodextrin chemistry: Preparation and application*. World Scientific; 2013.
18. Cid-Samamed A, Rakmai J, Mejuto JC, et al. Cyclodextrins inclusion complex: Preparation methods, analytical techniques and food industry applications. *Food Chemistry*. 2022;384:132467.
19. Michalska P, Wojnicz A, Ruiz-Nuño A, et al. Inclusion complex of ITH12674 with 2-hydroxypropyl-β-cyclodextrin: Preparation, physical characterization and pharmacological effect. *Carbohydrate Polymers*. 2017;157:94–104.
20. Tang X, Yang T, Yu D, et al. Current insights and future perspectives of ultraviolet radiation (UV) exposure: Friends and foes to the skin and beyond the skin. *Environment International*. 2024;185:108535.
21. Yam LY. A study on ultraviolet protection factor of cotton and coolmax based on plain knitted structure. *Thesis:* Pao Yue-kong Library, The Hong Kong Polytechnic University, Hung Hom, Kowloon, Hong Kong 2012:1–151.
22. Hatch KL, Osterwalder U. Garments as solar ultraviolet radiation screening materials. *Dermatologic Clinics*. 2006;24(1):85–100.
23. Sarkar AK. An evaluation of UV protection imparted by cotton fabrics dyed with natural colorants. *BMC Dermatology*. 2004;4:1–8.
24. Felton LA, Wiley CJ, Godwin DA. Influence of Cyclodextrin complexation on the in vivo photoprotective effects of Oxybenzone. *Drug Development and Industrial Pharmacy*. 2004;30(1):95–102.
25. Mori T, Tsuchiya R, Doi M, et al. Solubilization of ultraviolet absorbers by cyclodextrin and their potential application in cosmetics. *Journal of Inclusion Phenomena and Macrocyclic Chemistry*. 2019;93(1):91–96.
26. Kou X, Zhang Y, Su D, et al. Study on host-guest interaction of aroma compounds/γ-cyclodextrin inclusion complexes. *LWT*. 2023;178:114589.
27. Simeoni S, Scalia S, Benson H. Influence of cyclodextrins on in vitro human skin absorption of the sunscreen, butyl-methoxydibenzoylmethane. *International Journal of Pharmaceutics*. 2004;280(1–2):163–171.
28. Scalia S, Tursilli R, Bianchi A, et al. Incorporation of the sunscreen agent, octyl methoxycinnamate in a cellulosic fabric grafted with β-cyclodextrin. *International Journal of Pharmaceutics*. 2006;308(1–2):155–159.
29. Roy MN, Saha S, Kundu M, et al. Exploration of inclusion complexes of neurotransmitters with β-cyclodextrin by physicochemical techniques. *Chemical Physics Letters*. 2016;655:43–50.
30. Yuan L, Li S, Huo D, et al. Studies on the preparation and photostability of avobenzone and (2-hydroxy) propyl-β-cyclodextrin inclusion complex. *Journal of Photochemistry and Photobiology A: Chemistry*. 2019;369:174–180.

31. Molinari A, Tursilli R, Casolari A, et al. Influence of complexation with cyclodextrins on photo-induced free radical production by the common sunscreen agents octyl-dimethylaminobenzoate and octyl-methoxycinnamate. *Die Pharmazie-An International Journal of Pharmaceutical Sciences*. 2006;61(1):41–45.
32. Simeoni S, Scalia S, Tursilli R, et al. Influence of cyclodextrin complexation on the in vitro human skin penetration and retention of the sunscreen agent, oxybenzone. *Journal of Inclusion Phenomena and Macrocyclic Chemistry*. 2006;54:275–282.
33. Kundu M, Rahaman H, Roy MN. Physicochemical investigations on non-covalent interactions between Padimate O and cyclodextrin receptors in both solution and solid states. *Spectrochimica Acta Part A: Molecular and Biomolecular Spectroscopy*. 2019;218:9–14.
34. Scalia S, Molinari A, Casolari A, et al. Complexation of the sunscreen agent, phenylbenzimidazole sulphonic acid with cyclodextrins: Effect on stability and photo-induced free radical formation. *European Journal of Pharmaceutical Sciences*. 2004;22(4):241–249.
35. Sharma N, Baldi A. Exploring versatile applications of cyclodextrins: An overview. *Drug Delivery*. 2016;23(3):729–747.
36. Fenyvesi É, Vikmon M, Szente L. Cyclodextrins in food technology and human nutrition: Benefits and limitations. *Critical Reviews in Food Science and Nutrition*. 2016;56(12):1981–2004.
37. Loftsson T, Brewster ME. Pharmaceutical applications of cyclodextrins: Basic science and product development. *Journal of Pharmacy and Pharmacology*. 2010;62(11):1607–1621.
38. Ferreira L, Mascarenhas-Melo F, Rabaça S, et al. Cyclodextrin-based dermatological formulations: Dermopharmaceutical and cosmetic applications. *Colloids and Surfaces B: Biointerfaces*. 2023;221:113012.
39. Gadade DD, Pekamwar SS. Cyclodextrin based nanoparticles for drug delivery and theranostics. *Advanced Pharmaceutical Bulletin*. 2020;10(2):166.

14 Enhancing UVR Absorber Efficacy with Cyclodextrins in Cosmetic and Non-Cosmetic Materials

Maria Soledad Bueno[†], Belén Alejandra Mezzano[†], Micaela Ponce Ponte[†], Marcela Raquel Longhi, and Claudia Garnero

[†] *These authors contributed equally to this work.*

14.1 INTRODUCTION

Sunlight is essential for life, but solar ultraviolet (UV) radiation can have adverse effects on living organisms and damage materials like food, textiles, plastics, archaeological objects, and outdoor furniture [1, 2]. Prolonged exposure in natural environments can cause photodegradation, affecting appearance, mechanical integrity, and biodegradation in some cases, reducing quality and shelf life. This highlights the need for effective UV protection strategies.

Sunlight has benefits like vitamin D production and increased β-endorphin expression, but excessive exposure can cause short-term and long-term skin damage. Short-term effects involve inflammation and erythema, and posterior activation of melanocytes, which darkens the skin. The long-term effects include irreversible photoaging, loss of skin elasticity, increased pigment heterogeneity, and a higher risk of skin cancer [3, 4].

The solar UV radiation spectrum is divided into subregions: UVA (400–320 nm), UVB (320–290 nm), and UVC (290–100 nm). Sunlight predominantly consists of UVA (90–95%) and UVB (5–10%), while UVC radiation is absorbed by the ozone layer and does not have harm the skin. The main environmental risk to skin health are UVA and UVB radiation. The radiation energy is inversely proportional to its wavelength, while its penetration into the skin is proportional to the wavelength (Figure 14.1). UVA can penetrate deep into the skin, generate free radicals, and cause long-term and indirect DNA damage; while UVB can only penetrate the epidermis due to its shorter wavelength; however, it is 1000 times more effective at causing

DOI: 10.1201/9781003492757-14

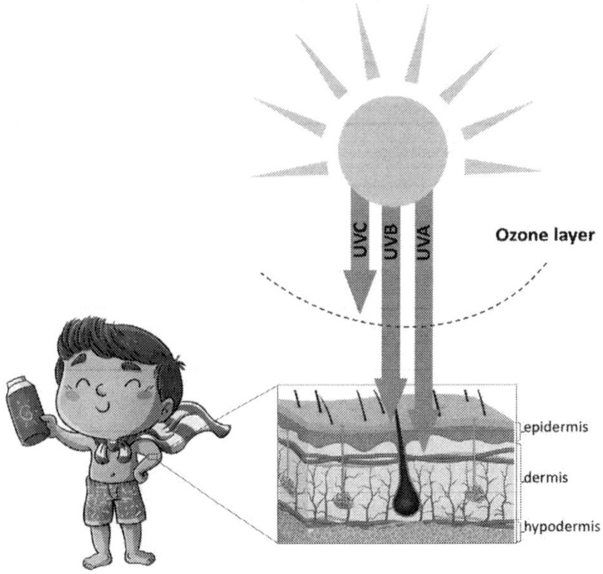

FIGURE 14.1 Wavelength-dependent penetration of UV radiation into human skin.

sunburn than UVA, has a higher energy, and can induce direct DNA damage and, as a result, skin cancer [3, 5, 6].

The first defense against the sunlight is melanin production, which absorbs dangerous UV radiation. However, the amount of melanin produced is insufficient for protection from UV exposure, making the use of UV filters in cosmetic formulations and photoprotective clothing the best approach [7, 8]. UV filters, the main active ingredient in sunscreen formulations, are organic and inorganic compounds that directly interfere with incident solar radiation by absorbing, reflecting, or dispersing radiation [4, 9].

An optimum formulation should provide broad protection against UVA and UVB radiation and prevent their penetration into the epidermis. The UV protection is indicated by the sun protection factor (SPF), which measures the UV filter efficiency in reducing solar UV radiation transmission. Better protection against sunburn is provided by a higher SPF. However, sunscreens can never be completely effective. The SPF value measures the formulation capacity to prevent erythema or sunburn caused by excessive UVB exposure. It is calculated as the ratio of the dose of solar-simulated UV radiation required to generate erythema with and without a UV filter on the skin of human volunteers [10, 11].

The UV textile protection is indicated by the UV protection factor (UPF). It is measured by placing the textile sample adjacent to human skin while UV radiation is focused onto the skin via the textile for various periods of time. The test is conducted on both unprotected and protected skins independently. The resulting sunburn from this exposure is then used to estimate the UPF value [12].

The SPF and UPF are calculated using the below equation:

$$SPF = UPF = \frac{MED_p}{MED_u} \tag{14.1}$$

where minimal erythema dose of protected skin (MEDp) is divided by minimal erythema dose of unprotected skin (MEDu) [11, 12].

Some conventional UV filters have limitations that affect their effectiveness, such as limited solubility [13] and the tendency to photodegrade upon exposure to UV radiation [14, 15], potential adverse effects due to skin penetration and entry into the blood circulation [16], as well as negative impacts on skin biology from prolonged use. In order to improve these limitations, which impact sun protection efficacy and constitute health risks to the users, the cosmetic and pharmaceutical sectors are focusing on research strategies to develop effective and safer photoprotective agents that improve formulation performance and reduce potential toxicity [5, 9, 17]. Therefore, were reported several nanoparticles [18, 19], microcapsules [20], liposomes [21], among others. Furthermore, cyclodextrin (CD) inclusion complexes appears as an innovative strategy [5, 6, 9, 17, 22] with significant potential for UV filter topical formulations.

CD are cyclic oligosaccharides composed by glucopyranose units linked by a 1,4-glycosidic bonds [23, 24]. The most used native CD are α-cyclodextrin (α-CD), β-cyclodextrin (β-CD), and γ-cyclodextrin (γ-CD) (Figure 14.2) [25].

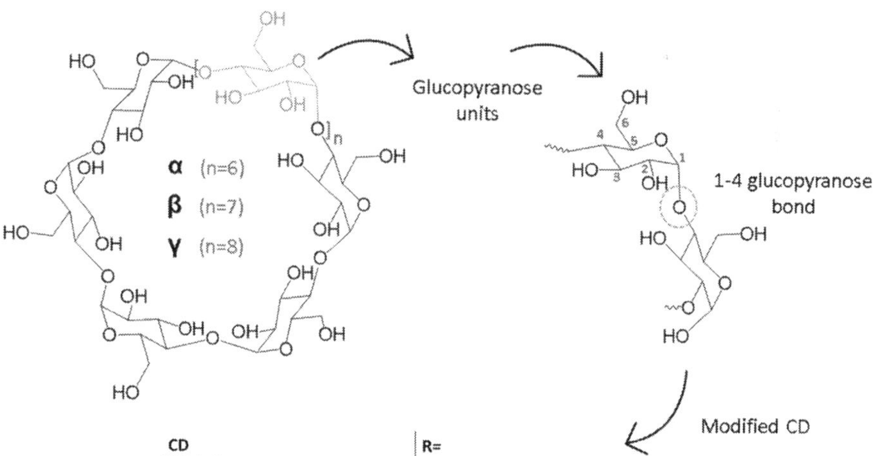

CD	R=
Methyl-cyclodextrin	$-CH_3$
Hydroxypropyl-cyclodextrin	$-CH_2CH(OH)CH_3$
Sulfobutylether-cyclodextrin	$-(CH_2)_4SO_3Na$

FIGURE 14.2 Molecular structure and conformation of α-CD, β-CD, γ-CD, and the substituted CD.

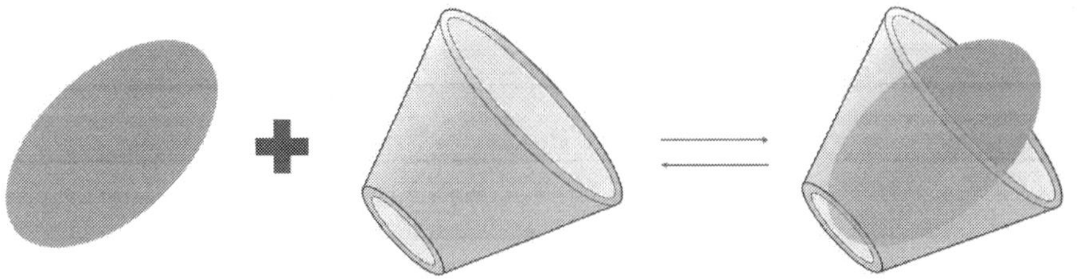

FIGURE 14.3 Schematic structure of an inclusion complex.

The number of glucopyranose units determines the CD solubility. Chemically modified CD have been developed to enhance their solubilizing and complexation efficiency by introducing functional groups like methyl, ethyl, hydroxyethyl, carboxymethyl, hydroxypropyl, and sulfobutylether through substitution reactions such as amination, etherification, and esterification of hydroxyl groups at the 2, 3, and 6 positions of the glucose molecule (Figure 14.2) [26, 27]. This selective CD functionalization enhances aqueous solubility, provides protection against degradative factors like light or oxygen, and enhances the guest molecule physical stability [6, 28].

Considering their cone-shaped structure with a hydrophobic interior and hydrophilic exterior, CD are highly valued for their ability to form inclusion complexes by encapsulating a variety of molecules (Figure 14.3) through hydrophobic interactions, hydrogen bonds, and van der Waals interactions, which protects the guest molecule within the cavity, mainly from hydrolysis, dehydration, and oxidation, improving the UV filters properties. In topical formulations, it protects them from photodegradation and oxidation while also limiting their permeation into deeper skin layers and systemic absorption, thus reducing potential undesirable side effects and restraining incompatibilities between the formulation ingredients [5, 15, 22]. And, CD complexes with UV filters incorporated into the cotton fabric provide UV protecting properties, improving the overall UV blocking capability [8, 29, 30].

In this context, this chapter will explore some important aspects of the CD complexes impact on several materials, including cosmetic and non-cosmetic, analyzing the modifications in UV filter properties and the resulting efficacy improvements evidenced by the UPF/SPF value.

14.2 UV FILTER CLASSIFICATION

The classification is based on their mechanism of action upon exposure to sunlight, which can be either organic or inorganic [4, 10, 31]. Recently, hybrid materials that combine properties of organic and inorganic compounds acquired interest [31]. Figure 14.4 illustrates a representative list of them.

ORGANIC

- UVA: benzophenones, ecamsule
- UVB: PABA derivatives, cinnamates, salicylates, octocrylene
- BROAD SPECTRUM: silatriazole, bisoctriazole, bemotrizinol, avobenzone

INORGANIC

- Zinc oxide
- Titanium dioxide
- Kaolim
- Calamine
- Talc

HYBRID

- Organic + Inorganic compounds

FIGURE 14.4 Classification of UV filters.

14.2.1 Organic UV Filters

These filters are chemical aromatic molecules with a carbonyl group that absorb UV radiation to convert it into heat energy, which is dissipated innocuously from the skin. The most important compounds include benzophenones, triazines, and derivatives of aminobenzoic acid, salicylic acid, cinnamic acid, dibenzoylmethane, benzylidene camphor, benzimidazole, and benzotriazole. They are categorized by protection range: UVB, UVA, and broad spectrum agents that cover UVA and UVB. These filters are frequently preferred for their cosmetic benefits, such as a lighter texture. However, some compounds are photolabile, resulting in a reduction in their effectiveness, and can be absorbed into the skin, which may cause systemic absorption, photoirritation, allergic contact reactions, acute sunburn, or chronic actinic damages. These filters are predominantly used in combination. However, it is important to select appropriate UVA/UVB filters to avoid adverse interactions. Certain filters, such as para-aminobenzoic acid (PABA), PABA derivatives, and benzophenones, can cause significant adverse effects.

14.2.2 Inorganic UV Filters

These filters are mineral compounds recognized as sunblocks, acting as a physical barrier by reflecting and scattering UV light, protecting across a broad UV spectrum. Titanium dioxide and zinc oxide are the most widely utilized. They have a minimal potential for allergic sensitization and exhibit high photostability, and are suitable for sensitive skin products. However, their reflective properties may have an excessive shine and a whitish appearance, restricting their application to specific formulations due to limited cosmetic acceptance.

14.2.3 Hybrid UV Filters

These filters are composed of two half-blended materials combining organic components (such as organic molecules or polymers) with inorganic components (like metal oxides, carbonates, and phosphates) [31] at the molecular or nanoscale level, resulting in hybrid materials with specific attributes that enhance UV filtering capacity, reduced toxicity, and better biocompatibility.

14.2.4 Antioxidants

Antioxidants are incorporated into formulations as an additional layer of protection. In contrast to UV filters, antioxidants must penetrate deeper into skin layers to find their site of action, resulting in a reduction in DNA damage, skin ageing, and diseases by various mechanisms of action. Enzymes, vitamins, flavonoids, carotenoids, phenolic acids, and polyphenols are some of the categorizations for antioxidants [6, 9].

14.3 APPLICATION OF CD COMPLEXES IN COSMETIC UV PROTECTIVE PRODUCTS

14.3.1 Effect of CDs on UV Filters Susceptible to Degradation upon Exposure to Sunlight and Oxygen

Sunscreens must remain stable on the skin layer during UV exposure for effectiveness. Photostable UV filters maintain UV protection without chemical or physical changes [32]. However, some UV filters exposed to light can generate toxic products and lose their protective capacity. Their UV radiation exposure generates reactive compunds like free radicals and reactive oxygen species (ROS) that can interact with sunscreen components or skin, reducing their effectiveness [33]. Photolabile UV filters can be stabilized by encapsulation or the addition of antioxidants. CD are effective [15, 34], and safe for use in cosmetic formulations and in sunscreen applications, preventing degradation and oxidation.

The ensulizole photostability, a UVB protecting agent that can generate ROS upon photoexcitation, damaging proteins, DNA, and lipids, was significantly improved through complexation with methyl-β-cyclodextrin (M-β-CD) [35] and β-CD [36], maintaining a constant absorption profile for at least 4 hours under solar radiation. Consequently, CD complexation provides a protective barrier against solar light-induced changes, demonstrating that it is a promising method to extend its sun protection life.

Avobenzone and dibenzoylmethanes also are susceptible to photodegradation, producing products that reduce photoprotective efficacy and can cause phototoxic, photosensitizing, or photoallergic reactions [31]. Encapsulation of avobenzone with 2-hydroxypropyl-β-cyclodextrin (2-HP-β-CD) improves photostability, reducing photolysis to 8% compared to 49% for avobenzone alone [37]. Moreover, the incorporation of avobenzone and octinoxate into the β-CD reduced their degradation, enhancing photostability and providing an eco-friendly sunscreen preparation [34].

Similarly, the complexation of isoamyl 4-(dimethylamino) benzoate with sulfobutylether-β-cyclodextrin (SBE-β-CD) reduced photolysis from 9.16% to 2.25% after 8 hours of UVB irradiation, suggesting the SBE-β-CD is safe for use in UVB sunscreen [38].

14.3.2 Effect of CDs on Retention, Absorption, and/or Penetration of the UV Filters in Cutaneous Layers

Human skin comprises three layers: the epidermis (50 to 100 μm), the upper layer of the skin and the main barrier to drug permeation; the dermis (3–5 mm), metabolically activate and essential for wound repair; and the hypodermis, containing blood vessels and connective tissue.

Retention of UV filters in the epidermis is crucial, as deeper penetration into the dermis reduces efficacy and may cause irritation or sensitization. Sunscreen formulations optimize UV filter retention in the outer layers, using excipients that prevent deep absorption into the skin [6].

Shokri et al. [39] demonstrated that complexes of avobenzone, oxybenzone, and ensulizol with β-CD in a cream formulation significantly reduced penetration into the stratum corneum and increased absorption lag time, with avobenzone: β-CD showing the greastest reduction, enhancing safety in sunblock products. Oxybenzone is absorbed systemically after topical application, Felton et al. [40] found that hydroxypropyl-β-cyclodextrin (HP-β-CD) concentration higher than 10% (w/w) reduced transdermal permeation and skin accumulation, forming a reservoir on the skin surface. While Simeoni et al. [41] demonstrated that SBE-β-CD complexes increased drug solubility by 1049-fold while maintaining photodecomposition, reducing penetration and retention in the stratum corneum, and minimizing toxicological risks.

Similarly, the ferulic acid:α-CD complex in oil/water (O/W) emulsion and gel reduced transdermal permeation due to slower release from the CD cavity, acting as a reservoir. Furthermore, the complex protects the molecule from photodegradation, enhancing photostability and ensuring prolonged protection [42].

Butyl methoxydibenzoylmethane, a UVA filter, undergoes photodegradation, producing free radicals, and permeates the stratum corneum and viable epidermis. Scalia et al. [43] found that the HP-β-CD complex incorporated into lipid microparticles in O/W emulsion, reduced penetration, retaining over 70% of the dose on human skin, enhancing protection and reducing toxicity. Simeoni et al. [44] demonstrated that butyl methoxydibenzoylmethane:SBE-β-CD complex significantly decreased epidermal penetration (0.66% of the applied dose), while no effect was observed with HP-β-CD complex, indicating SBE-β-CD reduces epidermal retention, limiting sunscreen contact with viable skin tissues.

14.3.3 Effect of CDs on UV Protection Efficacy of UV Filters

Different methodologies have been investigated to evaluate the UVA/UVB protection of sunscreens including SPF values [6]. Generally, *in vivo* methods using humans are employed. Another alternative is *in vitro* methods, like spectrophotometry. Although recognition and standardization of *in vitro* UVA protection tests, none suggested *in vitro* SPF methods has been officially designated as regulatory standards due to the absence of *in vivo*/*in vitro* SPF values equivalence [11]. UV filter efficacy is measured by SPF, with higher values providing greater skin burn protection [6].

Several ways to improve sunscreens' effectiveness were studied. UV filters complexes with CD are an extensively used alternative. Felton et al. [45] showed formulations with HP-β-CD that enhanced the oxybenzone photoprotective effects. The 20% (w/w) HP-β-CD preparation reduced sunburn cell formation, while 5% (w/w) HP-β-CD showed significantly less edema. This was attributed to UV filter accumulation on the skin surface, providing enhanced photoprotection during longer exposure. HP-β-CD formulations containing small amounts of a single UV filter exhibited similar photoprotection to that of a commercial SPF 30 sunscreen with three UV filters.

The sunscreen O/W emulsion containing a *trans*-ferulic acid:hydroxypropyl-γ-cyclodextrin complex presented a 1.4-fold higher SPF compared to trans-ferulic acid alone, likely due to increased solubility. Additionally, a 1.6-fold higher UVA protection factor was attributed to a bathochromic shift of the CD complex [46].

In vivo studies showed that formulations containing a higher HP-β-CD concentration (30% w/w) reduced the avobenzone skin penetration, generating a reservoir that enhanced photoprotection, as evidenced by minimal suburban cell formation and edema induction [47].

14.3.4 Patents

Several inventions involve CD use in UV protection products mentioned in patents. Some patents described compositions and properties.

WO2014165471A1 [48] and US10485748B2 [49] present CD as UV and IR radiation screening agent for topical formulations for skin or hair treatments in humans or animals. CD acts as a booster for filtering radiation with chemical and/or physical UV filters. The invention proposes the β-CD use as a physical UV absorber, providing effective general UV protection and robust UVA protection in an O/W emulsion for an SPF 30 sunscreen formulation.

BR102013007969A2 [50] relates to cosmetic formulations, often emulsions comprising chemical UV filters and CD as the only physical UV filter. These multifunctional cosmetics include anti-ageing products, for facial or body treatment, makeup, or shampoo. A comparative analysis with conventional formulations demonstrated that using CD as a physical UV blocker in the invention provides effective protection against UV radiation and high levels of UVA protection.

WO2016/161493A1 [51] discloses a topical formulation including UV filters and CD designed for application to wet skin and exhibits a non-whitening effect, and provides the same SPF on dry or wet skin.

14.4 APPLICATION OF CD COMPLEXES IN NON-COSMETIC UV PROTECTIVE MATERIALS

CDs are currently being investigated for their broad potential in non-cosmetic applications; they are used in industries, such as textiles, food technology, dyes, and environmental protection as UV protective agents due to their versatility, biodegradability, and ability to form stable complexes with a wide variety of molecules.

14.4.1 Protective Clothing

In recent decades, chemical treatments have enhanced the functionality of natural and synthetic fibres, improving UV blocking properties while addressing demands for durability, hygiene, and comfort. UV protective finishes are vital for protecting individuals and textiles from the harmful UV radiation effects [7, 52].

Textiles are categorized by their UPF according to fabric structure, fibre type, and post-processing treatments like dyeing and finishing. Fabric with a UPF ≥ 40 are excellent UV filters, while untreated cotton fabrics offers poor UV protection. Functional finishing treatments enhance UV absorption [30, 53, 54].

CD possess a key role in textile processing (Figure 14.5), due to their UV protection properties, environmental benefits, and versatility. CD are used in spinning,

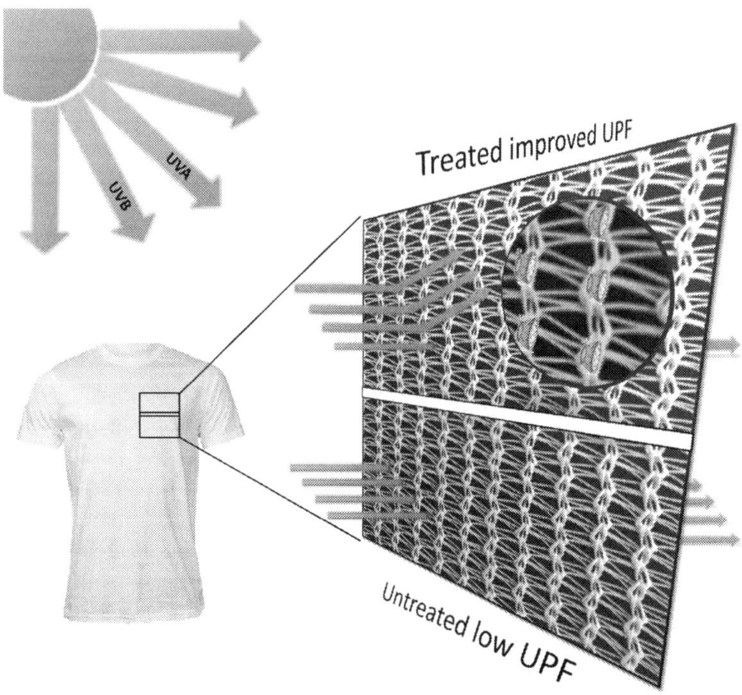

FIGURE 14.5 Comparison of the influence of CD on treated and non-treated fabrics in relation to the effects of UV radiation.

pretreatment, dyeing, and finishing, with notable applications in water treatment and eco-textiles [55, 56]. Their function as encapsulating agents is one of the most important applications in textiles.

β-CD and its derivatives are widely used for UV protection due to their ease of processing, affordability, non-irritating, and non-mutagenic properties [57, 58] by forming complexes with UV absorbers. CD can be bonded to natural fabrics, such as cotton, by using crosslinking agents, such as citric acid, to improve the UV absorption as it cannot form covalent bonds directly with textile fibres [8, 59]. Natural fabrics like cotton cellulose are modified with CD to create functional textiles. Modified CD with reactive groups, like monochlorotriazinyl-β-cyclodextrin (MCT-β-CD), form covalent bonds with cellulose, while itaconic acid or acryloyl derivatives enhance polymerization and improve CD adhesion to textiles.

MCT-β-CD complexes enhance disperse dyeing and UV protection in wool/polyester fabrics during printing. UV absorbers (oxalanilide and metal oxide nanoparticles) with MCT-β-CD improved color depth and UV shielding in wool (UPF > 40) [60]. The MCT-β-CD complexes with oils of peppermint and clove enhanced fragrance retention after washing and UV protection in cotton, with peppermint achieving the highest UPF (25.5) compared to untreated cotton (UPF = 17.8) [29].

To apply CD into fibers, CD derivatives and crosslinking agents, like citric acid, maleic anhydride, and others can be used [55]. Citric acid is preferred for its affordability and eco-friendliness [8, 56]. Nunes et al. [61] demonstrated that grafting β-CD and ZnO nanoparticles onto cotton using citric acid and tetracarboxylic acid butane increased UV protection by 17.5-fold compared to the control, which UV protection was retained after 5 wash cycles.

Liu et al. [8] demonstrated that grafting β-CD onto cotton with citric acid and forming complexes with flavonoids from broadleaf holly improved UV protection (UPF > 40), durability after 30 wash cycles, wrinkle recovery, and a slight reduction in strength.

14.4.2 Packaging Materials

Active packaging materials not only protect but also interact with contents like food to extend the shelf life and preserve product quality. CD are useful in UV protective packaging, encapsulating UV filters or sensitive compounds to prevent UV induced degradation. Additionally, CD improve product stability by slowing oxidation or microbial growth, and are biodegradable, making them safe and sustainable for packaging technology [62, 63]. For example, polylactic acid, a biodegradable material from crops like corn, is sensitive to UV radiation and temperature. The β-CD use with a self-crosslinking Schiff base in its cavity and surface modification by a metal-organic framework was proposed to enhance fire retardancy and UV resistance. Polylactic acid composites prepared with this material showed a 2% increase in UV absorbance, reducing the polylactic acid chains degradation and improving their UPF value from 4 to 7.3 [64].

A multifunctional chitosan film loaded with nitrogen-doped citric acid carbon dots that incorporated an HP-β-CD:geraniol complex increased UV protection, improved antioxidant and antimicrobial activities, and preserved pork meat for 7 days, demonstrating potential as smart bio-based packaging for food preservation and quality control [65].

14.4.3 Other Protective Materials

Srinivasan et al. [66] reported that complexes of dinitroaromatic compounds (dinitrophenol, dinitroaniline, and dinitrobenzoic acid) with β-CD enhance UV absorption in the solid phase, which used in ballpoint pen ink reduced photodegradation. The complex absorbs 15 times more UVA and 28 times more UVB radiation than free compounds, increasing SPF value by 1.24-fold. Additionally, the complex increases the stability under sunlight exposure by up to 19.8%.

Peng et al. [67] synthesized a self-healing UV shielding coating by combining β-CD grafted with TiO_2 nanoparticles and 2-hydroxyethyl-methacrylate-adamantaneto obtain a complex polymerized with 2-hydroxyethyl-methacrylate and butyl acrylate. This material demonstrated high UV absorption capacity, acting as a protective film and retarding UV accelerated aging.

14.5 CONCLUSION AND PERSPECTIVES

The specific chemical structures and different sizes of CD enable them to form reversible inclusion complexes with a variety of substances. Several studies evidence the advantages of CD complexes with UV filters, contributing to the efficiency of UV protection and demonstrating considerable progress in the development of innovative materials.

Due to the significant impact of solar UV radiation, both beneficial and negative, on human health, several materials and the environment, multiple researchs continue on non-toxic and biodegradable CD complexes, exploring their chemistry and responses to solar UV radiation for identify novel applications. This research will specifically explore the retention of protection factors and the potential use and impact of complexes in the environment, among others.

REFERENCES

1. Andrady, A.L., Heikkilä, A.M., Pandey, K.K., Bruckman, L.S., White, C.C., Zhu, M., Zhu, L. (2023) 'Effects of UV radiation on natural and synthetic materials'. *Photochem. Photobiol. Sci.*, Vol 22 (5), pp 1177–1202.
2. Ainali, N.M., Bikiaris, D.N., Lambropoulou, D.A. (2021) 'Aging effects on low-and high-density polyethylene, polypropylene and polystyrene under UV irradiation: An insight into decomposition mechanism by Py-GC/MS for microplastic analysis' *J. Anal. Appl. Pyrolysis*, Vol 158, pp 105207.
3. D'Orazio, J., Jarrett, S., Amaro-Ortiz, A., Scott, T. (2013) 'UV Radiation and the Skin' *Int. J. Mol. Sci.*, Vol 14 (6), pp 12222–12248.
4. Geoffrey, K., Mwangi, A.N., Maru, S.M. (2019) 'Sunscreen products: Rationale for use, formulation development and regulatory considerations' *Saudi Pharm. J.*, Vol 27, pp 1009–1018.

5. Blasi, P., Schoubben, A., Giovagnoli, S., Rossi, C., Ricci, M. (2011) 'The real value of novel particulate carriers for sunscreen formulation' *Expert Rev. Dermatol.*, Vol 6 (5), pp 509–517.
6. Dahabra, L., Broadberry, G., Le Gresley, A., Najlah, M., Khoder M. (2021) 'Sunscreens containing cyclodextrin inclusion complexes for enhanced efficiency: A strategy for skin cancer prevention' *Molecules*, Vol 26 (6), pp 1698.
7. Sfameni, S., Hadhri, M., Rando, G., Drommi, D., Rosace, G., Trovato, V., Plutino, M.R. (2023) 'Inorganic finishing for textile fabrics: recent advances in wear-resistant, UV protection and antimicrobial treatments' *Inorganics*, Vol 11, pp 19.
8. Liu, J., Ma, X., Shi, W., Xing, J., Ma, C., Li, S., Huang, Y. (2020) 'Anti-ultraviolet properties of β-cyclodextrin-grafted cotton fabrics dyed by broadleaf holly leaf extract' *Text. Res. J.*, Vol 90 (21–22), pp 2441–2453.
9. Jesus, A., Sousa, E., Cruz, M.T., Cidade, H., Sousa Lobo, J.M., Almeida, I.F. (2022) 'UV filters: Challenges and prospects' *Pharmaceuticals*, Vol 15 (3), pp 263.
10. Latha, M.S., Martis, J., Shobha, V., Sham Shinde, R., Bangera, S., Krishnankutty, B., Bellary, S., Varughese, S., Rao, P., Naveen Kumar, B.R. (2013) 'Sunscreening agents: A review' *J. Clinic. Aesthetic Dermatol.*, Vol 6 (1), pp 16–26.
11. Zou, W., Ramanathan, R., Urban, S., Sinclair, C., King, K., Tinker, R., Bansal, V. (2022) 'Sunscreen testing: A critical perspective and future roadmap' *Trends Anal. Chem.*, Vol 157, pp 116724.
12. Bashari, A., Shakeri, M., Shirvan, A.R. (2019) 'UV-protective textiles' In S. U. Islam, B. S. Butola (Eds.), *The impact and prospects of green chemistry for textile technology, The textile institute book series*, Cambridge, UK, Woodhead Publishing, pp 327–365.
13. Herzog, B., Schäfer, A., Quass, K., Giesinger, J. (2020) 'Synergistic effects of binary oil mixtures on the solubility of sunscreen UV absorbers' *Eur. J. Pharm. Sci.*, Vol 145, pp 105230.
14. Mandar, S., Wulandari, P., Savitri, V.F., Aziz, H., Refilda, N., Rahmawati, Y. (2023) 'Photostability of avobenzone in a commercial sunscreen SPF 50 with addition of quencher upon sun exposure' *Int. J. Sci. Rep.*, Vol 9 (7), pp 202–209.
15. Promkatkaew, M., Boonsri, P., Suramitr, S., Karpkird, T., Wolschann, P., Hannongbua, S. (2023) 'Stability improvement of UV-filter between methoxy cinnamic acid derivatives and cyclodextrins inclusion complexes based on DFT and TD-DFT investigations' *J. Mol. Graph. Model.*, Vol 125, pp 108619.
16. Rehfeld, A., Dissing, S., Skakkebæk, N.E. (2016) 'Chemical UV filters mimic the effect of progesterone on Ca^{2+} signaling in human sperm cells' *Endocrinology*, Vol 157, pp 4297–4308.
17. Gholap, A.D., Sayyad, S.F., Hatvate, N.T., Dhumal, V.V., Pardeshi, S.R., Chavda, V.P., Vora, L.K. (2023) 'Drug delivery strategies for avobenzone: A case study of photostabilization' *Pharmaceutics*, Vol 15(3), pp 1008.
18. Chavda, V.P., Acharya, D., Hala, V., Daware, S., Vora, L.K. (2023) 'Sunscreens: A comprehensive review with the application of nanotechnology' *J. Drug Del. Sci. Technol.*, Vol 86, pp 104720.
19. Santos, A.C., Marto, J., Chá-Chá, R., Martins, A.M., Pereira-Silva, M., Ribeiro, H.M., Veiga, F. (2022) 'Nanotechnology-based sunscreens—a review' *Mater. Today Chem.*, Vol 23, pp 100709.
20. Xu, C., Zeng, X., Yang, Z., Ji, H. (2021) 'Sunscreen enhancement of Octyl Methoxycinnamate microcapsules by using two biopolymers as wall materials' *Polymers*, Vol 13 (6), pp 866.

21. Bernasqué, A., Faure, C., Rezvani, H., Cario, M. (2024) 'A new eco-friendly and water-resistant sunscreen agent: Lecithin-based multilamellar liposomes' *J. Cosmet. Dermatol.*, Vol 23 (3), pp 918–925.
22. Ferreira, L., Mascarenhas-Melo, F., Rabaça, S., Mathur, A., Sharma, A., Giram, P. S., Pawar, K.D., Rahdar, A., Raza, F., Veiga, F., Mazzola, P.G., Paiva-Santos, A.C. (2023) 'Cyclodextrin-based dermatological formulations: Dermopharmaceutical and cosmetic applications' *Colloids Surf. B Biointerfaces*, Vol 221, pp 113012.
23. Jansook, P., Ogawa, N., Loftsson, T. (2018) 'Cyclodextrins: structure, physicochemical properties and pharmaceutical applications' *Int. J. Pharm.*, Vol 535 (1–2), pp 272–284.
24. Kurkov, S.V., Loftsson, T. (2013) 'Cyclodextrins' *Int. J. Pharm.*, Vol 453 (1), pp 167–180.
25. Duchêne, D. (2011) 'Cyclodextrins and Their Inclusion Complexes' In E. Bilensoy (Ed.), *Cyclodextrins in pharmaceutics, cosmetics, and biomedicine: Current and future industrial applications*. Wiley, Hoboken, NJ, USA, pp 1–18.
26. Kali, G., Haddadzadegan, S., Bernkop-Schnurch, A. (2024) 'Cyclodextrins and derivatives in drug delivery: New developments, relevant clinical trials, and advanced products' *Carbohydr. Polym.*, Vol 324, pp 121500.
27. Liu, J.Y., Zhang, X., Tian, B.R. (2020). 'Selective modifications at the different positions of cyclodextrins: a review of strategies' *Turk. J. Chem.*, Vol 44 (2), pp 261–278.
28. Wei, Y., Zhang, J., Memon, A. H., Liang, H. (2017) 'Molecular model and in vitro antioxidant activity of a water-soluble and stable phloretin/hydroxypropyl-β-cyclodextrin inclusion complex' *J. Mol. Liq.*, Vol 236, pp 68–75.
29. Khanna, S., Yadav, M., Singh, N. (2024) 'Functionalization of outdoor cotton textiles: combining fragrance and uv protection through β-cyclodextrin derivative inclusion complexes infused with peppermint and clove essential oils' *Cellul. Chem. Technol.*, Vol 58 (5–6), pp 603–615.
30. Krishnamoorthy, S., Allabasha, N., Mani, M.K., Sarkar, A.K. (2023) 'Effect of β-cyclodextrin on benzophenones and its impact on UVR filtration of cotton fabric through "guest absorbers-host enhancer" mechanism' *J. Incl. Phenom. Macrocycl. Chem.*, Vol 103, pp 131–145.
31. Nhu Ngoc, L.T., Van Tran, V., Moon, J.Y., Chae, M., Park, D., Lee, Y.C. (2019) 'Recent trends of sunscreen cosmetic: An update review' *Cosmetics*, Vol 6, pp 64.
32. Romanhole, R.C., Ataide, J.A., Cefali, L.C., Moriel, P., Mazzola, P.G. (2016) 'Photostability study of commercial sunscreens submitted to artificial UV irradiation and/or fluorescent radiation' *J. Photochem. Photobiol. B, Biology*, Vol 162, pp 45–49.
33. do Prado, R., Bertges, F.S., Figueiredo, S.A., Vieira Fonseca, M.J., Tavares, G.D., Pinto Vilela, F.M. (2020) 'Photostability of commercial sunscreens upon sunlight exposure' *HU Revista*, Vol 46, pp 1–9.
34. d'Agostino, S., Azzali, A., Casali, L., Taddei P, Grepioni, F. (2020) 'Environmentally friendly sunscreens: Mechanochemical synthesis and characterization of β-CD inclusion complexes of avobenzone and octinoxate with improved photostability' *ACS Sustain. Chem. Eng.*, Vol 8 (35), pp 13215–13225.
35. Rajamohan, R., Muthuraja, P., Govindasamy, C., Subramanian, S.A., Kim, S.J., Krishnan, M.M., Murugavel, K., Lee, Y.R. (2023) 'Enhanced photostability and biocompatibility of sunscreen formulation of 2-phenylbenzimidazole-5-sulfonic acid with methyl-beta-cyclodextrin' *J. Mol. Liq.*, Vol 390, pp 123013.
36. Rajamohan, R., Kamaraj, E., Govindasamy, C., Subramanian, S.S., Kim, S.K., Krishnan, M.M., Murugavel, K., Murugan, M., Lee, Y.R. (2024) 'Formulation of

Ensulizole with Beta-Cyclodextrins for improved sunscreen activity and biocompatibility' *J. Pharm. Sci.*, Vol 113 (6), pp 1536–1545.

37. Yuan, L., Li, S., Huo, D., Zhou, W., Wang, X., Bai, D., Hu, J. (2019) 'Studies on the preparation and photostability of avobenzone and (2-hydroxy)propyl-β-cyclodextrin inclusion complex' *J. Photochem. Photobiol. A*, Vol 369, pp 174–180.
38. Wang, X., Han, L., Hu, X., Li, S., Ma, W., Song, W. (2022) 'Photostability of the inclusion complex of isoamyl 4-(Dimethylamino) benzoate with sulfobutylether-β-cyclodextrin' *J. Photochem. Photobiol. A*, Vol 423, pp 113641.
39. Shokri, J., Hasanzadeh, D., Ghanbarzadeh, S., Dizadji-Ilkhchi, M., Adibkia, K. (2013) 'The effect of Beta-Cyclodextrin on percutaneous absorption of commonly used Eusolex® sunscreens' *Drug Res.*, Vol 63 (11), pp 591–596.
40. Felton, L.A., Wiley, C.J., Godwin, D.A. (2002) 'Influence of Hydroxypropyl-β-Cyclodextrin on the Transdermal permeation and skin accumulation of Oxybenzone' *Drug Dev. Ind. Pharm.*, Vol 28 (9), pp 1117–1124.
41. Simeoni, S., Scalia, S., Tursilli, R., Benson, H. (2005) 'Influence of Cyclodextrin complexation on the in vitro human skin penetration and retention of the sunscreen agent, oxybenzone' *J. Incl. Phenom. Molec. Recognit. Chem.*, Vol 54 (3–4), pp 275–282.
42. Monti, D., Tampucci, S., Chetoni, P., Burgalassi, S., Saino, V., Centini, M., Staltari, L., Anselmi, C. (2011) 'Permeation and distribution of ferulic acid and its α-Cyclodextrin complex from different formulations in hairless rat skin' *AAPS Pharm. Sci.*, Vol 12 (2), pp 514–520.
43. Scalia, S., Coppi, G., Iannuccelli, V. (2011) 'Microencapsulation of a cyclodextrin complex of the UV filter, butyl methoxydibenzoylmethane: In vivo skin penetration studies' *J. Pharm. Biomed. Anal.*, Vol 54 (2), pp 345–350.
44. Simeoni, S., Scalia, S., Benson, H. (2004) 'Influence of cyclodextrins on in vitro human skin absorption of the sunscreen, butyl-methoxydibenzoylmethane' *Int. J. Pharm.*, Vol 280 (1–2), pp 163–171.
45. Felton, L.A., Wiley, C.J., Godwin, D.A. (2004) 'Influence of Cyclodextrin complexation on the in vivo photoprotective effects of oxybenzone' *Drug Dev. Ind. Pharm.*, Vol 30 (1), pp 95–102.
46. Mori, T., Tsuchiya, R., Doi, M., Nagatani, N., Tanaka, T. (2018) 'Solubilization of ultraviolet absorbers by cyclodextrin and their potential application in cosmetics' *J. Incl. Phenom. Macrocycl Chem.*, Vol 93 (1–2), pp 91–96.
47. Yang, J., Wiley, C.J., Godwin, D.A., Felton, L.A. (2008) 'Influence of hydroxypropyl-β-cyclodextrin on transdermal penetration and photostability of avobenzone' *Eur. J. Pharm. Biopharm.*, Vol 69 (2), pp 605–612.
48. ISP Investments Inc., (2014) 'Composition comprising cyclodextrins as UV- and IR-radiation screen agent' Inventors: Perassinoto, N.L., Bartuccio Raponi, M.R., Yoshida Shitara, J.L., Miyashiro Kumayama, T. Date of Patent: Oct. 9, 2014. World Intellectual Property Organization WO2014/165471A1.
49. ISP Investments Inc., (2019) 'Composition comprising cyclodextrins as UV- and IR-radiation screen agent' Inventors: Perassinoto, N.L., Bartuccio Raponi, M.R., Yoshida Shitara, J.L., Miyashiro Kumayama, T. Date of Patent: Nov. 26, 2019. PCT Filed: Apr. 1, 2014. United States Patent US10485748B2.
50. ISP do Brasil LTDA (2014) 'Composicies cosmeticas e uso de ciclodextrina' Inventors: Yoshida Shitara, J.L., Bartuccio Raponi, M.R., Perassinoto, N.L., Miyashiro Kumayama, T. Date of Patent: Nov. 11, 2014. Data Filed: Apr. 3, 2013. República Federativa do Brasil BR1020130079693A2.

51. ISP do Brasil LTDA (2016) 'Topic sun-protection formulations' Inventors: Perassinoto, N.L., Raponi, M.R.B. Date of Patent: Oct. 13, 2016. World Intellectual Property Organization WO2016/161493A1.
52. Mavrić, Z., Tomšič, B., Simončič, B. (2018) 'Recent advances in the ultraviolet protection finishing of textiles'. *Tekstilec*, Vol 61 (3), pp 201–220.
53. Attia, N.F., Osama, R., Elashery, S.E.A., Kalam, A., Al-Sehemi, A.G., Algarni, H. (2022) 'Recent advances of sustainable textile fabric coatings for UV protection properties' *Coatings*, Vol 12, pp 1597.
54. Bernhard, A., Caven, B., Wright, T., Burtscher, E., Bechtold, T. (2022) 'Improving the ultraviolet protection factor of textiles through mechanical surface modification using calendering' *Text. Res. J.*, Vol 92 (9–10), pp 1405–1414.
55. Bezerra, F.M., Lis, M.J., Firmino H.B., Dias da Silva J.G., Curto Valle R.C.S., Borges Valle J.A., Scacchetti F.A.P., Tessaro A.L. (2020) 'The role of β-Cyclodextrin in the textile industry-review' *Molecules*, Vol 25 (16), pp 3624.
56. Radu, C.D., Parteni, O., Ochiuz, L. (2016) 'Applications of cyclodextrins in medical textiles – review' *J. Control. Release*, Vol 224, pp 146–157.
57. El-Sayed, E., Othman, H.A., Hassabo, A.G. (2021) 'Cyclodextrin usage in textile industry'. *J. Text. Color. Polym. Sci.*, Vol 18 (2), pp 111–119.
58. Ibrahim, N.A., El-Zairy, E.M., Allah, S.S.A., Emam, E.M. (2023) 'A Greener Facile approach to develop durable multifunctional Cellulosic pigment prints using β-CD, ZnO-NPs and several bioactive guest molecules' *Fibers Polym.*, Vol 24, pp 109–118.
59. Jajpura, L., Rangi, A., Khandual, A. (2020) 'Natural finishies, technologies and recet developments'. In Gekas, V.K. (ed.), *Sustainable technologies for fashion and textiles*, Cambridge, UK, Woodhead Publishing, pp 209–229.
60. Ibrahim, N.A., Khalil, H.M., Eid, B.M. (2015) 'A cleaner production of ultra-violet shielding wool prints' *J. Clean. Prod.*, Vol 92, pp 187–195.
61. Nunes, J., Lange De Aguiar, C.R., Krause Bierhalz, A.C. (2023) 'Functionalization of cotton fabric with β-cyclodextrin and nano-ZnO' *Cellul. Chem. Technol.*, Vol 57 (3–4), pp 427–436.
62. Puebla-Duarte, A.L., Santos-Sauceda, I., Rodríguez-Félix, F., Iturralde-García, R.D., Fernández-Quiroz, D., Pérez-Cabral, I.D., Del-Toro-Sánchez, C.L. (2023) 'Active and intelligent packaging: A review of the possible application of cyclodextrins in food storage and safety indicators' *Polymers*, Vol 15, pp 4317.
63. Szente, L., Fenyvesi, E, (2018) 'Cyclodextrin-enabled polymer composites for packaging' *Molecules*, Vol 23, pp 1556.
64. Zhang, J. Zhang, X., Wang, R., Wang, W., Zhao, H., Yang, S., Dong, Z., Wang, D., Pan., Y. (2024) 'Cyclodextrin-based host-guest hierarchical fire retardants: Synthesis and novel strategy to endow polylactic acid fire retardancy and UV resistance' *Carbohydr. Polymers*, Vol 341, pp 122313.
65. Ding, X., Luo, X., Shi, W., Cheng, D., He, X., Yue, L., Wang, J. (2024) 'Chitosan film loaded with carbon dots and geraniol/hydroxypropyl-β-cyclodextrin inclusion complex: Simultaneous freshness preservation and quality monitoring of pork' *Food Hydrocoll.*, Vol 157, pp 110326.
66. Srinivasan, K., Radhakrishnan, S., Stalin, T. (2014) 'Inclusion complexes of β-cyclodextrin-dinitro compounds as UV absorber for ballpoint pen ink' *SAA Mol. Biomol. Spectrosc.*, Vol 129, pp 551–564.
67. Peng, L., Lin, M., Zhang, S., Li, L., Fu, Q., Hou, J. (2019) 'A self-healing coating with UV-shielding property' *Coatings*, Vol 9 (7), pp 421.

Index

Made in the USA
Las Vegas, NV
19 September 2026

4f3bc4eb-785a-456c-9b42-80644598e2c1R01